Photobiology of
Higher Plants

Photobiology of Higher Plants

Maurice S. Mc Donald

National University of Ireland, Galway

WILEY

Copyright © 2003   John Wiley & Sons Ltd, The Atrium, Southern Gate, Chichester,
West Sussex PO19 8SQ, England

Telephone (+44) 1243 779777

Email (for orders and customer service enquiries): cs-books@wiley.co.uk
Visit our Home Page on www.wileyeurope.com or www.wiley.com

Other Wiley Editorial Offices

John Wiley & Sons Inc., 111 River Street, Hoboken, NJ 07030, USA

Jossey-Bass, 989 Market Street, San Francisco, CA 94103-1741, USA

Wiley-VCH Verlag GmbH, Boschstr. 12, D-69469 Weinheim, Germany

John Wiley & Sons Australia Ltd, 33 Park Road, Milton, Queensland 4064, Australia

John Wiley & Sons (Asia) Pte Ltd, 2 Clementi Loop #02-01, Jin Xing Distripark, Singapore
129809

John Wiley & Sons Canada Ltd, 22 Worcester Road, Etobicoke, Ontario, Canada M9W 1L1

Wiley also publishes its books in a variety of electronic formats. Some content that appears
in print may not be available in electronic books.

Library of Congress Cataloging-in-Publication Data

McDonald, Maurice S.
  Photobiology of higher plants / Maurice S. McDonald.
    p. cm.
  Includes bibliographical references.
  ISBN 0-470-85522-3 (cloth : alk. paper) – ISBN 0-470-85523-1 (pbk. : alk. paper)
  1. Plants, Effect of light on. 2. Photobiology. I. Title.

QK757.M393 2003
571.4′552–dc21   2002192414

British Library Cataloguing in Publication Data

A catalogue record for this book is available from the British Library

ISBN 0 470 85522 3 (ppc)
ISBN 0 470 85523 1 (paperback)

Typeset by Dobbie Typesetting Ltd, Tavistock, Devon
Printed and bound in Great Britain by Antony Rowe Ltd, Chippenham, Wilts
This book is printed on acid-free paper responsibly manufactured from sustainable forestry
in which at least two trees are planted for each one used for paper production.

To Evelyn and our family

Contents

Photobiology of Higher Plants. By M. S. Mc Donald.
© 2003 John Wiley & Sons, Ltd: ISBN 0 470 85522 3; ISBN 0 470 85523 1 (PB)

Chapter 3   CO$_2$ Fixation   74

Chapter 8   Selected Photobiological Responses   274

Chapter 9   Signal Transduction   302

Preface

Photobiology of Higher Plants has arisen out of the need for a single text on this topic and has developed out of 30 years of teaching plant biochemistry/physiology to undergraduate students. The current trend in undergraduate teaching appears to be moving away from long, one-year courses and towards concise modules. Traditional courses such as, say, Plant Biochemistry and Plant Physiology are being replaced by modules such as Plant Biotechnology, Plant Tissue Culture, Plant Growth and Development etc. – topics which have been components of existing, traditional one- or two-year courses. This new approach has evolved in order to give students a wider choice of specialized topics; it also facilitates the increasingly popular student-exchanges, especially for those who make one-semester visits.

There is no single text available to cover a course in the photobiology of higher plants. Various chapters, in a range of contemporary plant biochemistry/physiology texts, treat the individual topics of the subject with various emphases and pitched at varying levels. Hence the need to 'pull it all together' in a single text. This is achieved in *Photobiology of Higher Plants* which has been written with a balanced emphasis by a single author, thus ensuring a consistent writing style.

In writing the text I have consulted an exhaustive range of research literature on previous and ongoing research throughout the world, and reviews and opinions, and have distilled the basic concepts into a teacher-written, readable, student-friendly form. I have included many references with each chapter in order to document our sources of information and to provide a starting point for students who wish to learn more about the subject. A glossary is added to provide easy access to definitions of the technical terms used in the text.

The intended readership is specifically second/third-year undergraduate students in Botany, Plant Science, Agriculture, Horticulture and Forestry and in general for students of plant biochemistry and plant physiology. It should also form a component of Denominated Degree courses such as Earth Science, Environmental Science, Marine Botany and Plant Biotechnology which require a plant input and should prove to be a valuable source of information for post-graduate students beginning work in these fields.

Photobiology of Higher Plants. By M. S. Mc Donald.
© 2003 John Wiley & Sons, Ltd: ISBN 0 470 85522 3; ISBN 0 470 85523 1 (PB)

A word of acknowledgement is due to Patrick Cooke and Thomas Naughton, Department of Botany, National University of Ireland, Galway for their help with diagrams and layout and to a wide range of colleagues who reviewed the text and made helpful suggestions and criticisms.

Maurice S. Mc Donald

1
Light and Pigments

The Nature of Light

The English physicist Sir Isaac Newton (1642–1727) is credited with being the first person to propose a theory that attempted to explain the nature of light. Newton showed that white light, when passed through a prism, a droplet of water or a soap bubble separated into a spectrum of visible colours (Fig. 1.1). If passed through a second prism the colours recombined to form white light again.

Newton proposed that white light is composed of a spectrum of different colours ranging from violet at one end of the spectrum to red at the other. Their separation is possible because light is broken up (diffracted) into different colours which are deflected (refracted) at different angles when passing through the prism. Newton's explanation of the nature of light was called the 'corpuscular theory': light was considered to be made up of tiny particles (corpuscles) which travel in straight lines from the source of the light. Newton's corpuscular theory provided a satisfactory explanation for reflection: it predicted that when light was reflected, the angle of

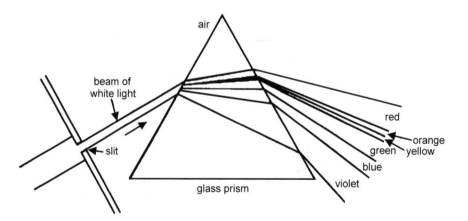

Figure 1.1 When passed through a glass prism white light is dispersed into its component colours by refraction

Photobiology of Higher Plants. By M. S. Mc Donald.
© 2003 John Wiley & Sons, Ltd: ISBN 0 470 85522 3; ISBN 0 470 85523 1 (PB)

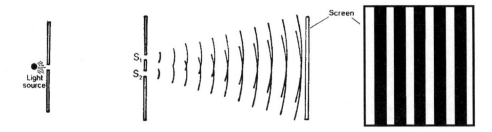

Figure 1.2   Young's experiment: light emerging from two narrow slits overlap and produce an interference pattern

incidence should be equal to the angle of reflection. The theory also offered an explanation for refraction: it predicted that light should have a greater speed in a denser medium. Light does, in fact, travel more slowly in a denser medium and so Newton's theory does not offer a satisfactory explanation of refraction; neither does it explain diffraction or interference. The corpuscular theory is no longer used but wasn't abandoned (probably due to Newton's reputation) until the beginning of the nineteenth century. In 1670, a Dutchman, Christiaan Huygens, proposed that the laws of reflection and refraction could better be explained on the basis of a wave theory of light. The wave theory was not immediately accepted, partly because it clearly opposed Newton's theory and partly due to the fact that Huygens offered no experimental proof. Thomas Young (1773–1829), an Englishman, designed experiments to show that light is diffracted when passed through a narrow orifice and that beams of light emerging from two narrow slits overlap and produce an interference pattern (Fig. 1.2). It was not until Young's published results appeared in 1817 that the wave theory of light was accepted and the corpuscular theory found to be inadequate.

In the 1860s, a Scottish mathematician named James Maxwell (1831–1879) showed that an oscillating electrical circuit could radiate electromagnetic waves. The velocity of the propagation of these waves was found to be 3×10^{10} cm s^{-1}, very close to the measured velocity of light. This concept supported the hypothesis for the existence of light as electromagnetic waves of very short wavelengths. The problem appeared to be solved. However, by 1900 it became apparent that the wave model of light was not adequate. One puzzling observation seemed to contradict the wave theory of light: a zinc plate exposed to ultraviolet light acquires a positive charge. This phenomenon, called *photoelectric emission*, by which the zinc becomes positively charged because the incident light energy dislodges electrons from the zinc atoms, can be produced in all metals. Every metal has a critical wavelength for the effect; the incident light must be of a particular wavelength or shorter (of higher energy) to cause the effect. The wave model of light suggests that the brighter the light, i.e. the stronger or more intense the beam, the greater the force available to dislodge electrons; in fact a weak beam of a critical wavelength displaces electrons from metal atoms whereas a stronger (brighter) beam of longer wavelength will not. In addition, whereas increasing the intensity of light increases the number of electrons dislodged, the shorter the wavelengths the greater the velocity at which the electrons are dislodged. With this evidence, Albert Einstein (1905) postulated that the energy in a light beam is concentrated in small particles called photons or quanta. Because of this quantization of radiation the photon can be considered as a particle. This particle model gave some validity to the corpuscular theory. However, since a photon is considered to have a frequency and the energy of a photon is believed to be proportional to this frequency, some of the wave theory is retained. Thus, in order to understand the nature of light it is necessary to appreciate its dual wave–particle nature.

The Wave Nature of Radiation

In the 1860s Maxwell showed that the visible light that Newton separated into a spectrum of colours occupies only a small portion of a much larger electromagnetic radiation with wavelengths in the range of 200 nm to 49 000 nm (1 nanometre (nm) = 10^{-7} cm = 10^{-9} m, i.e. it is 1 billionth of a metre) (Fig. 1.3). The term 'light' describes the narrow band (400–700 nm) within the total electromagnetic spectrum that causes the physiological sensation of vision. This range which we call light includes the colours violet, blue, green, yellow, orange and red; shorter wavelengths ($\lambda < 400$ nm) and longer wavelengths ($\lambda > 700$ nm), which our eyes cannot detect (although they may have significant biological effects), are called ultraviolet and infrared radiation, respectively. Light absorbed by higher plants and used in photobiological reactions is in the range 400–700 nm.

Electromagnetic radiation is actually composed of two waves – one electrical and one magnetic – which oscillate at 90° to each other and to the direction of propagation (Fig. 1.4).

Radiation may be described as a wave function whose length (λ) and frequency (v) are related to the speed of propagation:

$$c = \lambda v$$

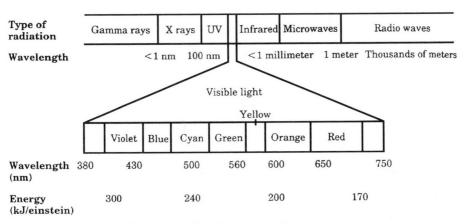

Figure 1.3   The electromagnetic spectrum

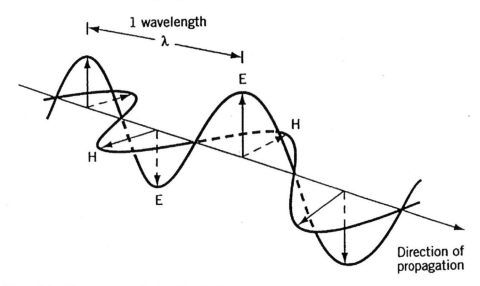

Figure 1.4   Wave nature of light. Electrical vectors (E) and magnetic vectors (H) oscillate at 90°
to each other

c = velocity, the distance travelled by a peak of radiant energy in a specified interval
of time. The velocity of all forms of radiant energy is the same in a vacuum and is
equal to $3 \times 10^8 \, m \, s^{-1}$ (300 000 km s^{-1} or 186 000 miles s^{-1}).

λ = wavelength, the distance between waves or crests of energy in electromagnetic
radiation. Wavelengths of radiant energy vary from less than the diameter of an
atom to several kilometres in length. Green light has a wavelength of about 500 nm
(5×10^{-7} m) whereas radio waves have wavelengths between about 10^{-3} and 10^4 m.

v = frequency, the number of wave crests passing a given point in a given interval
of time. Frequency is usually expressed in terms of energy crests (waves) per second.
Green light has a frequency of about 6×10^{14} pulsations s^{-1} whereas radio waves
have frequencies between about 10^4 and 10^{11} s^{-1}. The wavelengths most important
in photobiological reactions of plants fall into three distinct ranges: ultraviolet,
visible and infrared (Table 1.1).

Refraction

When a ray of radiant energy passes from one medium to another of different
density it will refract (bend) because of the change of velocity. If the radiation
passes from a less dense to a more dense medium (air to water, for example) it will
refract toward the normal; when passing from a more dense to a less dense medium
it will refract away from the normal. The bending or refraction is greater for short-
wave radiation than for long-wave radiation. The wave-like nature of light thus
helps to explain how wavelengths are separated into a spectrum when they pass
through a prism (Fig. 1.1). Light is refracted within leaves of plants as it passes
from air into a cell wall or the cytoplasm.

Table 1.1 Wavelengths of light important in biology. The wave nature of radiation explains the phenomena of refraction, diffraction, interference and polarization

Colour	Wavelength (nm)	Average energy (kJ mol^{-1})
Ultraviolet	100–400	
UV-C	100–280	471
UV-B	280–320	399
UV-A	320–400	332
Visible	400–700	
Violet	400–425	290
Blue	425–490	266
Green	490–550	230
Yellow	550–585	218
Orange	585–640	196
Red	640–700	184
Far red	700–740	160
Infrared	> 740	85

Interference and Diffraction

When two or more radiation waves occur at the same time in the same space they produce *interference*. If the two waves are superimposed on each other so that they coincide (are in phase) they reinforce each other to produce a new wave of exactly twice the amplitude in a phenomenon called constructive interference; if the two waves are 180° out of phase they dampen or cancel each other in what is called *destructive interference* (Fig. 1.5).

Plant physiologists often use two devices based on these principles. *Diffraction gratings*, which consist of fine lines ruled very close together on a transparent surface, separate a mixture of wavelengths into a spectrum similar to that produced by a prism. *Interference filters* use a thin layer of reflective medium on a glass surface; one wavelength is strongly reinforced by passing through the filter, while other wavelengths are reflected and cancelled. The study of certain physiological phenomena in plants sometimes requires light of narrow bandwidths, i.e. single colours. Such monochromatic light may be useful for studying seed germination, flowering or stomatal movement as functions of light quality. A glass prism for visible light or a quartz prism for ultraviolet light separates wavelengths by diffraction. Particular wavelengths can be obtained from the diffraction patterns produced by diffraction gratings. High-intensity monochromatic light can be produced by lasers. A ruby laser emits light with a wavelength of 694.3 nm; others can produce ultraviolet light. Alternate constructive and destructive interference results in diffraction and arises if a beam of light strikes a reflecting surface that has a regular array of scattering or reflecting centres. Such a surface is provided by the

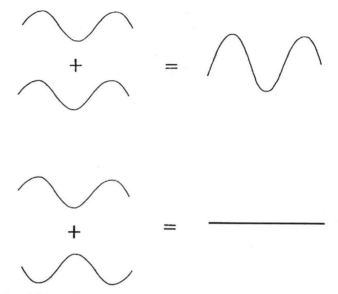

Figure 1.5 The addition of two waves of equal amplitude, in phase (upper) and out of phase (lower)

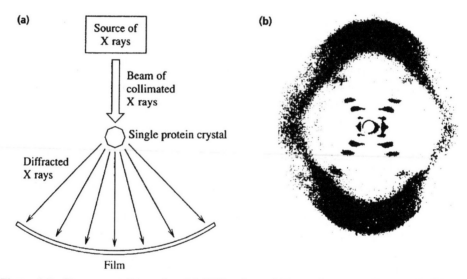

Figure 1.6 X-ray crystallography. (a) Diffraction of X-rays by a protein crystal. (b) X-ray diffraction photograph of the DNA molecule. (From (a) Horton *et al*. 1996; (b) Lehninger *et al*. 1993)

uniform lattice structure of a crystal and the crystalline structure can be determined by diffraction patterns. In the light microscope images are generated by focusing light from a point source on an object; the light waves are scattered by the object and then recombined by a series of lenses to generate an enlarged image of the object. The resolving power of the microscope is determined by the wavelength of light. Objects smaller than half the wavelength of the incident light cannot be resolved. Hence X-rays are used since their wavelengths are about 0.1–0.2 nm, short enough to resolve atoms. The X-rays are scattered by the electrons, the extent of scattering being proportional to the density of the electrons. The three-dimensional structure of an object can be deduced from the diffraction pattern of discrete spots that is produced when the scattered radiation is intercepted by photographic film (Fig. 1.6). Once the pattern of direction, wavelength, amplitude and phase of the radiation is determined the three-dimensional structure of the molecules of the crystal can be computed. X-ray diffraction patterns and subsequent computer analysis have been used to determine the structure of a variety of biological molecules. For example, X-ray analysis of sodium chloride crystals shows that Na^+ and Cl^- ions are arranged in a simple cubic lattice. The spacing of different kinds of atoms in complex organic molecules such as nucleic acids (Fig. 1.6) and proteins can be analysed by X-ray diffraction methods.

The Particulate Nature of Light

The dual nature of electromagnetic radiation (wave and particle) means that light behaves both as waves and as particles at the same time. While the wave nature of radiation fits well with the phenomena of refraction, interference and diffraction, it is best to consider radiation as discrete particles when it interacts with matter. These particles are packets of radiant energy and are called photons or quanta; they cannot be further subdivided and can be regarded as 'atoms' of energy. The two terms 'photon' and 'quantum' are often used interchangeably: a quantum is the unit of energy in the quantum theory whereas a photon is the quantum of energy at a particular wavelength of electromagnetic radiation.

In 1901 Max Planck concluded that radiation was not wave-like and that it exists as discrete particles with energies expressed by the equation

$$q \text{ (quantum)} = hv = \frac{hc}{\lambda}$$

where $h = $ Planck's constant, with a value of 6.6255×10^{-34} J s photon^{-1} (1.58×10^{-34} cal s photon^{-1} or 6.6255×10^{-27} erg s photon^{-1}). This equation is called Planck's law and shows that the energy in a quantum is related to wavelength and frequency. Thus, the energy of a quantum of radiation is inversely proportional to its wavelength (the shorter the wavelength the greater the energy (Table 1.2) or directly proportional to the frequency. Accordingly, a photon of half the wavelength of another has twice the energy of that other; a photon of one third the energy of another has three times the energy of that other (Table 1.2). The intensity (i.e. brightness) of radiation depends on the number of photons (i.e. the amount of energy) emitted from a source per unit time. When a pigment molecule absorbs photons of light, the molecule is excited, that is, electrons in the outer shells of the pigment molecules may be raised from their normal (ground) state to an excited state (higher energy level).

Table 1.2   Energy value of light quanta at different wavelengths

Wavelength (nm)	Colour of light	Energy of quanta $(kJ\,mole^{-1})$
350	Ultraviolet	342
450	Blue	266
550	Yellow	218
650	Red	184
750	Far red	160

According to the Stark–Einstein law of photochemical equivalence, only one molecule or atom can be excited by one photon – that is, one photon of light, regardless of its energy level, will only activate one molecule. Since the energy content of one photon is very small it is more practical to consider the amount of energy absorbed by one mole of a substance rather than the energy absorbed by one molecule. Thus the energy level of a single photon is multiplied by Avogadro's number ($N = 6.023 \times 10^{23}$, the number of molecules in one mole of a substance); i.e. 6.023×10^{23} photons $= 1$ 'mole-photon', or, more commonly, a mole-quantum. Therefore we would need N quanta to excite one mole (N molecules) of a substance. A mole of quanta is known as the *photochemical equivalent*. This quantity was originally called an einstein, but since the einstein is not an SI unit, 'mole of quanta' is most commonly used. The energy (E) contained in a mole of quanta can be calculated, for any wavelength, using the following equation:

$$E = Nh\nu$$

Substituting c/λ for ν we get:

$$E = \frac{Nhc}{\lambda}$$

For the energy content of blue light, with a representative wavelength of 435 nm ($\lambda = 4.35 \times 10^{-7}$ m; $\nu = 6.895 \times 10^{14}\,s^{-1}$) :

$$E = \frac{\overset{(N)}{(6.02 \times 10^{23}\text{ photons mol}^{-1})}\overset{(h)}{(6.62 \times 10^{-34}\text{ J s photons}^{-1})}\overset{(c)}{(3 \times 10^{8}\text{m s}^{-1})}}{\underset{(\lambda)}{4.35 \times 10^{-7}\text{ m}}}$$

$$= 274\,kJ\,mole^{-1}\ (Table\ 1.2)$$

The energy content of red light, with a representative wavelength of 660 nm ($\lambda = 660 \times 10^{-7}$ m; $\nu = 4.545 \times 10^{14}\,s^{-1}$) is given by $E = 181\,kJ\,mole^{-1}$ (Table 1.2). Thus, blue light ($\lambda = 435 \times 10^{-7}$ m; $\nu = 6.895 \times 10^{14}\,s^{-1}$) has a frequency 6.895/4.545 times that of red light ($\lambda = 660 \times 10^{-7}$ m; $\nu = 4.545 \times 10^{14}\,s^{-1}$) so its energy per photon is one and half times that of red light.

Table 1.3   Radiation data terminology. (From Hall and Rao 1994)

Term	Unit	Definition
Radiant energy	J	Energy in the form of electromagnetic radiation
Radiant flux	$W = J\,s^{-1}$	Radiant energy emitted or absorbed by a surface per unit time
Radiant flux density	$W\,m^{-2}$	Radiant flux (of a specific wavelength region) incident on a unit surface area
Irradiance	$W\,m^{-2}$	Radiant flux incident on a unit area of plane surface per unit time
Fluence	$mol\,m^{-2}$	Number of photons incident across a unit area of plane surface
Photon flux density (PFD)	$mol\,m^{-2}\,s^{-1}$	Photon flux per unit area
Photosynthetically active radiation (PAR)		Solar radiation from 400 to 700 nm
Photosynthetic photon flux density (PPF)	$mol\,m^{-2}\,s^{-1}$	Flux of 400–700 nm solar radiation

Historically light was measured in terms of *lumens* (lm), a lumen being defined as the luminous flux on a unit surface, all points of which are at unit distance from a uniform point source of one candle. Intensity of illumination (illuminance) was expressed either as *foot-candles* (lm ft^{-2}) or *lux* (lm m^{-2}); 10.76 lux = 1.0 foot-candle, ft[c]. The exact definition of the lux is rather complicated, being weighted in relation to the sensitivity of the human eye and has little or no revelance for plant physiology.

Nowadays, photobiologists prefer to measure light energy incident on a surface, i.e. *radiant flux density* or *irradiance* in terms of units of power as watts per square metre (W m^{-2}). Roughly 1.3 kW m^{-2} of radiant energy from the sun reaches the earth. Whereas joules are used to measure total energy, watts per square metre are the units of irradiance, the flux of radiant energy per unit time; thus 1.3 kW m^{-2} equals 1.3 kJ s^{-1} m^{-2}. Since photobiological reactions depend more on the number of photons (quanta) absorbed by a surface rather than on the energy content of these photons, it is more accurate to express the response to incident light in moles rather than in watts or joules, i.e. as the *photon flux density*. Energy fluxes, expressed in watts or joules, can be converted to photon flux densities, expressed in moles, when the precise wavelength of the light under consideration is known.

The most widely accepted measure of light quantity is based on the concept of *fluence*. Fluence is defined as the quantity of radiant energy falling on a unit surface area summed over a given time. Fluence refers to the moles of photons (*photon fluence*; the unit is mol m^{-2}) or amount of energy (energy fluence; the unit is J m^{-2}). Consequently we speak of *photon fluence rate* (mol m^{-2} s^{-1}) and energy fluence rate (J m^{-2} s^{-1} or W m^{-2}). The term 'irradiance' is frequently used interchangeably with energy fluence rate although in principle the two are not equivalent. Irradiance refers to the flux of energy on a flat surface rather than on a sphere. Photon fluence rate is measured in units of mol m^{-2} s^{-1} where the mol is 6.023×10^{23} (Avogadro's number) photons or quanta. Since the einstein (E) is defined as 6.023×10^{23} quanta, the photon fluence rate can be expressed as $E\,m^{-2}s^{-1}$. The unit einstein (or microeinstein), where 1 einstein equals 1 mole of light, is not, however, a unit in the SI system (Système Internationale), and moles (or micro-moles) should be used instead. A more practical unit to express photosynthetic photon flux density (PPF) is $\mu mol\,m^{-2}s^{-1}$ (PPF used to be called 'photosynthetic photon flux *density*' or PPFD, but because the term flux includes the concept of density, 'density' has been dropped). For example, the solar irradiance of full sunlight is approximately $1000\,W\,m^{-2}$ or 100 000 lux, in which the PPF (400 to 700 nm, the range of light broadly defined as photosynthetically active radiation) is $2000\,\mu mol\,m^{-2}s^{-1}$. Some of the terms used in plant physiology to express radiation are summarized in Table 1.3.

Absorption of Light

Before light can be effective it must be absorbed. A fundamental principle of photobiology, known as the Gotthaus–Draper law states that only light that is absorbed can be active in a photochemical process. A second basic principle of light absorption, the Stark–Einstein law, states that a molecule can absorb or be activated by only one photon at a time; this photon causes excitation of only one electron, i.e. absorption of one photon results in only one event. Photons that are absorbed produce electronic changes of great biological importance. It is the valence electrons of the outer shells of atoms that are most affected since they can be excited by the absorption of photons. The most stable states of atoms are those in which the valence electrons are in their low-energy orbitals – their *ground state*. In the ground state, the spins of paired electrons are in opposite directions (Pauli exclusion principle) and the total spin is zero (Fig. 1.7).

When a pigment molecule in the ground state absorbs a photon of the correct wavelength a valence electron is raised to a higher-energy orbital – an *excited state*.

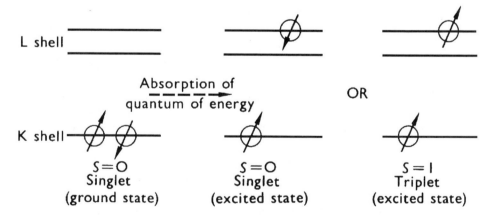

Figure 1.7 Energy levels of electrons in the helium atom. (From Hall and Rao 1994). Reproduced with permission of Cambridge University Press)

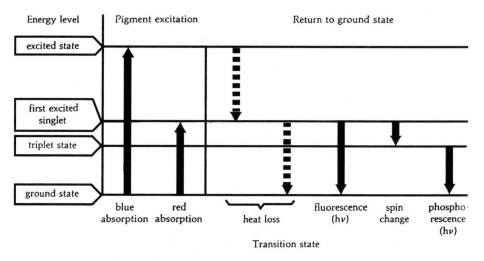

Figure 1.8   Absorption and dissipation of light energy by pigment molecules

This happens only if the energy of the photon absorbed is precisely equal to the difference in energy between the ground-state and the excited-state orbitals. The photon that has been absorbed no longer exists. Its energy, however, has not disappeared but has been captured by the absorbing molecule. The time taken to excite a pigment molecule with a photon of light is incredibly short, about a femtosecond (10^{-15} s). The fate of the energy of the absorbed photons may best be explained by reference to Fig. 1.8. If a pigment molecule absorbs a quantum of blue light, a valence electron is raised to a higher energy level (excited singlet state); where it may exist for a very short period (10^{-9} s) and then return to its ground state. Similarly, absorption of a quantum of red light will raise a valence electron to a lower-energy singlet state, with a return to the ground state after 10^{-9} s. The light energy absorbed in both instances does not disappear; if it does not interact with any other chemical, it will reappear as radiation and the excess energy may be dissipated in a number of ways:

(i)   *Thermal dissipation.* The energized electron will fall very quickly to the lowest excited singlet state, accompanied by heat loss. As the electron falls further to its ground state energy is further lost as heat.

(ii) *Fluorescence.* A second way in which molecules may lose excitation energy is by a combination of heat loss and fluorescence. Fluorescence is the re-emission of light as excited electrons decay rapidly to their ground state. Fluorescence emission can occur only after the electron has undergone heat dissipation to the lowest excited singlet state. Because of this heat loss the emitted photon has a lower energy content than the absorbed photon and emission occurs at a longer wavelength (Fig. 1.9). Chlorophyll *a* extracts, for example, absorb in the blue and red regions of the spectrum, but fluoresce only to the long wavelength side of the red absorption band (Fig. 1.9). The

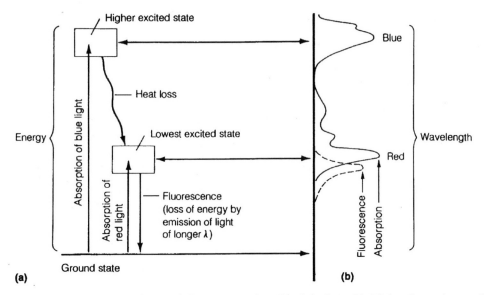

Figure 1.9 Light absorption and fluorescence by chlorophyll *a*. (a) Light absorption and dissipation; (b) absorption and fluorescence spectra showing the Stoke shift. (From Sauer 1975)

maximum intensity of fluorescence for a chlorophyll *a* molecule in solution occurs at 668 nm whereas the maximum absorption occurs at 663 nm. This is true regardless of whether the pigment is excited by blue light (450 nm) or red light (660 nm). Solutions of chlorophyll usually fluoresce in the red and the shift in wavelength, commonly called the *Stoke's shift*, is typically about ten nanometres; in the leaf, fluorescence is weak because the excitation energy is used in photosynthesis. Figure 1.9 may help to explain why blue light is less efficient than red light, in photosynthesis. After excitation by a blue photon the electron in a chlorophyll molecule decays very rapidly by heat loss to a lower energy level, a level that the lower-energy red light produces without heat loss when a red photon is absorbed.

(iii) *Phosphorescence*. There is always the possibility that an electron brought to a higher energy level (excited-singlet state) by the absorption of a quantum of light, may have its spin reversed (Fig. 1.7). Since two electrons cannot exist at the same energy level with parallel spins, the excited electron cannot return to its companion. The electron is said to be 'trapped' at a high energy level, and is in what is called the *triplet state*, which, due to transitions and slight loss of energy, is at a lower energy level than the excited-singlet state (Fig. 1.10). However, the electron may have its spin changed again, return from the triplet to the ground state and give off its excess energy as radiation. This delayed light emission is termed 'phosphorescence' and is of a longer wavelength than fluorescence. The major difference between fluorescence and phosphorescence is the amount of time for each process to occur after the initial absorption of a quantum of light of the correct wavelength. The half-life of the singlet state is 10^{-9} s and that of the triplet state is 10^{-3} s or longer. Because of its metastable state with a longer half-life, the triplet state seems most appropriate as the energy level where electron transfers in photochemical reactions can occur. At present, there is no conclusive evidence for the participation of chlorophyll triplet states in photosynthesis; surprisingly, it is the singlet state that is used in this process.

(iv) *Inductive resonance.* If pigment molecules lose energy by light emission (either fluorescence or phosphorescence) little can be achieved by way of useful work. The prevention of light emissions by interaction of excited pigment molecules with either solvent or other solute molecules is termed *quenching.* By means of quenching the ground state of pigment molecules can be re-established without loss of energy through heat or wasteful radiation. The energy of excited electrons may be transferred between accessory pigments and chlorophyll molecules by a process of *inductive resonance* (Fig. 1.10). By this means the excitation of any one of many pigment molecules in a chloroplast can be collected into a chlorophyll *a* reaction centre. This energy transfer, or sensitized fluorescence, can occur with high efficiency.

The process requires that the different pigment molecules involved be very close together in the thylakoid, and that the fluorescence band of the donor molecule must overlap the absorption band of the acceptor molecule

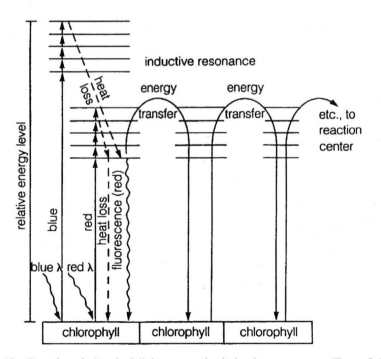

Figure 1.10   Transfer of absorbed light energy by inductive resonance. (From Salisbury and Ross 1992)

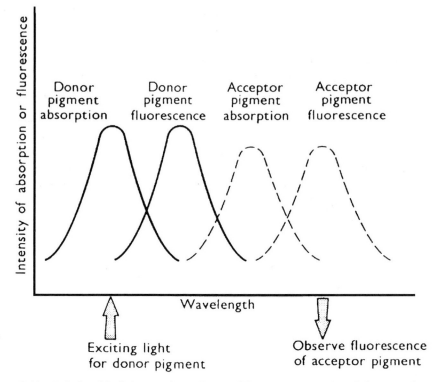

Figure 1.11 Relationship between absorption and fluorescence spectra of donor and acceptor pigments illustrating sensitized fluorescence. (From Hall and Rao 1994). Reproduced with permission of Cambridge University Press

(Fig. 1.11). Phycobilins, carotenoids and chlorophylls exhibit the necessary overlapping fluorescent and absorption characteristics. In addition, the molecular arrangement of the chloroplast is such that the proximity of the pigment molecules accommodates the phenomenon.

Pigments

Light must be absorbed by plants before it can be effective in physiological processes. The pigments that absorb light are called *photoreceptors*. The pigment molecules can process both the energy and information contained in the absorbed light. The principal photoreceptors present in plants are described here; their roles in the various physiological processes carried out by plants will be discussed in detail in later chapters.

Chlorophylls

Chlorophylls, the green pigments of plants, are the most important photoreceptors active in photosynthesis. We can distinguish four types of chlorophyll, designated

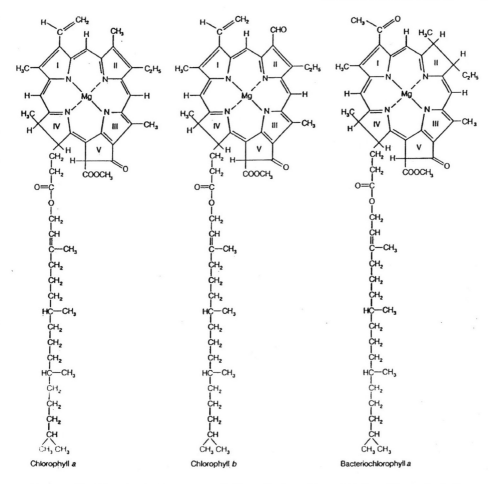

Figure 1.12   The chemical structure of chlorophyll *a*, chlorophyll *b* and bacteriophyll *a*

chlorophyll *a*, *b*, *c* and *d*. Chlorophyll *a* occurs in all photosynthetic eukaryotes and in the cyanobacteria. Chlorophyll *b* is an accessory pigment that broadens the range of light that can be used in photosynthesis; light energy absorbed by chlorophyll *b* is transferred to chlorophyll *a* which then transforms it into chemical energy in photosynthesis. In some groups of algae, especially in the brown algae and diatoms, chlorophyll *c* replaces chlorophyll *b* as an accessory pigment. Photosynthetic bacteria (other than cyanobacteria) contain bacterial chlorophylls: bacteriochlorophylls *a*, *b*, *c*, *d*, *e* and *g*. Photosynthetic bacteria (other than cyanobacteria) contain either bacteriochlorophyll (purple bacteria) or chlorobium chlorophyll (green sulphur bacteria).

The chlorophyll molecule is composed of a hydrophylic porphyrin head and a hydrophobic phytol tail (Fig. 1.12). Together these two features give chlorophyll molecules amphipathic properties. The porphyrin head contains four substituted pyrrole rings and a fifth ring which is not a pyrrole. The four inward-oriented

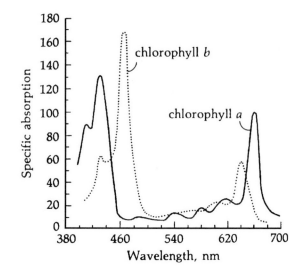

Figure 1.13   Absorption of ether extracts of chlorophylls *a* and *b*. (From Zscheile and Comar 1951)

nitrogen atoms of the pyrrole rings chelate a magnesium atom (Mg^{2+}) at the centre of the molecule. The porphyrin ring is similar to the haem prosthetic groups of haemoglobin and the cytochromes, except that these contain Fe^{2+} instead of Mg^{2+}. Thus chlorophyll is a magnesium porphyrin, whereas the cytochromes are iron porphyrins. Different chlorophylls are distinguished from each other by the chemistry of the side chain attached to the main part of the molecule. Chlorophyll *b* differs from chlorophyll *a* in having an aldehyde (—CHO) group instead of a methyl (—CH₃) group attached to one (II) of its pyrrole rings. The principal difference between chlorophyll *a* and chlorophyll *c* is that chlorophyll *c* lacks the phytol tail. Chlorophyll *d* (red algae) is similar to chlorophyll *a* except that a (—O—CHO) group replaces the (—CH=CH₂) group in ring I. The complex, heterocyclic, five-ring system of the porphyrin part of the chlorophyll molecule has a network of alternating single and double bonds. These structures are called *polyenes*; they show very strong absorption in the visible part of the spectrum and so make chlorophylls very effective photoreceptors. The peak molar absorption coefficients of chlorophylls *a* and *b* are higher than $10^5\,\mathrm{cm^{-1}\,M^{-1}}$, which puts them among the highest observed for organic compounds. The electron that is excited by absorbed light energy is in the porphyrin head, near the magnesium atom.

The phytol tail is a long hydrocarbon tail which is esterified to ring IV of the porphyrin. The phytol tail is hydrophobic and this facilitates the binding of chlorophyll to the hydrophobic lipid region of chlorophyll-binding proteins and to the thylakoid membrane. Chlorophyll is found exclusively in the lipid thylakoid membranes of the chloroplast; it is always found in association with protein and never in the free form *in vivo*. The absorption spectra of protein-bound chlorophyll in chloroplasts are markedly different from those of free chlorophyll in a solvent.

For example, chlorophyll in a chloroplast suspension has an absorption peak at 675 nm whereas an acetone extract of chlorophyll has a peak at 663 nm. In fact, the absorption spectra of chlorophylls differ with different solvents and peaks may differ by a few nanometres for chlorophyll extracts from different species. In addition to the minor differences in their molecular structure, chlorophylls *a* and *b* exhibit different absorption spectra (Fig. 1.13).

Both chlorophyll *a* and *b* show absorption maxima in the blue–violet region (400–500 nm) and the orange–red region (600–700 nm) of the visible spectrum. Because their structures are slightly different, the absorption spectra of chlorophyll *a* and chlorophyll *b* are not identical. Light that is not appreciably absorbed by chlorophyll *a*, for example, at 460 nm, is absorbed by chlorophyll *b* which has maximum absorption at this wavelength. Thus chlorophylls *a* and *b* complement each other and extend the range of wavelengths that chlorophylls absorb. The strong absorbency in the blue and the red, and transmittance in the green, is what gives chlorophyll its characteristic green colour. Most higher plants contain two or three times as much chlorophyll *a* as chlorophyll *b*.

Accessory pigments

Accessory pigments are molecules that absorb wavelengths not absorbed by chlorophyll *a*. Accessory pigments overcome the narrow absorption of chlorophyll *a* and broaden the action spectrum of photosynthesis (Fig. 1.14) (Table 1.4).

Light energy absorbed by accessory pigments is transferred to chlorophyll *a*. The most common accessory pigments found in land plants are chlorophyll *b* and the carotenoids (Table 1.4). Algae contain other accessory pigments such as phycobilins which play an important role in the absorption of light energy in red algae and cyanobacteria. One of the strongest pieces of evidence for the involvement of chlorophylls and β-carotene in photosynthesis is that the absorption spectra of these pigments is similar to the action spectrum of photosynthesis; the absorption of chlorophyll *a* alone does not match the action spectrum (Fig. 1.14).

Carotenoids

The second most abundant group of pigments in nature is the carotenoids. These are widely distributed in plants, both photosynthetic and non-photosynthetic, in red and green algae, fungi and photosynthetic bacteria. These pigments are largely responsible for the striking autumn colours of foliage and for the range of colours found in insects, crustaceans, fish and birds. The carotenoids of plants are usually yellow or orange, or occasionally red; those of animals are generally purple, violet, blue or green. Natural carotenoids can be considered as derivatives of lycopene (Fig. 1.15), a red pigment occurring in tomatoes and in many other plants. Lycopene, a tetraterpenoid, is a highly unsaturated, straight-chain hydrocarbon of 40 carbon atoms linked symmetrically with alternating single and double bonds; its empirical formula is $C_{40}H_{51}$. The molecule consists of two identical halves; each

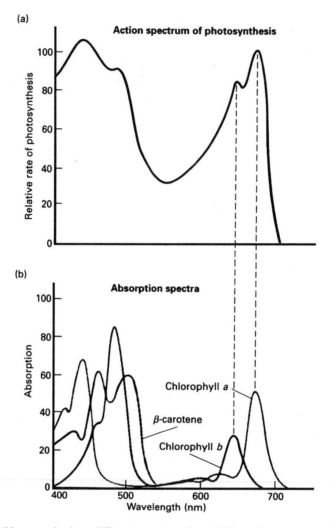

Figure 1.14   Photosynthesis at different wavelengths. (a) The action spectrum of photosynthesis in plants. (b) The absorption spectra of three photosynthetic pigments: chlorophyll a, chlorophyll b and β-carotene. A comparison of (a) and (b) suggests that photosynthesis at 650 nm is primarily due to light absorbed by chlorophyll b; at 680 nm, by light absorbed by chlorophyll a; and at shorter wavelengths, to light absorbed by chlorophyll b and by carotenoid pigments, including β-carotene. (From Lodish *et al.* 1995)

half-molecule is probably derived from four isoprene units (isoprene has the formula $CH_2\!=\!C(CH_3)\!-\!CH\!=\!CH_2$).

Thus carotenoids are composed of eight isoprene-like residues. Carotenoids absorb light because they contain networks of single and double bonds, that is, they are polyenes. Molecular structures of a number of carotenes are shown in Fig. 1.15. Since carotenoids are predominantly hydrocarbon they are lipid-soluble and are found in the thylakoid membranes of chloroplasts or in specialized plastids

Table 1.4 Pigment composition of oxygenic photosynthetic organisms

Organism	Chlorophyll				Carotenoids	Phycobilins
	a	*b*	*c*	*d*		
Plants	+	+	−	−	+	−
Green algae	+	+	−	−	+	−
Diatoms	+	−	+	−	+	−
Dinoflagellates	+	−	+	−	+	−
Brown and yellow algae	+	−	+	−	+	−
Red algae	+	−	−	+	+	+
Cyanobacteria	+	−	−	−	+	+

called chromoplasts. In chromoplasts the concentrations of carotenoids may be so high as to form crystals. Carotenoids divide into two groups – carotenes and xanthophylls. Carotenes are hydrogen carotenoids and consist exclusively of hydrogen and carbon, e.g. α-carotene, β-carotene and lycopene. Carotenes are mainly orange or orange–red pigments. The major carotenoid in plant tissues is the orange–yellow β-carotene, which is generally accompanied, in smaller amounts (0–35%), by α-carotene. In β-carotene and α-carotene, both ends of the molecule are cyclized to form ionine rings (Fig. 1.15). The difference between the two is that β-carotene has two β-ionine rings whereas α-carotene has one α- and one β-ionine ring (Fig. 1.16).

Xanthophylls contain oxygen, which occurs at the ends of the molecule as hydroxy, methoxy, aldehyde or carboxylic acid. For example, lutein (Fig. 1.17) and zeaxanthin are hydroxylated forms of α-carotene and β-carotene, respectively.

Generally the common names of different carotenes end with -*ene* and those used to describe the xanthophylls end with -*in*. In nature, the xanthophylls (Table 1.5) are more abundant than carotenes: in growing leaves the ratio may be 2:1.

The carotenoids have a limited number of functions. Most studies on their physiological role have centred on their relationship with vitamin A. Beta-carotene consists of two phenolic ring structures connected by a chain of eighteen carbon atoms joined alternately by double and single bonds (Fig. 1.18). When the double bond that links the two halves of β-carotene is broken, two molecules of vitamin A are produced. When vitamin A is oxidized in turn, retinal, the pigment involved in human vision, is produced. This may help to explain the connection between carrots (rich in β-carotene), vitamin A and vision. In plants, carotenoids play an important role in photosynthesis. However, their role is only secondary, since tissues rich in carotenoids and devoid of chlorophyll do not photosynthesize. The light energy absorbed by plant carotenoids depends largely on the number of double bonds in the molecules and is also influenced by the structures of the end groups. Carotenoids absorb typically in the blue–green part of the spectrum (Baskin and Iino 1987), where chlorophyll shows minimum absorption (Fig. 1.19).

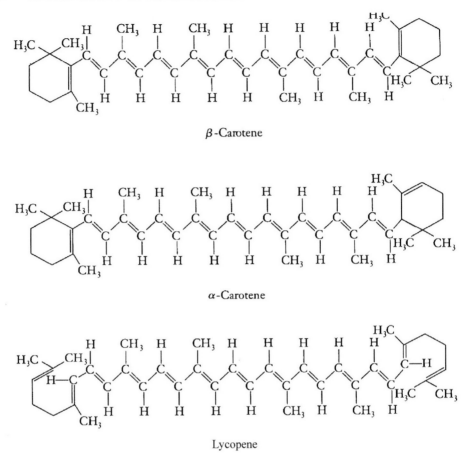

β-Carotene

α-Carotene

Lycopene

Figure 1.15   The chemical structures of representative carotenes

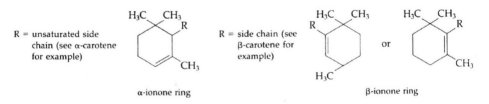

R = unsaturated side
chain (see α-carotene
for example)

α-ionone ring

R = side chain (see
β-carotene for
example)

or

β-ionone ring

Figure 1.16   Structure of α- and β-ionine rings

Light energy absorbed by carotenoids is transferred to chlorophyll *a* and used in photosynthesis. In support of this it has been shown that light absorbed by carotenoids (450 nm) results in chlorophyll fluorescence (680 nm). Carotenoids also play an important role in photoprotection (Demmig-Adams and Adams 1996). When chlorophyll is activated by the absorption of light it is generally returned to its original state as a result of its participation in photosynthesis. Under conditions of high light flux more light is absorbed than can be used immediately in

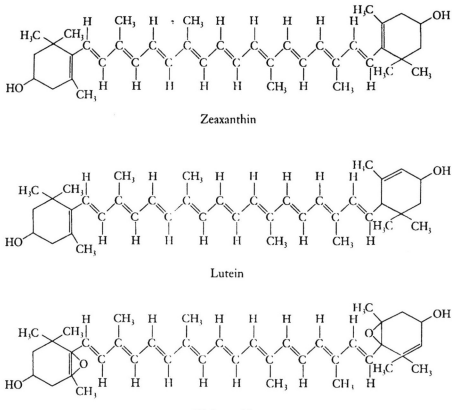

Zeaxanthin

Lutein

Violaxanthin

Figure 1.17   The chemical structure of representative xanthophylls

Table 1.5   The major xanthophylls found in green leaves

Pigment	Structure	Relative amount (%)
Cryptoxanthin	3-hydroxy-β-carotene	4
Lutein	3,3-dihydroxy-α-carotene	40
Zeaxanthin	3,3-dihydroxy-β-carotene	2
Violaxanthin	5,6,5′,6′-diepoxyzeaxanthin	34
Neoxanthin	$C_{40}H_{56}O_4$	19

photosynthesis. Absorption of excess light can destroy chlorophyll in a process of photo-bleaching termed *photo-oxidation*. If the light-excited state of chlorophyll is not quenched either by photochemistry or resonance transfer, the energy is passed on to oxygen to form an excited state of oxygen known as *singlet oxygen*. These extremely reactive singlet oxygen species exist as *free radicals* (molecules with one or several unpaired electrons) such as superoxide (O_2^-), which can oxidize (bleach) organic compounds such as chlorophyll. This transfer of electrons to oxygen is

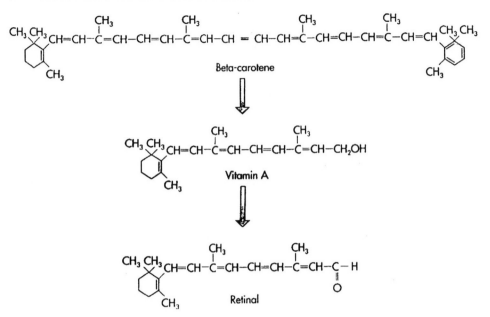

Figure 1.18   The chemical structures of β-carotene, vitamin A and retinal

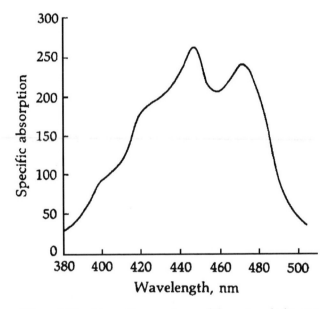

Figure 1.19   Absorption spectrum of β-carotene in hexane

most marked in illuminated chloroplasts because of their high concentration of molecular oxygen. Photo-oxidation can be prevented by either removing superoxide or by preventing its formation. Removal of superoxide is effectively achieved by superoxidase dismutase (SOD), and enzyme found in most cellular compartments, including chloroplasts. SOD can scavange and inactivate superoxide radicals by forming hydrogen peroxide and molecular oxygen:

$$O_2^- + O_2^- + 2H^+ \longrightarrow H_2O_2 + O_2$$

The H_2O_2 in turn is reduced to water by sequential reduction with ascorbate, glutathione and NADPH.

Alternatively, formation of the superoxide radical can be prevented from forming by trapping and dissipating excess excitation energy before it reaches the reaction centre and enters the electron transport chain. An important link has been established between energy dissipation and the presence of the xanthophyll zeaxanthin (Demmig-Adams and Adams 1996). Zeaxanthin is formed from violaxanthin by a process known as the *xanthophyll cycle* (Fig. 1.20). Violaxanthin is a diepoxide, containing two epoxy groups, which are reactive electrophilic groups that are rapidly hydrolysed into hydroxyl groups and eventually form water. Under conditions of excess light, violaxanthin is converted to zeaxanthin through the removal of these oxygens (de-epoxidation). The stepwise removal of the first oxygen generates an intermediate monoepoxide, antheraxanthin. De-epoxidation is also induced by low pH in the lumen, which is a normal consequence of electron transport under high light conditions. In the dark, in the reverse reaction, zeaxanthin is epoxidated to form violaxanthin again. While it has been established that the xanthophyll cycle plays a key role in photoprotection of the chloroplast, the molecular mechanism of the process is not clear. It is known that carotenoids readily lose energy in the form of heat. It has been shown that zeaxanthin with its increased number of conjugated carbon–carbon double bonds can accept the downhill transfer of energy from excited chlorophyll. These observations form the

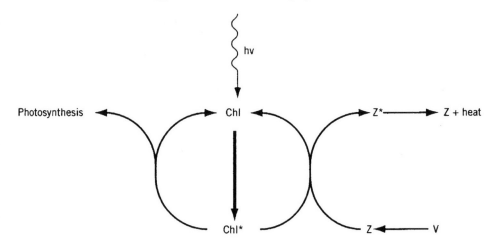

Figure 1.20   The xanthophyll cycle – energy dissipation by zeaxanthin. Under low light, the energy of excited chlorophyll (Chl*) is preferentially allocated to photosynthesis. As irradiance increases, zeaxanthin (Z) is formed by de-epoxidation of violaxanthin (V) and excitation energy is transferred to zeaxanthin to be dissipated as heat. (From Hopkins 1999, reproduced with permission)

basis of the hypothesis that zeaxanthin in the antenna complexes (Chapter 2) receives excess energy directly from chlorophyll, dissipating it harmlessly as heat (Demmig-Adams and Adams 1996). The xanthophyll cycle thus operates as an effective switch, generating zeaxanthin whenever dissipation of excess energy is required but removing zeaxanthin under conditions of low irradiance when more energy is required for photosynthesis (Fig. 1.20).

Mutants that lack carotenoids are susceptible to such radicals, which explains why they are bleach-white when grown in light and soon die. In fact, many so-called chlorophyll mutants (lacking chlorophyll) are actually carotenoid mutants (lacking carotenoids). This phenomenon suggests that carotenoids protect against chlorophyll destruction. Many herbicides kill plants by blocking the synthesis of carotenoids, thereby causing the plant to die from photo-oxidation. Carotenoids, due to their double bonds, quench the excited state of chlorophyll by absorbing the activated oxygen molecule. They provide photoprotection by acting as preferred substrates in photosensitized oxidations and by absorbing excess radiation (especially blue). The carotenoid excited state does not have sufficient energy to form singlet oxygen and so it decays rapidly back to its ground state, dissipating energy as heat. In addition, carotenes are possibly oxidized to xanthophylls in a light-mediated conversion. Carotenoids are also responsible for the colours of petals and fruits. In these tissues carotenoids occur in modified (degenerate) chloroplasts called chromoplasts. Fruit carotenoids are based mainly on α-carotene and generally occur, in these tissues, in association with anthocyanins.

Phycobilins

Phycobilins are a group of accessory pigments found in the red algae (Rhodophyta) and in the prokaryotic blue-green algae (Cyanophyta). Neither of these two algal groups contain chloroplasts; they have an open photosynthetic apparatus in which the thylakoid membranes carry particles, of approximately 40 nm diameter, in regular patterns. These structures are called *phycobilisomes*. Phycobilisomes are large assemblies of many phycobiliprotein subunits, each containing attached bilin prosthetic groups and linker polypeptides.

Phycobilins are linear tetrapyrroles that have the extended polyene system found in chlorophylls, but lack their cyclic structure and central Mg^{2+}. They also differ from chlorophyll in that their tetrapyrrole group is covalently linked to a protein that forms part of the molecule. A pigment that contains a protein as an integral part of the molecule is termed a *chromoprotein*. The protein part is called the *apoprotein*. The tetrapyrrole part of the molecule is a chromophore, which is the part responsible for light absorption. Also, unlike the carotenoids, the phycobilins are water-soluble. Examples are the red phycoerythrin, the blue phycocyanin (Fig. 1.21) and the blue allophycocyanin. Phycobilins are useful in the process of light harvesting for photosynthesis because they absorb light energy effectively in the green (270–650 nm) part of the spectrum where chlorophyll does not absorb light quanta (Fig. 1.22).

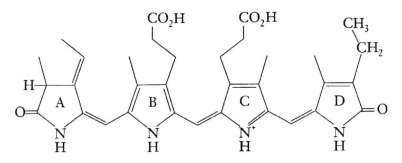

Figure 1.21 The open-chain tetrapyrrole chromophore of phycocyanin

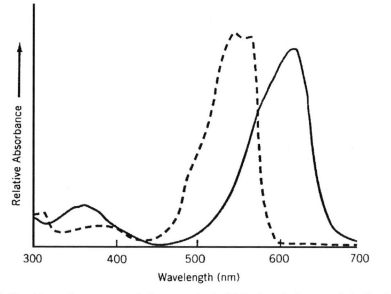

Figure 1.22 Absorption spectra of phycocyanin (solid line) and phycoerythrin (broken line) in dilute buffer

Red and blue-green algae grow in deep water, below the zones of green and brown algae, to which predominantly green quanta penetrate. The red algae, for example, appear almost black because the chlorophyll and phycoerythrin together absorb almost all of the visible spectrum used in photosynthesis. Phycobilins enable algae to occupy ecological niches that would not support organisms relying solely on chlorophyll for the trapping of light.

A fourth phycobilin is phytochrome, which is found in most, if not all, photosynthetic higher plants. Phytochrome is the photoreceptor in higher plants which is responsible for the processes of growth and development which are regulated by light – *photomorphogenesis*. Its chromophore structure (Fig. 1.23) and absorption spectrum (Fig. 1.24) are very similar to that of allophycocyanin. Phytochrome is unique in that it exists in two, photoreversible forms – P660 or P_r

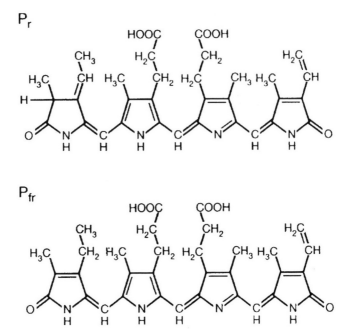

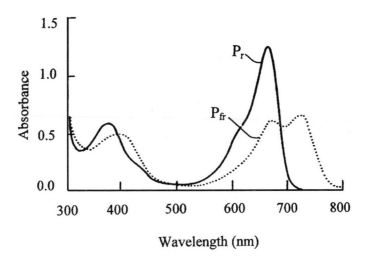

Figure 1.23   Light-induced changes in the phytochrome chromophore

Figure 1.24   The absorption spectrum of phytochrome

and P735 or P_{fr}. The P_r form absorbs maximally at 660 nm and is thus converted into the second form, P_{fr}, which itself absorbs maximally at 735 nm. Absorption of far-red light by P_{fr} converts it back to the red-absorbing P_r form (Fig. 1.23).

P_{fr} is considered to be the physiologically active form of the pigment that is capable of mediating a wide range of morphogenic responses. Phytochrome will be

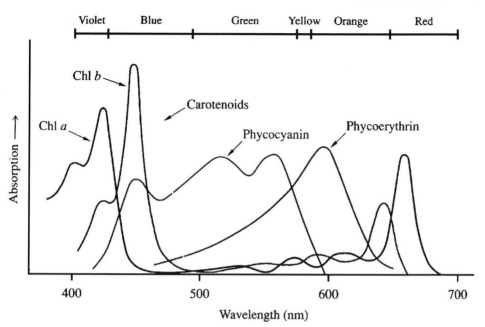

Figure 1.25   The absorption spectra of the major photosynthetic pigments. (From Govindjee and Govindjee 1974)

discussed in more detail in Chapter 5. The absorption spectrum of the major photosynthetic pigments (Fig. 1.25) shows how, collectively, the radiant energy of the whole visible spectrum is absorbed.

Flavonoids

Flavonoids form one of the largest groups of phenolic compounds found in plants (Swain 1965). They are phenyl propane derivatives with a basic C_6—C_3—C_6 (flavone) composition, in which the C_3 link forms a heterocyclic pyrone ring (Fig. 1.26). Complex phenolics that have a three-carbon side-chain are termed phenylpropanoids (phenyl- refers to the ring and propan refers to the three-carbon side-chain) and are derived from tyrosine and phenylalanine which are phenylpropanoid amino acids (Beggs *et al.* 1986).

Flavonoids are classified into different groups based mainly on the degree of oxidation of the three-carbon bridge. Three groups of flavonoids of particular interest in plant physiology are anthocyanins, flavonols and flavones (Fig. 1.27).

Anthocyanins are the most widespread of the pigmented flavonoids and are responsible for most of the red, pink, purple and blue colours of plant parts. They occur as glycosides, usually containing glucose or galactose units attached to the hydroxyl group in the central ring. When the sugars are removed, the remaining part of the molecule (the chromophore) is termed anthocyanidin. Anthocyanidins

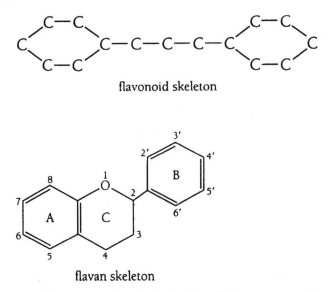

flavonoid skeleton

flavan skeleton

Figure 1.26   Flavonoid structure. Top: basic C_6—C_3—C_6 flavone composition; bottom: the C_3 link of the flavone forms a heterocyclic pyrone ring

are usually named after the particular plant from which they were first extracted; the most widely occurring anthocyanidins and their sources are listed in Table 1.6.

Anthocyanins are distinguished by the pattern of hydroxyl (—OH) or methoxyl (—OCH_3) groups on the B ring, which in fact determines their colour. In general, hydroxylation increases the blueness of the pigment (delphinidin), whereas replacement of the hydroxyl with a methoxyl group gives a redder colour (peonidin). Unlike chlorophylls, anthocyanins are water-soluble pigments and are found mainly in the vacuolar sap. The colour of anthocyanins is related to pH, so that they behave as natural indicator dyes. Most anthocyanins are reddish in acidic solutions and change through violet to blue as the pH is raised. Anthocyanins absorb strongly in the UV-B region (280–320 nm) and in the blue–green (475–560 nm) part of the spectrum (Fig. 1.28) and transmit both blue and red light.

The difference between flavones, flavonols and anthocyanidins is based on the substitution of the C-ring of the flavan skeleton (Fig. 1.29). Flavones and flavonols are ivory-coloured pigments that occur in flower parts and are also widely distributed in leaves of higher plants. They absorb in the UV-B region (280–320 nm) of the spectrum. It is thought that flavones and flavonols thus protect cells from excessive UV radiation, while letting the visible, photosynthetically active, wavelengths pass through to the palisade layer of cells.

In addition there are many other pigments found in biological systems, particularly in fungi, lichens and invertebrates, which have not been recorded in higher plants. The pigments of higher plants have different functions: light-collecting function – absorption and transfer of light energy for photosynthesis (green chlorophyll); signal function – communication between plants and insects or animals (red anthocyanin); light-filtering function – absorption of unwanted, excess

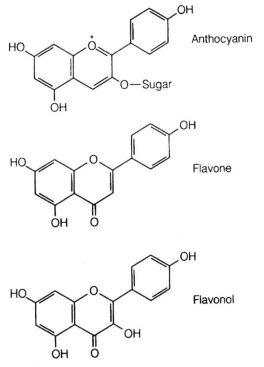

Figure 1.27   Carbon skeletons of the three major groups of flavonoids

Table 1.6   Commonly occurring plant anthocyanidins

Anthocyanidin	Glycoside=	Anthocyanin	Source
Pelargonidin	3,5-diglycoside	Pelargonin (*orange/red*)	Geranium petals
Cyanidin	3,5-diglycoside	Cyanin (*purplish red*)	Red rose petals
Delphinidin	3-glucoside	Violanin (*bluish purple*)	Violet flowers
Malvidin	3-glucoside	Oenin (*blue*)	Blue grapes
Peonidin	3,5-glycoside	Peonin (*rosy red*)	Peony petals

radiation to prevent photo-oxidative effects (yellow carotenoids). For these functions a relatively high concentration of the pigments is required and they are therefore termed mass pigments. In addition to the mass pigments, higher plants also contain photosensor pigments, which, in low concentrations, control photomorphogenesis.

There are three main groups of photosensor pigments which divide broadly according to the wavelengths of radiation that they absorb:

1. *Phytochrome* (> 520 nm) – absorbs most strongly in the red and far-red part of the spectrum and absorbs blue light;

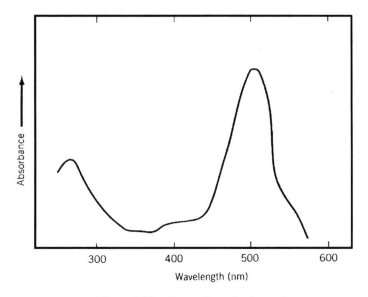

Figure 1.28 Absorption of pelargonin

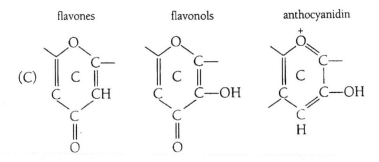

Figure 1.29 Flavones, flavonols and anthocyanidins

2. *Cryptochrome* (340–520 nm) – a group of similar, mainly unidentified pigments that absorb blue light and long-wave ultraviolet wavelengths (UV-A region, about 300–400 nm) (Ahmad and Cashmore 1996; Cashmore 1997);

3. *UV-B photosensor* (280–320 nm) – a number of unidentified compounds that absorb ultraviolet radiation in the 280–320 nm region.

Summary

Newton (1642–1727) first demonstrated that white light is made up of a spectrum of different colours ranging from violet to red. He proposed the 'corpuscular theory' – light is made up of tiny particles which travel in straight lines from their source. At that time this theory satisfactorily explained the laws of reflection and refraction. The wave theory of light, published by Young in 1817, replaced Newton's corpuscular theory. About one hundred years later it was established that light has a dual nature – it exists as both particles and waves at the same time.

The term 'light' describes the narrow band of the visible range ($\lambda = 400–700$ nm) within the larger electromagnetic radiation which ranges from gamma rays ($\lambda < 1$ nm) to radiowaves ($\lambda = 10^3$ m). Only the visible range of light is absorbed and used in photobiological processes by higher plants; some wavelengths in the ultraviolet ($\lambda < 400$ nm) and in the infrared ($\lambda > 700$ nm) may have significant biological effects.

Light must be absorbed to be effective in photobioligical processes. When a pigment molecule absorbs a photon of light of the correct wavelength, an electron within the pigment molecule is raised to a higher energy level. This energy is absorbed by photoreceptors (pigments) which process both the energy and information contained in the various wavelengths of the absorbed light. The pigments of higher plants include mass pigments which are required in relatively high concentrations and function in light-collecting (chlorophyll), light-signalling (anthocyanins) and light-filtering (carotenoids). In addition, higher plants contain photosensor pigments, which, in low concentrations, control photomorphogenesis. The three main groups of photosensor pigments are phytochrome (520 nm), cryptochrome (340–520 nm) and UV-B (280–320 nm).

References

Ahmad, M. and Cashmore, A.R. (1996). Seeing blue: the discovery of cryptochrome. *Plant Molecular Biology* **30**: 851–861.

Baskin, T.I. and Iino, M. (1987). An action spectrum in the blue and ultraviolet for phototropism in alfalfa. *Photochemistry and Photobiology* **46**: 127–136.

Beggs, C.J., Wellman, E. and Griesebach, H. (1986). Photocontrol of flavonoid biosynthesis. In: *Photomorphogenesis in Plants*, R.E. Kendrick and G.H.M. Kronenberg (eds), pp. 467–499. Martinus Nijhoff, Boston.

Cashmore, A.R. (1997). The cryptochrome family of photoreceptors. *Plant, Cell and Environment* **20**: 764–767.

Demmig-Adams, B. and Adams, W.W. (1996). The role of xanthophyll cycle carotenoids in the protection of photosynthesis. *Trends in Plant Science* **1**: 21–26.

Govindjee and Govindjee, R. (1974). The absorption of light in photosynthesis. *Scientific American* **231**: 68–82.

Hall, D.O. and Rao, K.K. (1994). Importance and role of photosynthesis, *Photosynthesis*, 5th edn, pp. 1–21. Cambridge University Press, Cambridge.

Hopkins, W.G. (1999). Bioenergetics and the light-dependent reactions of photosynthesis, *Introduction to Plant Physiology*, pp. 163–187. Wiley, New York.

Horton, H.R., Moran, L.A., Ochs, R.S., Rawn, J.D. and Scrimgeour, K.G. (1996). Proteins: three-dimensional structure and function, *Principles of Biochemistry*, pp. 79–119. Prentice-Hall, Upper Saddle River, NJ.

Lehninger, A.L., Nelson, D.L. and Cox, M.M. (1993). Nucleotides and nucleic acids, *Principles of Biochemistry*, 2nd edn, pp. 324–357. Worth, New York.

Lodish, H., Baltomore, D., Berk, A., Zipursky, S.L., Matsudaira, P. and Darnell, J. (1995). Photosynthesis, *Molecular Cell Biology*, 3rd edn, pp. 779–808. Scientific American Books, New York.

Salisbury, F.B. and Ross, C.W. (1992). Photosynthesis: chloroplasts and light, *Plant Physiology*, pp. 207–224. Wadsworth, Belmont, CA.

Sauer, K. (1975). Primary events and the trapping of energy. In: *Bioenergetics of Photosynthesis*, Govindjee (ed.), pp. 116–181. Academic Press, New York.

Swain, T. (1965). Nature and properties of flavonoids. In: *Chemistry and Biochemistry of Plant Pigments*, T.W. Goodwin (ed.). Academic Press, London and New York.

Zscheile, F. and Comar, C. (1951). Influence of preparative procedure on the purity of chlorophyll components as shown by absorption spectra. *Bot. Gaz.* **102**: 463.

2
Photosynthesis

Introduction

Plants are autotrophs: they need only carbon dioxide from the air and water and minerals from the soil, as nutrients. More specifically plants are photoautotrophs – they use sunlight as a source of energy to make food molecules from the carbon dioxide and water, in the process of photosynthesis. This is the most important chemical process on earth, providing the major supply of food (energy) for virtually all living organisms – animals, plants, protists, fungi and bacteria.

A major feature of most photosynthetic organisms is that they release oxygen from water. Life, as we know it today, is therefore sustained not only by the food supply from photosynthesis but also by the oxygen that it releases. During the evolution of life on earth fermentation probably developed first. This was most likely followed by the evolution of photosynthesis which released large amounts of oxygen into the previously CO_2-filled atmosphere. The oxygen allowed the evolution of aerobic respiration – a source of energy in the form of ATP. Once oxygen-dependent cellular respiration was established a wide range of organisms evolved – multicellular (higher) plants and animals.

Ancient Greeks reasoned that plants obtained all their nutrition directly from the soil: plant and animal debris in the soil was readily converted to food that plant roots absorbed. This concept went unquestioned for centuries. In the early 1600s a Dutch physician named Jan-Baptista van Helmont (1579–1644) carried out a simple but very important experiment. He grew a weighed willow seedling in a known weight of soil and provided it with rainwater only, over a growing period of five years. At the end of this period the loss of weight of the soil was negligible while the weight of the plant had increased from 2 kg to 75 kg. Helmont attributed the weight gain totally to the water. About fifty years later an English botanist, James Woodward, stated that plants required more than water. He grew sprigs of mint in water from various sources and concluded that 'Vegetables are not formed of water but of a certain peculiar terrestrial matter. It has been shown that there is considerable quantity of this matter contained in rain, spring and river water' (Loomis 1960). Since the chemistry of carbon dioxide was, as yet, unknown and the

Photobiology of Higher Plants. By M. S. Mc Donald.
© 2003 John Wiley & Sons, Ltd: ISBN 0 470 85522 3; ISBN 0 470 85523 1 (PB)

nature of light only beginning to be worked out, it is understandable that their effects on plant growth received little attention. Nevertheless, Stephen Hales (1677–1761), often referred to as the 'father of plant physiology', wrote in 1727 that plants probably drew through their leaves some part of their nourishment from the air and suggested that sunlight absorbed by leaves contributes to plant growth. In 1772, Joseph Priestley, the discoverer of oxygen, noted that air, within a closed space, 'contaminated' by the burning of a candle could not support the life of a mouse. He showed, however, that a sprig of mint grown in this contaminated air could 'purify' it, enabling it to support a living mouse again. He also noted that sprigs of mint flourished in this so-called 'contaminated' air. While Priestley understood the difference in gas exchange between plants and animals he did not recognize the role of either carbon dioxide or light in photosynthesis. In 1779, Jan Ingenhousz, a Dutch physician and contemporary of Priestley's, reported that plants purified the air only in the presence of light. He also noted that only the green parts of plants produced the purifying agent (O_2) while non-green parts, in a similar manner to animals, contaminated (with CO_2) the air. Ingenhousz thus established the importance of light and gave a hint that green was also important. In 1782, a Swiss scientist, Jean Senebier, claimed that 'fixed air' (CO_2) produced by animals and by plants in darkness, stimulated production of 'purified' air (O_2) by plants in light. Antoine Lavoisier (1743–1794) showed, in the late 1700s, that the two gases involved were CO_2 and O_2. Ingenhousz subsequently suggested that plants do not simply exchange 'good air' for 'bad air' but rather that they absorb the carbon from the carbon dioxide and release the oxygen into the atmosphere. This concept was extended in 1894 by the Swiss botanist Nicholas de Saussure, who noted that approximately equal volumes of CO_2 and O_2 are exchanged in light. He also proposed that water was involved and explained plant growth and gas exchange by the equation

$$\text{carbon dioxide} + \text{water} + \text{light} \longrightarrow \text{organic material} + \text{oxygen}$$

In 1883, T.W. Englemann studied both the light requirements and the biochemistry of photosynthesis in *Spirogyra*, a green alga that has a long, strap-like, spiral chloroplast in each cell. He placed the alga in a drop of water-containing, oxygen-requiring bacteria and used a prism to illuminate the cells with different colours of light. He observed, under the microscope, that the bacteria clustered around parts of the chloroplast illuminated by red and blue light. He concluded that red and blue light are most effective in producing oxygen during photosynthesis. In 1905, Frederick F. Blackman (1866–1947), an English plant physiologist, demonstrated that photosynthesis is not only a photochemical reaction but also a biochemical reaction. We now know that the photochemical or light reaction is very rapid and requires light energy. In contrast, the biochemical or 'dark' reactions (carbon dioxide fixation) proceed at a relatively slow rate. These so-called dark reactions can occur in light or darkness and can be more correctly termed *carbon dioxide fixation* reactions. Ingenhousz's suggestion that the oxygen released during photosynthesis came from carbon dioxide was still accepted up to the early 1900s. In the 1920s C.B. van Niel, at Stanford University, studied the source of the oxygen released during photosynthesis. He worked with photosynthetic bacteria that use H_2S as a source of electrons and deposit sulphur as a by-product. Photosynthesis in these bacteria occurs as follows:

$$CO_2 + 2H_2S + light \longrightarrow CH_2O + H_2O + 2S$$

Van Niel (1941) claimed that the oxygen released during photosynthesis comes from water, not from carbon dioxide. His conclusions were based on analogies between the roles of H_2S and H_2O and of S and O_2:

Sulphur bacteria:

$$CO_2 + 2H_2S + light \longrightarrow CH_2O + H_2O + 2S$$

Green plants:

$$CO_2 + H_2O + light \longrightarrow CH_2O + H_2O + O_2$$

General equation:

$$CO_2 + 2H_2X + light \longrightarrow CH_2O + H_2O + 2X$$

Van Niel's hypothesis was tested by Samuel Ruben and co-workers (1941). They exposed the green alga *Chlorella* to H_2O labelled with $^{18}O_2$, a non-radioactive isotope of oxygen. Their results showed that the oxygen came from the decomposition of water and not from CO_2:

$$CO_2 + 2H_2{}^{18}O + light \longrightarrow CH_2O + H_2O + {}^{18}O_2$$

These results confirmed van Niel's hypothesis. Further confirmation came from the findings of Robin Hill (1937) who discovered that isolated chloroplasts in the presence of light, water and a suitable electron acceptor (Hill used potassium ferricyanide as an artificial electron acceptor) could evolve oxygen in the absence of carbon dioxide. This light-driven splitting of water (*Hill reaction*) is evidence that oxygen evolution resulting from photochemical reactions is not linked directly to CO_2 reduction in photosynthesis.

As early as 1905, studies by Blackman suggested that photosynthesis is a two-step process, consisting of a fast photochemical or 'light' reaction and a slower non-photochemical or 'dark' reaction. This result led to studies of the photochemistry of photosynthesis (light reactions) and of carbon fixation (dark reactions). It had long been thought that only the light reactions occurred in the chloroplast and that CO_2 reduction occurred in the cell cytoplasm. In 1951, Arnon reported that a natural plant constituent nicotinamide adenine dinucleotide phosphate (NADP) could act as a Hill reagent. In the presence of light isolated chloroplasts could reduce NADP to $NADPH_2$ by accepting electrons from water and evolve O_2 simultaneously. In 1954, Arnon and his co-workers demonstrated photophosphorylation (light-induced synthesis of ATP from ADP and P_i), and the simultaneous assimilation of CO_2 and evolution of O_2 by isolated spinach chloroplasts. Therefore the enzymes involved in CO_2 reduction and the assimilatory power (reducing power) $NADPH_2$

needed to accomplish this assimilation must be present, and perhaps produced, within the chloroplast. Arnon and co-workers distinguished the 'light' and 'dark' phases of photosynthesis by separating chloroplasts into granal and stromal fractions. They demonstrated ATP and $NADPH_2$ formation in the grana in light and enzyme-controlled reduction of CO_2 in the stroma in the dark.

Using monochromatic light of different wavelengths, Robert Emerson (1957) and his co-workers measured the quantum yield (number of oxygen molecules released per light quantum absorbed) of photosynthesis over the range of the visible spectrum. They observed a decrease in quantum yield at wavelengths greater than 680 nm; because of its location in the red part of the spectrum the observed decrease is referred to as the *red drop effect*. They also showed that the efficiency could be restored by the simultaneous application of a shorter wavelength. The effect of the two superimposed beams was found to exceed the sum effect of the two beams applied separately; this is referred to as the *enhancement effect*. To explain the enhancement effect Emerson and Rabinowitch (1960) proposed that two sets of light reactions occurred in photosynthesis. In the same year Hill and Bendall proposed the Z-scheme (zigzag) for photosynthesis in which two distinct photosystems interact in photosynthetic electron transport and photophosphorylation.

With the method for the production of ATP and reduced NADP by photochemical reactions established, the pathway of CO_2 fixation and reduction to carbohydrates remained to be elucidated. The first theory on carbon reduction in photosynthesis is attributed to Justus Liebig, who suggested that plant acids are intermediate compounds in the reduction process of CO_2 to carbohydrate. He developed his theory purely from observation (for example, ripening fruit is first sour and later becomes sweet) but did not provide any supporting experimental evidence. Work using radioisotopes to trace the assimilation on CO_2 gained momentum after the Second World War. Calvin and Benson (1948) studied the incorporation of $^{14}CO_2$ into the photosynthesizing green alga *Chlorella*. From this study they mapped the path of carbon in photosynthesis and identified the intermediates in the assimilation of CO_2. The first stable organic intermediate produced in the Calvin–Benson cycle is the 3-carbon phosphoglyceric acid; plants in which this cycle uses CO_2 directly from the air are called C3 plants. Melvin Calvin received the Nobel Prize in Chemistry for his work, in 1961.

In 1965, Kortschak *et al.* reported that when $^{14}CO_2$ was incorporated into sugar cane leaves, the first stable products of photosynthesis are the 4-carbon organic acids malate and aspartate. These results were confirmed and extended by Hatch and Slack (1966) in Australia. They exposed photosynthesizing sugarcane leaves to $^{14}CO_2$ and found most of the radioactivity in the 4-carbon dicarboxylic acids malic, aspartic and oxaloacetic. The C4 pathway of CO_2 fixation was thus established and is referred to as the Kortschak–Hatch–Slack pathway; plants in which this pathway occurs are termed C4 plants.

It had been known for many years that the energy-transducing membrane systems in bacteria, chloroplasts and mitochrondria are able to link electron transport with ATP synthesis. The mechanism, however, was not understood until

Peter Mitchell proposed his chemiosmotic hypothesis in 1961. Although it was not readily accepted in the beginning, Mitchell's hypothesis, which includes the concept of both proton and electron carriers in the membranes, is now firmly supported by experimental results. Mitchell received the Nobel Prize in Chemistry, in 1978, for his pioneering work.

Biological Oxidation–Reduction

Redox reactions

Photosynthesis is an electrochemical process in which electrons are transferred from one compound to another in a series of oxidation/reduction reactions – *redox reactions*. Oxidation is the loss of electrons from an atom or molecule; reduction is the gain of electrons. Since electrons are neither created nor destroyed during a chemical reaction then if one reactant is oxidized then the other is reduced. Oxidation and reduction occur simultaneously. For example,

$$Fe^{2+} + Cu^{2+} \longrightarrow Fe^{3+} + Cu^{+}$$

Thus, Fe^{2+} (ferrous ion) loses an electron and becomes oxidized to Fe^{3+} (ferric ion); Cu^{2+} (cupric ion) gains an electron and is reduced to Cu^{+} (cuprous ion). In this oxidation–reduction reaction, the electron-donor molecule, Fe^{2+}, is the reducing agent (reductant) while the electron-accepting molecule Cu^{2+}, is the oxidizing agent (oxidant). Although oxidation and reduction occur together, it is convenient to describe the electron transfers as occurring in two half-reactions:

$$1. \ Fe^{2+} \longrightarrow Fe^{3+} + e^{-}$$
$$2. \ Cu^{2+} + e^{-} \longrightarrow Cu^{+}$$

The iron cations in the above reaction (Fe^{2+}, the electron donor, and Fe^{3+}, the electron acceptor) behave as a conjugate reduction–oxidation pair, designated a *redox couple*. While oxidation/reduction reactions may be shown as a direct transfer of electrons, as in the half-reactions shown above, electron transfer may also be shown in the form of protons:

$$AH_2 \longrightarrow A + 2H^{+} + 2e^{-}$$

Here AH_2 is the hydrogen (or electron) donor. AH_2 and A constitute a redox couple which can reduce another compound B by the transfer of hydrogen atoms:

$$AH_2 + B \longrightarrow A + BH_2$$

The neutral term 'reducing equivalent' is commonly used to designate a single electron equivalent participating in an oxidation–reduction reaction, irrespective of whether this equivalent be in the form of an electron or a proton.

Redox potentials

In an electrical circuit, electrons supplied by a battery will flow through an electric wire connecting two chemical species that differ in their affinity for electrons. Because of this difference, electrons flow through the circuit driven by a force proportional to the difference in electron affinity – *the electromotive force*. This electromotive force can do work if an energy transducer (a motor) is placed in the circuit. In an analogous biological 'circuit' the source of electrons is a photoexcited chlorophyll *a* pigment molecule. As chlorophyll is reoxidized the displaced electrons flow through a series of carriers which have increasing affinity for electrons. The energy released due to the electron transfer is used by membrane-bound transducer molecules which couple the electron flow to the production of a transmembrane pH difference. The resultant proton gradient (Prince 1985) has a potential energy sometimes called *proton-motive force*. In the thylakoid membranes of chloroplasts, transducer protein molecules use this proton-motive force to synthesize ATP from ADP and P_i.

The tendency of the redox couples in an electron transport chain to accept electrons or to donate electrons to another redox couple is termed *redox potential*. Redox potential, which is thus a measure of the relative affinity of an electron-acceptor for electrons, is an electrochemical concept and is defined against an arbitrary standard. Electrochemists have chosen the hydrogen half-cell as a standard of reference for redox potential. In the half-reaction

$$2H^+ + 2e^- \longrightarrow H_2$$

hydrogen is bubbled over a platinum electrode immersed in $1\,M\ H^+$, pH 0. The electrode at which this half-reaction occurs is arbitrarily assigned a standard reduction potential (E_0) of 0.00 V. Redox potentials, measured in volts, can be determined by connecting a second half-cell, in which the oxidized and reduced species are present at standard concentration (each solute at 1 M, each gas at 1 atm), through a voltmeter (Fig. 2.1). Electrons will flow through the external circuit from the half-cell of lower standard reduction potential to the half-cell of higher standard reduction potential. If H_2 loses electrons more readily than the test electron-donor then electrons will flow from the reference to the test half-cell, thereby oxidizing the H_2 in the reference half-cell and reducing the test electron-acceptor. By convention, the half-cell with the greater affinity for electrons is assigned a positive value for E_0. Thus, a negative reduction potential means that a substance has a lower affinity for electrons than H_2.

The reduction potential of a half-cell depends not only on the chemical species present but also on their concentrations. Over a hundred years ago, Walther Nernst derived an equation (Wood 1985) that relates standard reduction potential (E_0) to reduction potential (E) for any concentration of oxidized and reduced species in the cell:

$$E = E_0 + RT/nF = \ln \frac{[\text{electron acceptor}]}{[\text{electron donor}]}$$

where $R = $ gas   constant $1.987 \, cal \, deg^{-1} mol^{-1}$ (or $8.28 \, J \, deg^{-1} mol^{-1}$);
$T = °C + 273 °K$; $F = $ Faraday's constant $23.062 \, kcal \, mol^{-1} V^{-1}$ $(96.5 \, kJ \, V^{-1} mol^{-1})$;
$n = $ number of electrons; $\ln = $ natural logarithm. This expression reduces to

$$E = E_0 + 2.303 \, RT/nF \, \log \frac{[electron \, acceptor]}{[electron \, donor]}$$

(2.303 is the conversion factor from natural to common logarithms).

At $25 °C$, the term $2.303 \, RT/nF$ has a value of 0.059 volts for a one-electron ($n = 1$) transfer and since protons are nearly always involved in biochemical reactions, the redox potential is pH-dependent. In other words, the redox potential at $25 °C$ becomes more negative at the rate of $0.059 \, V$ per pH unit. Thus the hydrogen electrode at pH 7 is $7 \times (-0.059 \, V) = -0.413 \, V$ more negative than the standard hydrogen electrode at pH 0. For all pH-dependent redox systems (AH→A $+ H^+ + e^-$) the following applies:

$$E_0' \, (pH \, 7) = E_0 - 0.413 \, V$$

In biological systems a pH of zero is seldom encountered. Biochemists define the standard state for oxidation–reductions as pH 7 $(10^{-7} H^+)$ and express reduction potential as E_0', the standard reduction potential at pH 7. E_0' values for a number of biologically important redox couples are listed in Table 2.1.

The apparatus for measuring the difference in E_0' values for two redox couples is shown in Fig. 2.1. Since at the neutral pH of 7 and a ratio of the reduced form to the oxidized form equal to 1, a redox couple is 50% reduced, the redox potential is known as the *midpoint potential* and is designated E_m (the symbol E_m is always used in place of E_0' if the system is reduced to 50%). The fact that most chromophore redox systems, e.g. haem-proteins, ferredoxins and NADPH, show characteristic changes in their absorption spectra when the reduction state alters allows redox systems to be identified spectrophotometrically, even *in vivo*. By titration with a suitable redox substance of known potential the midpoint potential E_m can be easily measured (Dutton 1978).

Table 2.1   Standard reduction potentials ($25 °C$, pH 7) for some half-reactions important in photosynthesis

	Half-reactions		E_0' (V)
Ferredoxin $Fe^{3+} + e^-$	$\longrightarrow$	Fe^{2+}	-0.43
$2H^+ + 2e^-$	$\longrightarrow$	H_2O (pH 7)	-0.41
$NADP^+ + H^+ + 2e^-$	$\longrightarrow$	NADPH	-0.32
$PQ + 2H^+ + 2e^-$	$\longrightarrow$	PQH_2	$+0.05$
Cyt b_6 $Fe^{3+} + e^-$	$\longrightarrow$	Fe^{2+}	$+0.08$
Cyt f $Fe^{3+} + e^-$	$\longrightarrow$	Fe^{2+}	$+0.36$
$\frac{1}{2}O_2 + 2H^+ + 2e^-$	$\longrightarrow$	H_2O	$+0.82$

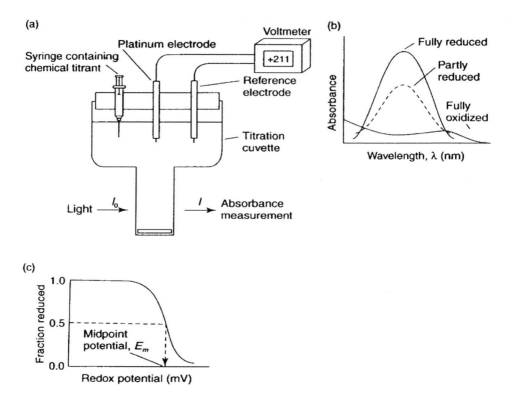

Figure 2.1 (a) Experimental set-up for redox titrations, including platinum and reference electrodes, a voltmeter and a light beam to detect absorbance. (b) Spectral absorbance band of a redox-active species. (c) Redox titration of a molecule. The midpoint potential is the potential where the species being investigated is half oxidized and half reduced. (From Blankenship 2002, reproduced with permission)

Again by convention, the more negative potentials are assigned to couples that have a tendency to donate electrons. The direction of electron transfer between redox couples can be predicted by comparing their midpoint potentials. In general, electron transfer occurs from couples with the more negative (less positive) redox potential to those with the less negative (more positive) redox potential. Thus, for the hydrogen ion itself, the midpoint potential for the $H_2/2H^+$ couple of -0.413 V indicates a high tendency to donate electrons. On the other hand, the O_2/H_2O redox couple with an $E'_0 = +0.82$ V, has a high tendency to accept electrons and thus to oxidize other substances. However, since none of the common biological oxidants have a higher E'_0 than O_2/H_2O (Table 2.1), there is very little tendency for water to become oxidized to O_2. In photosynthesis, however, H_2O is split by the absorption of photons of light.

Redox potentials and free energy change

The transfer of electrons or protons plays a central role in cell metabolism. In chloroplasts, chains of redox couples occur attached to the thylakoid membranes. In photosynthesis, carbon is converted from its maximally oxidized state of CO_2 to a strongly reduced state as carbohydrate, $(CH_2O)_n$. In the chloroplast, nicotine adenine dinucleotide phosphate, NADP, is the most important transport molecule. If two redox systems of different redox potentials (E) are coupled together, then the more negative donates electrons to the more positive system until dynamic equilibrium is established. The tendency to donate electrons is proportional to the difference in reduction potential, ΔE, between the two systems. By comparing the potential values of redox couples it is possible to predict the direction of electron transfer between them. In any oxidation–reduction reaction, the total voltage change (change in electric potential) ΔE, is the sum of the voltage changes (reduction potentials) of the component oxidation and reduction steps. Since redox potential describes the capacity to perform electrochemical work for each electron transferred during a chemical reaction and since all forms of energy are interconvertible, then ΔE can be expressed as the change in the Gibbs (after the American chemist Josiah W. Gibbs, 1839–1903) free energy, ΔG. Since biological systems usually exist at constant temperature and pressure, it is possible to predict the direction of a chemical reaction by calculating the potential energy of the system – the free energy or G. Gibbs showed that, under standard conditions, as generally found in biological systems, 'all systems change in such a way that free energy is minimized'. Redox potentials thus permit the calculation of the Gibbs chemical free energy for oxidation–reduction reactions.

This free energy change is given by:

$$\Delta G_0' = -nF\Delta E_0'$$
$$= -nF[E_0' \text{ acceptor} - E_0' \text{ donor}] \text{ volts}$$
$$= -n(96\,500)(E_0' \text{ acceptor} - E_0' \text{ donor}) \text{ volts}$$

The Faraday constant (F) is the charge on one mole (6×10^{23}) of electrons and has a value of 96 500 coulombs (96 500 J/volt) and n is the number of electrons transferred; the factor 4.184 may be used to convert joules into calories, when G is expressed in cal mol^{-1}.

Reduction potential is used to describe the change in electric energy that occurs when an atom or molecule gains an electron (reduction). The voltage change in a complete oxidation–reduction reaction, in which one molecule is reduced and another oxidized, is simply the sum of the oxidation and reduction potentials for the two half-reactions. Consider the reaction in which acetaldehyde is reduced, under standard conditions, by NADPH.

$$\text{Acetaldehyde} + \text{NADPH} + H^+ \longrightarrow \text{Ethanol} + \text{NADP}^+$$

The half-reactions are:

$$\text{Acetaldehyde} + 2H^+ + 2e^- \longrightarrow \text{Ethanol} \qquad E_0' = -0.197\,\text{V}$$
$$NADP^+ + 2H^+ + 2e^- \longrightarrow NADPH + H^+ \quad E_0' = -0.324\,\text{V}$$
$$E_0' = -0.197\,\text{V} - (-0.324\,\text{V})$$
$$= +0.127\,\text{V}$$
$$\Delta G_0' = -nF\Delta E_0'$$
$$= -2(96\,500\ \text{J V}^{-1}\,\text{mol}^{-1})\,(0.127\,\text{V})'$$
$$= -24.51\,\text{kJ mol}^{-1}$$
$$= -5.86\,\text{kcal mol}^{-1}$$

Note that in this example, a positive E_0' ($+0.127\,\text{V}$) gives a negative $\Delta G_0'$ ($-5.86\,\text{kcal mol}^{-1}$). Thus, under standard conditions, this reaction will occur spontaneously as written, from left to right. If, however, the $\Delta G_0'$ is calculated for the reverse reaction (the oxidation of NADPH by acetaldehyde), then

$$\Delta E_0' = E_0'\ \text{acceptor} - E_0'\ \text{donor}$$
$$= -0.324\,\text{V} - (-0.197\,\text{V})$$
$$= -0.127\,\text{V}$$
$$\Delta G_0' = +5.86\,\text{kcal mol}^{-1}$$

Such a reaction is not favoured in the direction written and would proceed only with the input of energy. Many chemical reactions in the cell have a positive ΔG; they are energetically unfavourable and will not proceed spontaneously. One example is the transfer of electrons to NADP in photosynthesis, an endergonic reaction which is driven by the absorption of light energy.

Overview of Photosynthesis

Photosynthesis occurs in chloroplasts

In higher plants photosynthesis occurs in specialized organelles called chloroplasts (Fig. 2.2). Chloroplasts are generally round to oval in shape; they are somewhat larger than mitochondria and have a diameter of 4–6 μm. They occur at about fifty per typical plant cell and can be easily seen under the light microscope; their finer structure, however, requires the electron microscope (Fig. 2.2).

Chloroplasts have three membranes (Staehlin and Arntzen 1986). They are bounded by an envelope of two membranes with an intervening space. Neither of the membranes contains chlorophyll and so they do not participate directly in photosynthesis. This double membrane is highly permeable to CO_2 and differentially permeable to other metabolites. The outer membrane contains a non-specific pore protein that freely permits water and a variety of ions and low-molecular weight metabolites (of sizes up to approximately 10 kDa) to pass into the intermembrane space, the aqueous compartment between the membranes. Among the polypeptides localized in the outer membrane are enzymes involved in

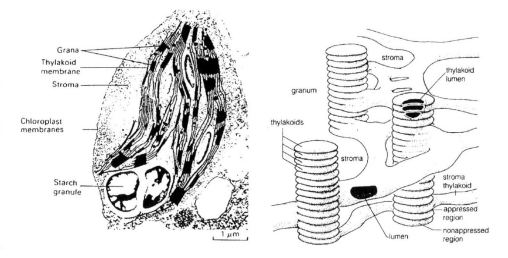

Figure 2.2   The chloroplast: electron micrograph of spinach-leaf chloroplast (left); schematic drawing of thylakoid membranes (right). (From Salisbury and Ross 1992)

galactolipid metabolism. The inner membrane presents a permeability barrier and contains carrier compounds that regulate the movement of metabolites, including small uncharged molecules such as O_2 and NH_3, into and out of the chloroplast. The inner membrane also contains enzymes involved in the assembly of thylakoid membrane lipids. The inner membrane encloses an aqueous matrix called the *stroma*. This gelatinous matrix contains ribosomes, DNA and the enzymes that convert CO_2 to carbohydrates. Chloroplasts, however, are not autonomous – they also contain proteins encoded by nuclear DNA.

The most prominent internal feature of chloroplasts is the third membrane which forms an extensive network of internal membranes called *thylakoids*, suspended in the stroma. The thylakoid membrane carries the photosynthetic pigments and is the site of the light reactions (Anderson 1986). Thylakoids are flattened sacs: the membrane may be considered to be a single, continuous, highly folded structure that encloses a single, continuous space, the thylakoid *lumen* (Fig. 2.2). Thus chloroplasts have three, different membranes – outer, inner and thylakoid membranes and three separate spaces – intermembrane, stroma and thylakoid lumen. In places the thylakoid membranes stack into piles of disc-like structures called *grana* (singular: granum). The stacked or appressed regions are known as *grana lamellae* while the membranes that do not stack but appear to interconnect the grana are called *stroma lamellae*.

Overall Reaction for Photosynthesis

Phytosynthesis is a redox (oxidation–reduction) reaction. The overall reaction for photosynthesis may be written as

$$6CO_2 + 12H_2O \longrightarrow C_6H_{12}O_6 + 6H_2O + 6O_2$$

This equation can be interpreted as a simple, redox reaction – the reduction of CO_2 to carbohydrate where H_2O is the reductant and CO_2 the oxidant. As indicated in the equation, when water molecules are split apart yielding O_2, they are actually oxidized, that is, they lose electrons along with the hydrogen ions (H^+). At the same time CO_2 is reduced to carbohydrate as the electrons and H^+ ions are added to it. In the redox reactions of photosynthesis electrons are caused to flow in an energetically uphill direction. As H_2O is oxidized and CO_2 reduced during photosynthesis the light energy captured by chlorophyll is converted to chemical energy and stored in the sugar molecules that are synthesized.

Photosynthesis Consists of Both Light and 'Dark' Reactions

The net reaction for photosynthesis is:

$$CO_2 + H_2O \xrightarrow{\text{light}} (CH_2O) + O_2$$

where (CH_2O) represents carbohydrate. Photosynthesis occurs in two separate, interdependent stages, each stage taking place in a particular location in the chloroplast. The light reactions occur in the stacked thylakoid membranes to produce ATP and NADPH which, in the stroma, are used to convert CO_2 to carbohydrate, in a series of reactions called the Calvin cycle.

Light reactions

The initial step in photosynthesis is the absorption of light by chlorophyll molecules attached to proteins in the thylakoid membranes of the chloroplast. The light energy absorbed is used to remove electrons and protons from water (an unwilling electron donor) to produce oxygen (Govindjee and Coleman 1990):

$$H_2O \xrightarrow{\text{light}} \tfrac{1}{2}O_2 + 2H^+ + 2e^-$$

The electrons released are transferred to a primary electron-acceptor and are subsequently passed through a non-cyclic chain of electron carriers until they reach the final acceptor, $NAPD^+$, reducing it to NADPH (Arnon 1991) (Fig. 2.3). The transport of electrons is coupled to the movement of protons from the stroma across the membrane into the thylakoid lumen (Fig. 2.18). This establishes a proton-motive force across the membrane that is used to synthesize ATP. Protons move down their own concentration gradient from the thylakoid lumen back into the stroma via a group of transport proteins. This results in the formation of ATP (Fig. 2.18).

The overall equation for the light reactions may be written as:

$$H_2O + ADP + P_i + NADP \xrightarrow{\text{light}} ATP + NADPH + H^+ + O_2$$

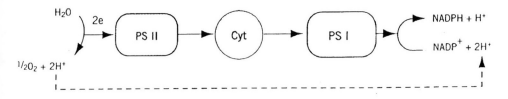

Figure 2.3 Summary of the photosynthetic electron transport system in chloroplasts

Thus, in the photochemical process, protons derived from water are used to synthesize ATP while a hydrogen atom, released from water due to photolysis, reduces NADP to $NADPH_2$.

'Dark' reactions

The so-called 'dark' reactions, which are not confined to darkness and can take place equally in the light, use the ATP and $NADPH_2$ produced in the light reactions to convert CO_2 into carbohydrates. The reaction for this fixation of CO_2 may be written as

$$CO_2 + ATP + NADPH_2 \longrightarrow (CH_2O) + ADP + P_i + NADP$$

Six molecules of CO_2 are converted into two molecules of triose phosphate (glyceraldehyde-3-phosphate) in a series of reactions called the Calvin cycle, which occur in the stroma of the chloroplast. Phosphoglyceraldehyde transported to the cytosol is converted to fructose-1,6-diphosphate and ultimately to the disaccharide sucrose. The reactions of CO_2 fixation are referred to as 'dark' reactions; they are not, however, confined to darkness and normally occur simultaneously with the light-dependent reactions.

Close-up on Light-dependent Reactions

Photosystems

The discovery by Emerson that the drop in the photosynthetic rate due to long wavelengths ($\lambda > 680$ nm) could be restored and enhanced by the application of shorter wavelengths ($\lambda < 680$ nm) led to the concept that photosynthesis requires the interaction of two sets of light reactions (Duysens *et al.* 1961). In fact, photosynthesis depends on the interaction of two separate photosystems. These are complex molecular structures designated photosystem I (PSI) and photosystem II (PSII). Both systems contain chlorophyll *a* and other pigments associated with several different proteins. The two photosystems act in series, linked by a third multiprotein complex called the cytochrome complex (Fig. 2.3) which acts as an electron carrier in a non-photochemical reaction.

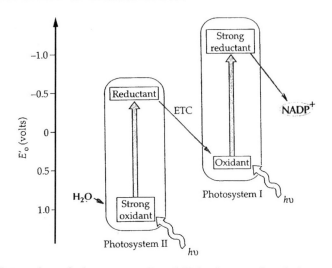

Figure 2.4 Interaction of photosystems I and II in the transfer of electrons from water to $NADP^+$. In the light, PSII produces a strong oxidant capable of oxidizing water and a reductant; illuminated PSI generates a strong reductant which can reduce $NADP^+$ and a weak oxidant. (Malkin and Niyogi 2000). Reproduced with permission of American Society of Plant Biologists

The effect of this chain is to remove electrons from water – an unwilling donor. The light energy trapped by PSI raises the energy level of these electrons to produce a strong reductant A_0^- ($E'_0 = -1.1$ volts) which leads to the formation of NADPH. Shorter wavelengths of light absorbed by PSII produce a strong oxidant ($P680^+$) which leads to the formation of O_2 (Fig. 2.4). In addition, the interaction of the two photosystems generates a transmembrane proton gradient which results in the synthesis of ATP.

Reaction centres

The initial energy transformations in photosynthesis take place at specific sites on the thylakoid membrane, known as *reaction centres*. Here, chlorophyll (Chl) absorbs a photon and subsequently transfers an electron to an acceptor:

$$Chl\ A \longrightarrow Chl^*A \longrightarrow Chl^+A^-$$

This primary charge separation is the only photosynthetic reaction that directly involves light. The isolation of the reaction centre of the photosynthetic bacterium *Rhodobacter sphaeroides* in the 1960s paved the way for the isolation of reaction centre complexes from a large number of photosynthetic organisms, including plants. The first reaction centre structurally resolved was that of the purple bacterium *Rhodopseudomonas viridis* (Deisenhofer *et al.* 1985). This was soon followed by the elucidation of several other purple bacterium structures and more recently by the appearance of detailed reaction centre structures from oxygenic systems, notably that of photosystem I (Krauss *et al.* 1996). Characterization of

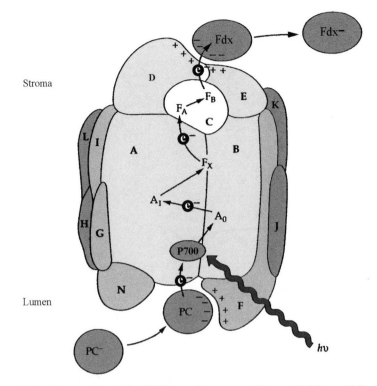

Figure 2.5   Structural model of PSI reaction centre. (From Malkin and Niyogi 2000)

these complexes supports the concept that the reaction centre exists as a discrete protein complex within the thylakoid membrane and contains all the essential elements for carrying out the primary charge separation reaction of photosynthesis. A structural model of the PSI reaction centre is shown in Figure 2.5. The PSI reaction centre contains two major proteins, designated A and B. Electrons are transferred from P700 to a chlorophyll molecule A_0, then on to the A_1 electron acceptor, phylloquinone. Electrons are then transferred through a series of Fe–S centres, called *Rieske proteins* after their discoverer, designated F_X, F_A and F_B and ultimately to the soluble iron–sulphur protein, ferredoxin (Fdx). Several PSI subunits such as F, D and E are involved in the binding of soluble electron transfer substrates to the PSI complex. The reaction-centre contains two chlorophyll *a* molecules called the *special pair*, which are distinct from the other chlorophyll molecules in that all the light energy absorbed by the antenna pigments is funnelled into them. The reaction-centre chlorophyll *a* thus constitutes an energy sink; it is the longest-wavelength and lowest-energy-absorbing chlorophyll in the system.

The reaction centres of PSI and PSII are designated P700 and P680 (P for pigment) respectively and named according to the absorption maxima of their chlorophyll *a* molecules. The pigments of the two reaction centres are actually identical chlorophyll *a* molecules but their association with different proteins in the thylakoid membrane accounts for their different light absorption maxima and light

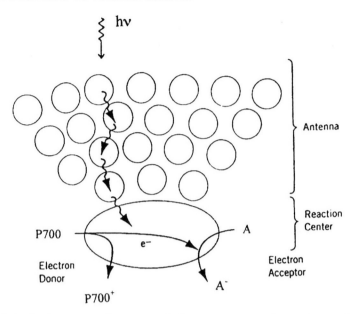

Figure 2.6   A photosystem is composed of an antenna complex and a reaction centre

absorption properties which are subtly different from those of free chlorophyll. When isolated chlorophyll *a* molecules *in vitro* are excited by light, the absorbed energy is rapidly released as fluorescence and heat, but when chlorophyll in intact leaf tissue is illuminated little or no fluorescence is observed. Instead, the energy from the excited antenna pigment molecules is transferred by resonance transfer to the reaction-centre special pair.

Each photosystem consists of a light-gathering *antenna complex* – a cluster of chlorophyll *a*, chlorophyll *b* and carotenoid molecules and a reaction centre (Glazer and Melis 1987) (Fig. 2.6). The antenna pigment molecules are associated with specific proteins to form chlorophyll–protein (CP) complexes (Hunter *et al.* 1989). Because the molecular structure of of PSII is not known, a great deal of structural insight has been obtained by comparison with the structure of the bacterial photosynthetic reaction centre. A structural model of the reaction centre core of PSII, based on the bacterial reaction centre complex of *Rhodopseudomonas viridis* is shown in Figure 2.7.

The structure is dominated by the two reaction-centre proteins D1 and D2. Building on this structure are two transmembrane chlorophyll-binding proteins, CP43 and CP47, products of the *psb*I gene (approx. 4 kDa); three extrinsic polypeptides, of 17, 23 and 33 kDa, are associated with the luminal side of the membrane (MSP, membrane-spanning proteins). Electrons are transferred from P680 to pheophytin (Pheo) and subsequently to two plastoquinone molecules, Q_A and Q_B. P680$^+$ is reduced by Z, a tyrosine residue in the D1 subunit.

The bulk of the chlorophyll in each photosystem functions as antenna chlorophyll; depending on the organism, an antenna complex may contain between

Stroma

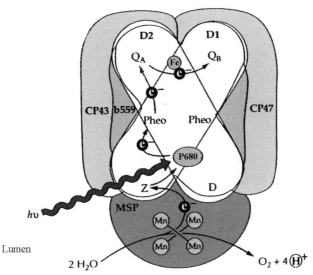

Lumen

Figure 2.7   Structural model of the PSII reaction centre including the electron-transfer cofactors and indicating the oxidation of water by the Mn cluster. (From Malkin and Niyogi 2000). Reproduced with permission of American Society of Plant Biologists

100 and 2000 pigment molecules. The antenna pigment molecules extend the range of wavelengths of light that can be absorbed; they do not participate directly in photosynthesis but they enhance the efficiency of the process by passing the absorbed light to the reaction centres. While the chlorophyll *a* molecules in the reaction centres absorb light directly and initiate photosynthesis, they do not, at normal light intensities, absorb sufficient light energy for the plant's photosynthetic needs. The antenna complexes overcome this limitation. Plants growing under low light conditions augment their antenna systems with even more chlorophyll in order to ensure efficient light harvesting. The augmented antennae are chlorophyll–protein combinations and are termed *light-harvesting complexes* (LHCs) (Zuber 1986); LHC I is associated with PSI and LHCII with PSII (Fig. 2.8).

Together, these LHCs, which are not considered to be part of the photosystem, contain more than half the chlorophyll of the chloroplast, including nearly all of the chlorophyll *b*. The light absorbed by the antenna pigments is funnelled, by resonance transfer, to the reaction centre, where it is used in photosynthesis (Fig. 2.6). In the process, photons are not simply emitted by one molecule and reabsorbed by another; rather the excitation energy is passed from one pigment molecule to another.

The process is very efficient and results in more than 95% of the photons absorbed by the antenna pigments passing their energy on to the reaction centre. The actual site of photochemistry is the reaction centre, which is composed of chlorophyll *a* and associated proteins (Fig. 2.6). Sharing the reaction centre with chlorophyll *a* is a specialized molecule called the primary electron-acceptor. In the

STROMA

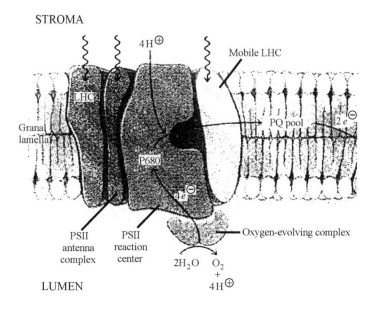

Figure 2.8   Organization of the light-harvesting complexes of PII. (From Horton *et al.* 1996)

reaction centre the light energy is converted to chemical energy in a redox reaction. Light raises an electron from the chlorophyll *a* molecule to a higher energy level and the primary electron-acceptor traps this energy before the excited electron falls back to its ground state in the chlorophyll molecule. The spatial arrangement of PSI and PSII and the cytochrome complex (Fig. 2.3) shows how the physical separation of the two photosystems is important, since electrons are transferred from PSII to PSI during photosynthesis. The orientation of each complex and its individual components is not random; specific polypeptide regions are oriented toward the lumen or stroma as appropriate. Such a vectorial arrangement is characteristic of all energy-transducing membranes. One important consequence of this is to allow the movement of protons between stroma and lumen (Malmstrom 1989) and thus provide the proton gradient necessary for ATP synthesis. Another consequence of the vectorial arrangement is that the oxidation of water and the reduction of NADP occur on opposite sides of the thylakoid membrane. Water is oxidized and protons accumulate on the lumen side of the membrane where they contribute to the proton gradient which drives ATP synthesis; NADPH and ATP are produced in the stroma (Fig. 2.18). The two photosystems are distinct both in structure and function. Thylakoid membranes can be selectively extracted to yield fractions composed of predominantly PSI or PSII complexes. PSII is located mainly on the stacked membranes of the grana while PSI is located mainly on the non-stacked lamellae (Fig. 2.9).

The third multiprotein system, the cytochrome *b/f* complex is found in both the stacked and unstacked regions of the thylakoid membrane. This complex connects two mobile electron carriers, plastoquinone (a quinone) and plastocyanin (a

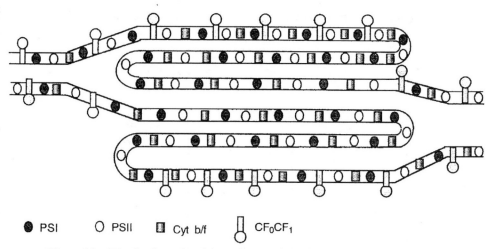

● PSI   ○ PSII   ▨ Cyt b/f   ᕼ CF₀CF₁

Figure 2.9   Distribution of multiprotein complexes in the thylakoid membrane

protein). Plastoquinone diffuses along the membrane and transports electrons from PSII to the cytochrome b/f complex. Plastocyanin, a soluble electron carrier in the lumen, transports electrons from the cytochrome b/f complex to PSI.

Charge Separation across the Thylakoid Membrane

The initial event of the light reactions of photosynthesis is the absorption of light by either the LHC or the antenna complex of PSII. The order of electron flow is from PSII to PSI (Fig. 2.3) and not from PSI to PSII, as might be expected; the designations PSI and PSII reflect the order of discovery of the two photosystems. The absorption of a photon of red light raises an electron of the antenna pigment molecule to the first excited state; the energy of one mole of photons is considerable, being equal to 42 kcal/mol for chlorophyll a. The energy of the excited pigment molecule is transferred rapidly (10^{-15} s) to other pigment molecules in the antenna system until it is eventually funnelled into the special pair of chlorophyll a molecules of the reaction centre. The special pair chlorophylls have a higher reduction potential (less negative) than the other chlorophyll molecules in the antenna complex and so energy transfer between pigments towards the special pair is energetically downhill. When the special pair is thus energized the reaction-centre Chl a is converted to an excited state Chl a^* (*denotes excited state), a strong reductant. Electron transfer between membrane-bound electron carriers in the chloroplast is based on oxidation/reduction. Electrons are transferred from carriers of lower reduction potential (more negative, stronger reductants) to carriers of higher reduction potential (more positive, stronger oxidants).

In photosynthesis the absorption of light energy lowers the reduction potential of the reaction centres P680 and P700 so that the excited states P680* and P700* respectively have lower reduction potentials than their ground-state equivalents (Fig. 2.10). The excited state of the reaction-centre chlorophyll is used to promote a

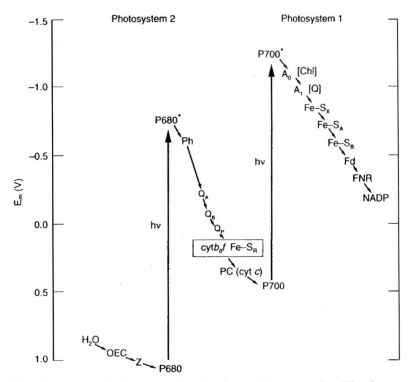

Figure 2.10   The current Z-scheme showing E_m values of electron carriers. The placement of each electron carrier of the non-cyclic electron chain corresponds to the midpoint of its redox potential. (From Blankenship 2002, reproduced with permission)

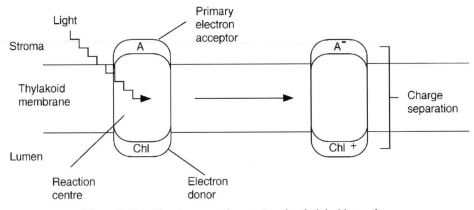

Figure 2.11   Charge separation across the thylakoid membrane

charge separation across the thylakoid membrane. An electron is transferred from the chlorophyll *a* special pair to a primary electron-acceptor (A) on the stromal surface of the membrane, leaving a positive charge on the photo-oxidized chlorophyll on the luminal surface (Fig. 2.11). The reduced primary electron-acceptor (A⁻) becomes a powerful reducing agent and can pass on electrons to another acceptor molecule. The positively charged chlorophyll a (Chl$^+$) is a strong oxidizing agent and will accept electrons from an electron donor on the luminal surface of the membrane.

Thus, the initial photochemical act results in the formation of a negative charge (A$^-$) on the stromal surface and a positive charge (Chl$^+$) on the luminal surface (Fig. 2.11). This charge separation is stabilized in two ways. First, the reduced primary electron-acceptor A$^-$ rapidly (within picoseconds) becomes reoxidized by passing electrons on to a series of acceptors. The initial charge separation is further stabilized because the strong oxidant Chl$^+$ is rapidly reduced to Chl, which is then ready for a further excitation. The formation of these strong biological oxidants and reductants in the green plant represents the actual conversion of light energy to chemical energy which will be needed to drive all the subsequent reactions of photosynthesis–electron transport and NADPH formation, ATP synthesis and CO_2 fixation.

Interaction of PSI and PSII

Photosystem II and the photolysis of water

The initial photochemical reaction is the excitation of P680, the reaction-centre chlorophyll of PSII, by the absorption of wavelengths of light < 680 nm (Fig. 2.10). The excited form of P680, designated P680*, a much stronger reductant than the ground state, is photo-oxidized within picoseconds (10^{-12} s) as it transfers its electron to pheophytin (Ph). Pheophytin, which is considered to be the primary electron-acceptor of PSII, is a form of chlorophyll a in which the magnesium ion is replaced by two hydrogens. With the loss of its electron, P680* is transformed into a cation radical, designated P680$^+$. This photochemical reaction results in the formation of P680$^+$ and Ph$^-$, a charge separation. P680$^+$ is the strongest biological oxidant known, and is capable of oxidizing H_2O. Ph$^-$ very rapidly passes its extra electron to a protein-bound quinone, plastoquinone (Q$_A$) which, in turn, passes its extra electron on to another, more loosely bound quinone, Q$_B$. When Q$_B$ has acquired two electrons in two such transfers from Q$_A$ and two protons from water, it is fully reduced to its quinol form QH$_2$ (Fig. 2.14). In its quinol form it has a lowered affinity for its protein and is released from the reaction centre to be replaced by another molecule of plastoquinone. Eventually, the electrons in QH$_2$ are transported through a chain of membrane-bound carriers to NADPH$^+$. P680$^+$ is also rapidly reduced (within picoseconds), thus returning to its ground state P680 and is again ready for excitation. The electrons for the reduction of the P680$^+$ come from the water molecules on the luminal side of the thylakoid membrane (Fig. 2.8). The protons released from the photolysis of water molecules remain in the lumen and contribute to the pH gradient that is a major part of the proton-motive force. In the process two molecules of water are split to yield four electrons, four protons and molecular oxygen:

$$\text{PSII: } 2H_2O \longrightarrow 4H^+ + 4e^- + O_2$$

$$\text{PSI: } 2NADP \longrightarrow 2NADP$$

The incorporation of two reaction centres in series during the evolution of plant and algal photosynthesis represents a brilliant strategy for using an inexhaustible supply of water in the unlikely role of a reductant, without sacrificing the ability to use photons in the red ($\lambda > 600 \, \text{nm}$) part of the spectrum. For a single reaction centre to oxidize water and reduce NADP, photons of about 500 nm would be required to span the entire redox-potential range between O_2 and NADPH, the two products of the reaction (Hillier and Babcock 2001). Using two photoreactions in series, this energy requirement is relaxed and photons in the longer-wavelength region (680 nm PSI, 700 nm PSII) become useful.

Herbicides can block electron transport during photosynthesis (Ashton and Crafts 1981). The herbicide DCMU (dichlorophenyl dimethylurea), a hydrophobic proton carrier, competes with Q_B for the Q_B-binding site in photosystem II and thus can uncouple phosphorylation from electron transport (Fig. 2.12). Other herbicides, such as paraquat, compete with the electron-acceptors between ferredoxin and NADP; they absorb the electrons and then react with oxygen to form superoxide, O_2^-. This free radical causes damage to chloroplast components, especially lipids, and thus causes loss of chloroplast activity.

Oxygen-evolving complex

A single photon of light does not possess enough energy to bring about the photolysis of water; four photons are needed. The four electrons released from water do not pass directly to P680$^+$, which can only accept one electron at a time. An oxygen-evolving complex (OEC) (Ghanotakis and Yocum 1990) passes four

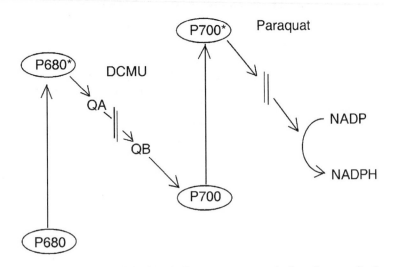

Figure 2.12 Herbicides block electron transport during photosynthesis

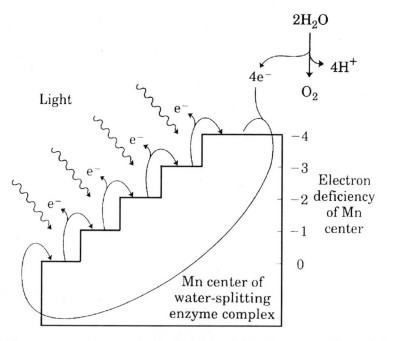

$2H_2O$

$4e^-$

$4H^+$

O_2

Light

e^-

e^-

e^-

e^-

−4

−3

−2

−1

0

Electron
deficiency
of Mn
center

Mn center of
water-splitting
enzyme complex

Figure 2.13 Oxygen-evolving complex: the sequential absorption of four photons with consequent loss of four single electrons, in turn, from the Mn^{2+} cluster, to produce an oxidizing agent that can split water and produce O_2. (From Lehninger *et al.* 1993)

electrons, one at a time, to $P680^+$. This water-splitting complex (Fig. 2.13) contains four manganese (Mn^{2+}) ions as well as bound chloride (Cl^-) and calcium (Ca^{2+}) ions (Debus 1992).

The oxidation of two molecules of water to yield O_2 requires the removal of four electrons. Since the absorption of a photon by PSII releases only one electron, the reaction centre (PSII) must be excited four times in order to oxidize two molecules of water to produce one molecule of oxygen, four protons and four electrons. The OEC has the capacity to store charges. Each photon absorbed by P680 releases one electron from a cluster of four manganese ions within the OEC; the residual positive charges are stored within the complex. When four electrons have been released (and four positive charges accumulated) the OEC oxidizes two molecules of water to yield one molecule of oxygen. The structure of the Mn^{2+} cluster is not fully known and details of its function remain to be clarified. The immediate electron-donor to $P680^+$ is a tyrosine (Tyr) residue (usually designated Z), on one of three proteins associated with PSII. This Tyr residue regains its electron by oxidizing the cluster of four Mn^{2+} in the OEC. With each individual electron transfer the Mn^{2+} cluster becomes more oxidized. Manganese can exist in stable oxidation states from $+2$ to $+7$ and so a cluster of four Mn^{2+} can easily donate four electrons. The pathway of electron flow from H_2O to QH_2 is shown in terms of redox potentials in Figure 2.10. QH_2 has a more negative (less positive), lower potential (0.1 V) than H_2O (0.82 V), i.e. QH_2 is a stronger reductant. This uphill movement of electrons is achieved through the absorption of photons of light by photosystem II. A photon of wavelength 680 nm contains 1.82 electronvolts, which is more than sufficient to change the potential of an electron by 0.72 volts (from 0.82 V to 0.1 V).

Cytochrome b_6/f Complex and the Q-cycle

The reactions of the cytochrome b_6/f complex are best explained by the Q-cycle (Fig. 2.14) based largely on a mechanism known as the *modified Q-cycle* (see below) which accounts for most of the observations of the mechanism of electron and proton flow through the cytochrome bc_1 complex (Fig. 2.15) (Crofts and Berry 1998). According to this model (Fig. 2.14) the cytochrome b/f complex contains a

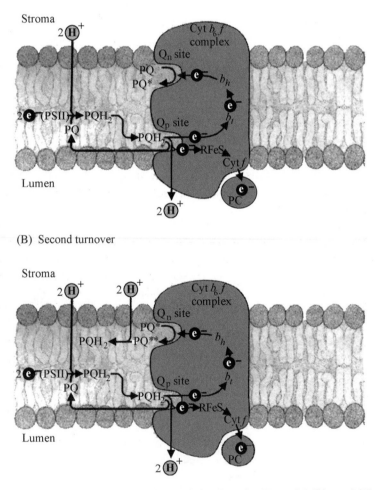

Figure 2.14 The cytochrome b_6/f complex and the Q-cycle. (From Malkin and Niyogi 2000). Reproduced with permission of American Society of Plant Biologists

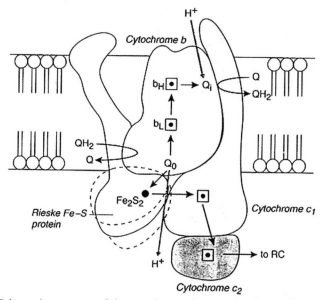

Figure 2.15 Schematic structure of the cytochrome bc_1 complex from mitochondria. The dashed lines indicate movement of the Rieske protein. (From Blankenship 2002, reproduced with permission)

quinol-binding site (Q_p) on the luminal side and a quinone-binding site (Q_n) on the stromal side of the membrane (Malkin and Niyogi 2000). At the Q_p site, a plastoquinol molecule (PQH_2) is oxidized and the electron released is passed through high-potential electron carriers, the Rieske Fe–S protein and cytochrome f to the electron-acceptor, plastocyanin (PC), on the luminal side of the membrane. The other electron passes through the two cytochrome b_6 haems (b_l and b_h) to a quinone-binding site on the stromal side of the membrane, Q_n, where it reduces a quinone molecule to plastosemiquinone (PQ*). Protons released from this oxidation reaction pass into the lumen. This cycle repeats itself to oxidize a second plastoquinol molecule. The pathways of the electron are identical to that of the first oxidation, except that the second electron is transferred to the Q_n site to produce a fully reduced plastoquinol molecule (PQH_2). The net result of this cycle is that a plastoquinol molecule is oxidized to a quinone (PQ) at the Q_p site, two electrons are transferred to plastocyanin (PC) and four protons are transferred from the stroma to the lumen of the chloroplast:

$$PQH_2 + 2PC_{ox} + 2H^+_{stroma} \longrightarrow PQ + 2PC_{red} + 4H^+_{lumen}$$

Electrons from PSII together with two protons from the stroma reduce Q to QH_2 (plastoquinol). QH_2 diffuses laterally through the thylakoid membrane towards the cytochrome b_6/f complex (Malkin 1992) where it donates its two electrons at a site on the luminal side of the membrane (Fig. 2.14). Since the cytochrome b_6/f complex can accept only electrons and not protons from QH_2 (Hope 1993), Q-cycling by Q and the cytochrome b_6/f complex contributes to the proton concentration gradient.

Each complete Q-cycle turnover results in the net oxidation of one molecule of QH_2 to one molecule of Q and the transfer of four protons into the lumen. In the lumen, the transported protons, along with those released from water, generate a proton concentration (pH) gradient across the thylakoid membrane; a pH gradient, rather than a membrane potential, is the principal component of the proton-motive force. The electrons are passed through the cytochrome b_6/f complex to a Cu-containing electron transport protein, plastocyanin. Plastocyanin is a small, soluble protein that diffuses freely in the thylakoid membrane from the cytochrome b_6/f complex to PSI (Fig. 2.14).

Cytochrome bc_1 Complex

The cytochrome b_6/f complex of oxygenic photosynthetic organisms is generally similar in both structure and function to the cytochrome bc_1 complex found in many non-photosynthetic bacteria and in mitochondria of eukaryotic cells. The cytochrome bc_1 complex consists of a minimum of three protein subunits – cytochrome b, cytochrome c_1 and the Rieske iron–sulphur protein. The structure of the bc_1 complex from mitochondria (Fig. 2.15) has been determined by X-ray diffraction (Xia et al. 1997; Iwata et al. 1998; Zhang et al. 1998). Much of the current discussion of the structure and mechanism of the oxygenic photosynthetic complex is based on these complexes. Recently available structural information coupled with a wealth of kinetic and inhibitor data, suggests a detailed mechanism that is being tested and elaborated (Blankenship 2002). In this mechanism, two molecules of ubiquinol are oxidized and one is reduced. The protons resulting from the quinone oxidation are sent to the periplasmic side of the membrane and those taken up during the reduction reaction come from the cytoplasmic side. In addition, electrons are transferred to cytochrome c_2 and eventually are used to reduce the oxidized special pair in the reaction centre. The overall reaction for two turnovers of the cytochrome bc_1 complex is given by:

$$2UQH_2 + 2cyt\ c_{2ox} + UQ + 2H^+ \longrightarrow 2UQ + 4H^+2cyt\ c_{2red} + UQH_2$$

One remarkable feature of the proposed mechanism is the large-amplitude swing of the Rieske protein back and forth from the cytochrome b to the cytochrome c_1. This aspect neatly solves a serious problem that would otherwise significantly reduce the proton pumping efficiency of the complex.

After the first electron is transferred to the Rieske complex it could rapidly react with the cytochrome c_1 and then extract a second electron from the ubisemiquinone. This would short-circuit the cycle and would reduce the proton-to-electron ratio from two to one, thereby cutting energy storage by half. The reduced Rieske protein cannot accept a second electron from the quinone so that the electron must go to haem b_L (Fig. 2.15). The Rieske protein then moves and delivers a single electron to cytochrome c_1 with essentially no chance of a double delivery. The 'swinging door' or 'ratchet' mechanism is thought to be essential in preventing this loss of energy storage in bacterial photosynthesis and in other electron transport chains using similar complexes (Blankenship 2002). By this mechanism, the electron flow connecting the acceptor side of the reaction centre to the donor side gives rise to a proton-motive force across the membrane, due to H^+ concentration differences between the two sides of the membrane. This proton-motive force is used to power the synthesis of ATP, as will be discussed later in this chapter.

PSI Generates NADPH

As described for photosystem II, the primary event at the photosystem I reaction centre is the separation of charge. Within photosystem I, P700 is excited by the absorption of a photon of $\lambda > 680$ nm. The excited reaction-centre P700* loses an electron to a primary electron-acceptor A_0, a chlorophyll molecule which is functionally analogous to pheophytin in photosystem II.

This electron transfer results in a charge separation that produces the strong oxidizing agent $P700^+$ and a strong reducing agent A_0^-. The oxidized chlorophyll $P700^+$ is re-reduced to P700 by an electron passed from PSII via reduced plastocyanin. A_0^- passes its electron through a chain of carriers and ultimately to NADP (Fig. 2.10). The first carrier in this series is phylloquinone (A_1 or vitamin K_1) which passes the electron through a cluster of iron–sulphur proteins designated Fe–S. From here the electron passes to ferredoxin (Fd), another iron–sulphur protein loosely associated with the thylakoid membrane on the stromal side. The fourth electron carrier in the chain is a flavoprotein ferredoxin-NADP-oxido-reductase (Fig. 2.16).

Ferredoxin-NADP-oxidoreductase transfers two electrons from two molecules of reduced ferredoxin to NADP, reducing the latter and oxidizing the ferredoxin:

$$2Fd^{2+} + 2H^+ + NADP^+ \longrightarrow 2Fd^{3+} + NADPH + H^+$$

Two molecules of NADPH are formed for each O_2 molecule released from H_2O. The reduction of NADP to $NADPH^+$ occurs on the stromal side of the membrane. The formation of NADPH completes a linear electron flow from H_2O through PSII, the cytochrome b_6/f complex and PSI. For each electron transferred from H_2O to NADP, two protons are absorbed, one by each photosystem. The release of one molecule of O_2 requires the transfer of four electrons from two molecules of H_2O to two molecules of $NADP^+$:

$$2H_2O + 2NADP^+ \longrightarrow O_2 + 2NADPH + 2H^+$$

The energy changes of the complete electron transport system are shown in Figure 2.10. The overall effect of the complete electron transport system is to create a continuous flow of electrons from H_2O through photosystems II and I and their connecting cytochrome system, to $NADP^+$:

$$PSII \xrightarrow{light} PSI \xrightarrow{light} 2NADPH + 2H^+$$
$$2H_2O + 2NADPH \longrightarrow O_2 + 4H^+ + 2NADP^+$$

STROMA

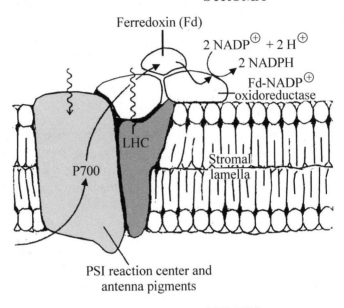

Figure 2.16   PSI complex and Fd-NADP-oxidoreductase. (From Horton *et al.* 1996)

This flow leads to the formation of a proton-motive force across the thylakoid membrane. In the process, electrons are removed from H_2O, a very weak reductant (a poor electron donor, $E'_0 = 0.82$ V) and raised to the energy level of ferredoxin, a very strong reductant and a good electron donor ($E'_0 = 0.42$ V). Ferredoxin, in turn, reduces $NADP^+$ to NADPH, also a strong reductant ($E'_0 = 0.32$ V). To raise the energy of electrons derived from H_2O to the energy level required to reduce $NADP^+$ to NADPH, each electron must be boosted twice by photons absorbed by PSII and PSI, i.e. one photon per electron per photosystem. The uphill transfer of electrons from P680 ($+1.1$ V) to P680* (-0.7 V) and from P700 ($+0.4$ V) to P700* (-1.2 V) is achieved by the absorption of photons of appropriate wavelength. After each excitation, the high-energy electrons flow energetically 'downhill' via electron carrier systems (Fig. 2.10). The downhill transfer of electrons between P680* and P700 represents a negative free-energy change. Some of this energy is conserved in the products of PSII (O_2 and QH_2); the balance is used to establish a proton gradient which, in turn, drives ATP synthesis. This linear electron flow is unidirectional: electrons are continuously supplied from H_2O and drawn off in the formation of NADPH. The process is referred to as *non-cyclic electron transport*; formation of ATP associated with this process is called *non-cyclic photophosphorylation*.

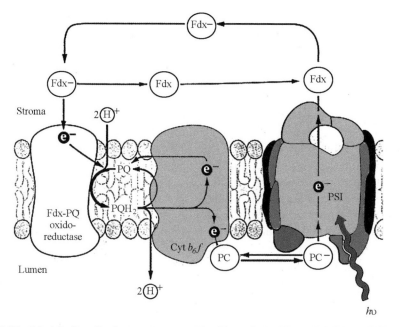

Figure 2.17 Model of cyclic electron transport in chloroplasts. (From Malkin and Niyogi 2000). Reproduced with permission of American Society of Plant Biologists

PSI and Cyclic Electron Transport

Electrons excited within photosystem I may be transported independently of PSII in a series of reactions called *cyclic electron transport* (Fig. 2.17). Chloroplasts can carry out a cyclic electron transport process that involves only PSI and yields ATP as its only product. The cyclic pathway requires a cofactor and there is experimental evidence to suggest that ferredoxin serves as the native cofactor. In a model that describes the pathway, PSI reduces ferredoxin in the light. It is proposed that instead of transferring an electron to $NADP^+$, reduced ferredoxin (Fdx_{red}) interacts with an Fdx-plastoquinone oxidoreductase, thus allowing for the transfer of electrons into the quinone pool (Fig. 2.17). Plastoquinol can then be oxidized by the cytochrome b_6/f complex, allowing for proton translocation across the membrane, possibly via a Q-cycle (Malkin and Niyogi 2000). According to this model, cyclic electron transport would not be expected to be inhibited by DCMU but would be expected to be inhibited by inhibitors of the cytochrome b_6/f complex.

While the cyclic electron pathway can be demonstrated *in vitro*, its role *in vivo* remains questionable. The biochemical evidence for the cycle is still incomplete. In particular, Fdx-plastoquinone-oxidoreductase, the putative key enzyme in the process, has not been fully characterized but is believed to be an NADH-plastoquinone-oxidoreductase that is present in the thylakoid membrane.

Cyclic electron transport and cyclic photophosphorylation are thought to occur when the plant already has a high ratio of NADPH to NADP but needs additional ATP for other metabolic purposes. By regulating the supply of electrons for either

NADP reduction or cyclic photophosphorylation a plant can adjust the ratio of NADPH to ATP produced according to its needs. Operation of the Q-cycle during cyclic electron transport boosts the proton-motive force and increases the synthesis of ATP without producing NADPH. The relative location of the different complexes on the thylakoid membrane admirably influences their functions (Cramer *et al.* 1991). As already mentioned, photosystem I and ATP synthase are located almost exclusively on the unstacked regions of the membrane, whereas photosystem II occurs mainly on the stacked regions (Fig. 2.9). The cytochrome b_6/f complex is distributed uniformly and the mobile electron carriers plastoquinone and plastocyanin are positioned on different regions. A common internal thylakoid space enables protons released from photosystem II, on the stacked regions, to be used by the ATP synthase located on the unstacked regions (Fig. 2.17). If both photosystems were present on the same region of the membrane a high proportion of the photons absorbed by photosystem II would be transported to photosystem I because the electron voltage jump from P680* ($-0.7\,eV$) to P700 ($+0.4\,eV$) is shorter (Fig. 2.9) than from P680* to its own ground state, P680 ($+1.1\,eV$). The separation of the two photosystems on the membrane circumvents this problem by positioning P680* about 10 nm away from P700. The location of photosystem I on the unstacked membranes gives it direct access to the stroma for the reduction of NADP.

Photosystem II, situated on the stacked regions, is in an ideal position to interact with H_2O, a small, polar electron donor, and with plastoquinone, a highly lipid-soluble electron carrier.

Chemiosmosis and ATP Synthesis

In 1954, Daniel Arnon demonstrated that when spinach leaves are illuminated, ATP is synthesized from ADP and P_i during electron transfer. This process is called photophosphorylation. In illuminated chloroplasts the synthesis of ATP is coupled to the energy released as high-energy electrons are transferred down an electron system from photoexcited photosystem II to electron-deficient photosystem I. The effect of the electron flow is to generate and maintain a proton gradient across the thylakoid membrane, thereby providing the energy for ATP synthesis. The synthesis of ATP in illuminated chloroplasts can be coupled to two types of electron flow – cyclic and non-cyclic, which are associated with PSI and PSII respectively (Bennett 1991).

While it had been known for many years that ATP synthesis was linked to electron transport in cell membranes, the mechanism was not clearly understood until Peter Mitchell (1961) proposed a hypothesis of *chemiosmosis*. His hypothesis was not readily accepted in the 1960s but is now well supported by experimental evidence. Mitchell's hypothesis broadly states that the transport of electrons laterally through the thylakoid membrane is accompanied by the pumping of protons across the membrane, from the stromal to the luminal side (Fig. 2.18). This results in a proton and pH gradient across the membrane: the lumen becomes acidic

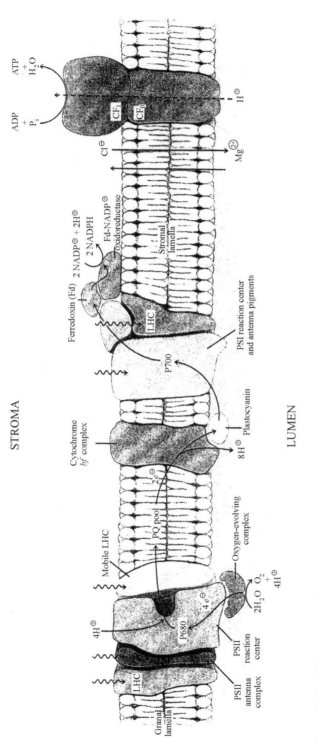

Figure 2.18   Light capture, electron transport and proton translocation in photosynthesis generating NADPH and a proton gradient that drives the formation of ATP. (From Horton *et al.* 1996)

(pH 5) and the stroma alkaline (pH 8). The electrochemical energy generated by this difference in proton concentration and separation of charge across the membrane is designated *proton-motive force*. This proton-motive force is subsequently used to synthesize ATP, as the protons flow passively back from lumen to stroma via ATP synthase. The hypothesis requires that the reaction centres, electron carriers and ATP-synthesizing enzymes are positioned asymmetrically in the thylakoid membrane and that the membrane is impermeable to protons. The effect of this is that some carriers can both transport electrons laterally through the membrane and protons vectorially across the membrane against a proton concentration gradient.

These carrier molecules are membrane proteins which constitute *electrogenic pumps*; in chloroplasts they are *proton pumps* which pump protons across the thylakoid membrane from the stroma into the lumen. The proton pumps thus generate a voltage (a charge separation) across the membrane. The voltage that results is designated *membrane potential*, and, like a battery, is an energy source and favours the passive transport of cations inwards towards the lumen and anions outwards towards the stroma. Thus, two forces, a chemical force due to the ionic concentration gradient and an electrical force due to the effect of the membrane potential, combine to drive protons across the membrane. Together these two forces constitute an electrochemical gradient down which protons move. This electrochemical gradient is the proton-motive force and is expressed as:

$$\Delta \mu H \quad = \quad \Delta \Psi \quad - \quad 2.3 n R T \Delta pH / F$$

electrochemical H^+ gradient	membrane potential	pH gradient

The re-entry of protons into the stroma via ATP synthase generates a free-energy change $(-\Delta G)$. The *chemical* contribution to the free-energy change is given by

$$\Delta G_{chem} = n R T \frac{[H^+]_{in}}{[H^+]_{out}}$$

where n is the number of protons, R is the universal gas constant ($8.315\,J\,mol^{-1}$ $^\circ K^{-1}$) and T is the temperature in degrees Kelvin.

Since $pH = -\log [H^+]$, the above equation can be written

$$\Delta G_{chem} = -2.303\, n R T\, (pH_{in} - pH_{out})$$

The electrical contribution to the free-energy change is $\Delta \Psi$, the membrane potential, which is the difference in charge across the membrane ($\Delta \Psi = \Psi_{in} - \Psi_{out}$). Thus,

$$\Delta G_{elect} = z F \Delta \Psi$$

where z is the charge on the transported ion and equals 1 for each proton transported; F is Faraday's constant ($96.5\,kJ\,V^{-1}\,mol^{-1}$). The total free-energy change for the transport of one proton from the stroma to the lumen is therefore given by

$$\Delta G = nF\Delta\Psi - 2.303\ nRT\Delta\text{pH}$$

Dividing across by nF,

$$G/nF = 2.303\ nRT\ \text{pH}/F$$

The term G/nF is the proton-motive force; at $25\,°C$, $2.303\ RT/F = 0.059\ V$
Thus,

$$\Delta\text{pH} = \Delta\Psi - (0.059\ V)\Delta\text{pH}$$

The electron-transferring molecules of the electron transport chain are oriented asymmetrically in the thylakoid membrane so that the photoinduced electron flow results in the pumping of protons across the membrane from the stroma into the lumen. The proton gradient (ΔpH) thus generated across the membrane may be quite substantial. When chloroplasts are illuminated, the pH in the lumen drops to about 5 while in the stroma the pH rises to about 8. Since a difference of one pH unit represents a ten-fold difference in H^+ concentration, a gradient of three pH units corresponds to a 10×3 difference in H^+ concentration. At $25\,°C$ a proton gradient (ΔpH) of 3 establishes a potential difference of $3\times 0.059\ V$ ($0.177\ V$). To pump protons into the lumen against a proton-motive force of this magnitude would require a large amount of energy; this energy is provided by the free-energy change (ΔG) of electron transport. In photosynthesis, the required energy is provided by the absorption of photons of light.

Some of the earliest experimental evidence to support Mitchell's model for the chemiosmotic synthesis of ATP came from the work of Jagendorf and his colleagues, who were studying ATP synthesis in washed thylakoids. Hind and Jagendorf (1963) reported that illuminated chloroplasts took up protons, thus causing an increase in pH of the bathing solution; they also showed a correlation between the extent of proton uptake and the amount of ATP synthesized. Jagendorf and Uribe (1966) showed that a pH gradient across the thylakoid membrane could supply the driving force needed to generate ATP. They soaked washed chloroplasts in a pH 4 buffer solution, in the dark (Fig. 2.19). At equilibrium, the pH of the lumen would be the same as that of the bathing solution. The chloroplast suspension was then transferred, in the dark, to a basic solution, pH 8, to which ADP and P_i had been added. This resulted in a transient pH gradient (4 pH units) across the membrane. As protons were channelled out of the lumen into the stroma there was a surge of ATP synthesis. Since this ATP synthesis occurred in the absence of light, this experiment demonstrated that a pH gradient across the membrane is a high-energy state that can mediate the transduction of energy from electron transfer into the chemical energy of ATP, in accordance with Mitchell's predictions. The combined effect of the electron transport chain and the ATP-synthase complex is to establish a proton circuit (Fig. 2.20).

The electron transport system pumps the protons from the stroma into the lumen, thus providing a transmembrane proton gradient. The ATP synthase allows the protons to return to the stroma. As the protons are funnelled through the CF_0–CF_1 complex, the free energy is made available for the synthesis of ATP (see

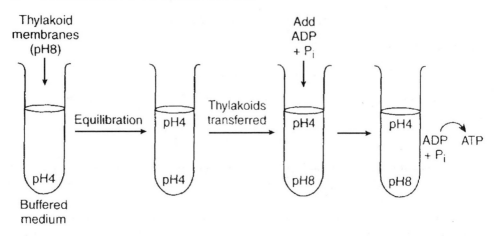

Figure 2.19 Jagendorf's experiment demonstrating chemiosmotic synthesis of ATP by isolated chloroplast thylakoids. Isolated thylakoids, pH 8, were placed in a pH 4 buffer solution, and allowed to equilibrate. The equilibrated thylakoids were rapidly transferred to a pH 8 buffer, to which ADP and P_i have been added

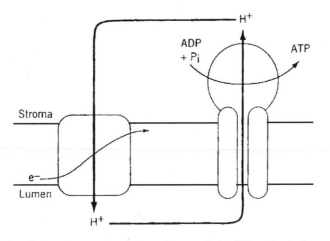

Figure 2.20 The electron transport system with coupled ATP synthesis forms a proton circuit

below). Likewise, NADPH formed by photosystem I is released into the stroma. Thus, ATP and NADPH, the products of the light reactions, are ideally located in the stroma for the dark reactions that convert CO_2 to carbohydrate.

Chloroplast ATP Synthase

The thylakoid ATP synthase consists of two segments: a transmembrane segment called CF_0 (F_0 in mitochondria) and a hydrophilic segment on the stromal surface,

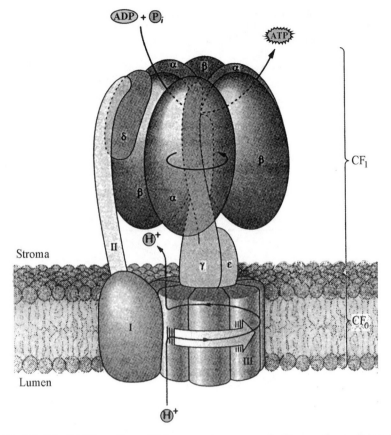

Figure 2.21 Model of ATP synthase. The enzyme is composed of two major regions: an integral membrane protein (CF_0) which functions as a channel for protons passing through the membrane, and an extrinsic portion (CF_1) which contains the catalytic sites for ATP synthesis. CF_1 consists of five different subunits α, β, γ, δ and ε; CF_0 contains at least three different subunits I, II and III, multiple copies of III being present in the membrane. (Malkin and Niyogi 2000)

called CF_1 (F_1 in mitochondria). The terms are historical: F_0 stands for digomycin-sensitive factor and F_1 stands for factor 1. CF_0 forms a H^+-permeable conduit across the membrane, allowing H^+ ions to diffuse spontaneously from the lumen into the stroma; CF_1 is the catalytic part of the enzyme and is involved in the actual conversion of ADP and P_i into ATP. The enzyme is often referred to as the CF_0–CF_1 complex (Fig. 2.21). CF_1, which readily dissociates from the transmembrane segment, can hydrolyse ATP *in vitro*. The regulation of CF_1 is accomplished both by the protein-gradient itself and by the light-driven electron transport providing reducing equivalents that are used to reduce an intramolecular disulphonic bridge in CF_1. If CF_1 can be reduced in the presence of a proton gradient when there is no ADP or P_i to make ATP, then it can act as an ATPase, hydrolysing available ATP. This hydrolysis, however, is accompanied by proton

Table 2.2   Subunits of chloroplast ATP synthase

Protein	Gene	Mol. mass (kDa)	Function
CF_1			
α subunit	atpA	55	catalytic
β subunit	atpB	54	catalytic
γ subunit	atpC	36	proton gating
δ subunit	atpD	20	binding of CF_1 to CF_0
ε subunit	atpE	15	ATPase inhibition; proton gating
CF_0			
I	atpF	17	binding of CF_0 to CF_1
II	atpG	16	binding of CF_0 to CF_1
III	atpH	8	proton translocation
IV	atpI	27	binding of CF_0 to CF_1

pumping to the thylakoid lumen, so that illuminated chloroplasts should be much less effective in causing ATP hydrolysis, since light is already establishing a proton gradient. The proton-motive force established across the thylakoid membrane in the light activates the ATP synthase and inhibits the wasteful ATPase activity of the enzyme. Chloroplast ATP synthase is a 400 kDa enzyme that contains nine different subunits and includes both chloroplast- and nuclear-encoded products (Table 2.2).

Structure of CF_1–CF_0 Complex

CF_1 contains the catalytic and regulatory portions of the enzyme whereas CF_0 is the membrane-embedded proton channel and the point of atachment of CF_1 to the thylakoid membrane. CF_1 consists of five different subunits: α, β, γ, δ and ε (Fig. 2.21). There are three copies of each of the large α and β subunits which are symmetrically arranged like the segments of an orange and enclose a central cavity. There is one copy each of the smaller γ, δ and ε subunits to give an overall stoichiometry of $\alpha_3\beta_3\gamma\delta\varepsilon$. The α and β subunits are almost identical. They have a central domain containing the nucleotide binding sites; they bind ADP and P_i and catalyse the formation of ATP. These catalytic sites communicate to the external solution via a water-filled conical funnel, allowing the phosphate and ribose moieties of ADP and ATP access and permitting diffusion of the substrates and products.

The γ subunit is essential for ATP hydrolysis by CF_1; it is part of a regulatory mechanism mediated by the ferredoxin/thioredoxin system which enhances ATP synthase in the light and deactivation of the enzyme in the dark, thereby preventing wasteful hydrolysis of ATP at night. The δ subunit is required for the binding of CF_1 to CF_0 and helps stabilize the $\alpha_3\beta_3\gamma$ core. The ε subunit has a dual role – it is required for proton gating and also blocks catalysis in the dark, preventing the

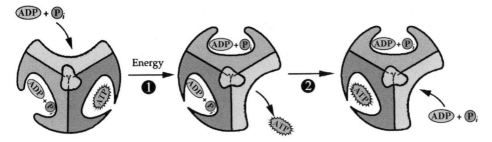

Figure 2.22 Model of the 'binding change mechanism' of ATP synthesis. (From Malkin and Niyogi 2000). Reproduced with permission

breakdown of ATP. The presence of the small subunits results in the overall asymmetry of CF_1 because the small, single-copy subunits cannot participate in the apparent three-fold symmetry of the $\alpha_3\beta_3$ structure.

The CF_0 complex is composed of polypeptide subunits I, II and III in the ratio 1:1:9–12. CF_0 is embedded in the thylakoid membrane with subunit II connecting it to CF_1; the δ subunit anchors subunit II to the α–β dimer. The 'rotor' complex of CF_0 is made up of subunit III, forming a disc which is linked to the asymmetric γ subunit by ε. When hydrogen ions flow through the membrane from lumen to stroma, the disc of the subunit III rotates anticlockwise as a consequence, causing the γ subunit to rotate within the central cavity of the CF_1 complex. Since the γ subunit is asymmetric it compels the β subunits to undergo structural changes, altering the structure of the catalytic sites sequentially to synthesize ATP (Fig. 2.21). In summary, a rotary mechanism is employed to synthesize ATP by CF_1–CF_0 in which the rotation of the central unit is driven by a flux of protons from the lumen to the stroma.

The most widely accepted mechanism of ATP synthesis is the so-called *binding change mechanism* (Fig. 2.22) proposed by Boyer (1993, 1997). Boyer's conformational model for proton-driven ATP synthesis postulates that ATP is synthesized by F-type ATPases through a process of rotational catalysis. According to this mechanism, the energy stored in the proton gradient is not directly used to drive the synthesis of ATP in the classical sense but rather is used to release the tightly bound form of ATP from its catalytic binding site on the enzyme.

Each α–β hexamer of CF_1 has three distinct nucleotide-binding sites: one site is open (O), another binds loosely (L) and the third binds tightly (T). At any one time, all three states are present in the CF_1 complex, each being associated with one of the three catalytic sites of the enzyme. According to the binding change mechanism, ADP and P_i initially bind to a vacant site in the open state. As a consequence of protons moving from the lumen to the stroma through the CF_0 channel, the γ subunit of CF_1 is forced to rotate and this rotation causes conformational changes in the three nucleotide binding sites. The T site, which contains bound ATP, is converted to the O site as the ATP is released, while the L site, which contains bound ADP and P_i, is converted to the T site: this facilitates ATP synthesis. Each individual site undergoes three binding changes (and three conformational changes) in the synthesis of one molecule of ATP. Thus as 3 ATPs are synthesized per rotation and 12 H^+s are needed to drive the cycle by one full turn of the γ subunit, 4 H^+s produce 1 ATP.

Crystallographic studies carried out by John Walker (Walker *et al.* 1991) of the $\alpha_2\beta_3\gamma$ core of the F_1 complex from mitochondria, demonstrated that three adenine nucleotide-binding sites on the complex (located primarily on the three β subunits) show three distinct conformations. Walker's work clarified the structural conditions of the enzyme's molecular machinery and thereby verified Boyer's binding change mechanism. As a result of their work, Boyer and Walker were co-recipients of the Nobel Prize in Chemistry in 1997.

Summary

Photosynthesis uses the energy of sunlight to synthesize organic molecules from CO_2 and H_2O. In higher plants photosynthesis occurs in chloroplasts.

The photochemical reactions take place in two distinct photosystems, PSI and PSII, located on the thylakoid membranes. During photosynthesis photons of light are absorbed by chlorophylls *a* and *b* and by ancillary pigments in an antenna system which is made up of several light-harvesting complexes (LHCs) associated with each photosystem. Light energy absorbed by the antenna pigments is transferred to a 'special pair' of chlorophyll *a* molecules, P680 in PSII and P700 in PSI.

PSI and PSII operate in series. Photoexcitation at the reaction centres results in a charge separation across the thylakoid membrane which produces a strong oxidant and a strong reductant. Photoexcitation of PSII releases an electron from P680; the resultant $P680^+$ is a powerful oxidant which can oxidize H_2O to O_2. This photolysis of water is catalysed by an Mn-containing protein complex on the luminal surface of the thylakoid membrane. In the process the Mn ion passes through five different oxidation states (S_0–S_4), releasing O_2 at state S_4. The resulting protons remain in the lumen and generate part of a proton-motive force. The electron released from P680 is transferred via pheophytin to plastoquinone, Q, on the stromal surface of the membrane. The quinone picks up two electrons and two protons from the stroma to form QH_2. The $P680^+$ is reduced as the PSII reaction centre regains electrons from H_2O. Reduced plastoquinone transfers electrons to the cytochrome b_6/f complex and releases its protons into the lumen, thus adding to the proton-motive force. The electrons are then transferred from the cytochrome b_6/f complex by plastocyanin, PC, in the lumen, to PSI.

Electrons released by the photoexcitation of PSI can follow one of two possible routes. The electrons may be transported through a series of carriers to ferredoxin, a powerful reductant, which then reduces $NADP^+$ to NADPH. Alternatively, the electrons from ferredoxin may flow back through the cytochrome b_6/f complex to PSI. This process, termed cyclic photophosphorylation, leads to the generation of a proton gradient; PSII is not involved, no NADPH is formed and no O_2 is evolved. A pH gradient (pH 5 in the lumen and pH 7.8 in the stroma) rather than an electrical potential is the main component of the proton-motive force. ATP synthesis is achieved through chemiosmosis whereby protons flow from the luminal to the stromal surface of the thylakoid membrane through a CF_0–CF_1 complex.

References

Anderson, J.M. (1986). Photoregulation of the composition, function and structure of thylakoid membranes. *Ann. Rev. Plant Physiol.* **37**: 93–136.

Arnon, D.I. (1951). Extracellular photosynthetic reactions. *Nature* **167**: 1008.

Arnon, D.I. (1991). Photosynthetic electron transport: emergence of a concept, 1949–59. *Photosyn. Res.* **29**: 117–131.

Arnon, D.I., Whatley, F. and Allen, M. (1954). Photosynthesis by isolated chloroplasts. II. Photosynthetic phosphorylation, the conversion of light into phosphate bond energy. *J. Am. Chem. Soc.* **76**: 6324.

Ashton, F.M. and Crafts, A.S. (1981). *Mode of Action of Herbicides*, 2nd edn. Wiley, New York.

Bennett, J. (1991). Phosphorylation in green plant chloroplasts. *Ann. Rev. Plant Physiol. Plant Mol. Biol.* **42**: 281–311.

Blackman, F. (1905). Optima and limiting factors. *Ann. Bot.* **19**: 281.

Blankenship, R.E. (2002). Electron Transfer Pathways and Components. In: *Molecular Mechanisms of Photosynthesis*, pp. 124–156. Blackwell Science, Oxford.

Boyer, P.D. (1993). The binding change mechanism for ATP synthesis – some probabilities and possibilities. *Biochem. Biophys.* Acta **1140**: 215–250.

Boyer, P.D. (1997). The ATP synthase – a splendid molecular machine. *Ann. Rev. Biochem.* **66**: 717–749.

Calvin, M. and Benson, A.A. (1948). The path of carbon in photosynthesis. *Science* **107**: 476.

Cramer, W.A., Furbacher, P.N., Szczepaniak, A. and Tae, G.-S. (1991). Electron transport between photosystem II and photosystem I. *Current Topics in Bioenergetics* **16**: 179–217.

Crofts, A.R. and Berry, E.A. (1998). Structure and function of the cytochrome bc_1 complex of mitochondria and photosynthetic bacteria. *Curr. Opin. Struct. Biol.* **47**: 501–509.

Debus, R.J. (1992). The manganese and calcium ions of photosynthetic oxygen evolution. *Biochem. Biophys. Acta* **1102**: 269–352.

Deisenhofer, J., Epp, O., Miki, K., Huber, R. and Michel, H. (1985). Structure of the protein subunits in the photosynthetic reaction centre of *Rhodopseudomonas viridis* at 3 Å resolution. *Nature* **318**: 618–624.

Dutton, P.L. (1978). Redox potentiometry: determination of mid-point potentials of oxidation–reduction components of biological electron-transfer systems. *Methods Enzymol.* **54**: 411–435.

Duysens, L.N.M., Amez, J. and Kamp, B.M. (1961). Two photochemical systems in photosynthesis. *Nature* **190**: 510–511.

Emerson, R. and Rabinowitch, E. (1960). Red drop and role of auxilary pigments in photosynthesis. *Plant Physiol.* **35**: 477.

Emerson, R., Chalmers, R. and Cederstrand, C. (1957). Some factors influencing the long-wave limit of photosynthesis. *Proc. Natl. Acad. Sci. USA* **43**: 133.

Ghanotakis, D.F. and Yocum, C.F. (1990). Photosystem II and the oxygen-evolving complex. *Ann. Rev. Plant Physiol. and Plant Mol. Biol.* **41**: 255–276.

Glazer, A.N. and Melis, A. (1987). Photochemical reaction centres : structure, organization and function. *Ann. Rev. Plant Physiol.* **38**: 11–45.

Govindjee and Coleman, W.J. (1990). How plants make oxygen. *Sci. Amer.* **262**: 50–58.

Hatch, M.D. and Slack, C.R. (1966). Photosynthesis by sugarcane leaves. A new carboxylation reaction and the pathway of sugar formation. *Biochem. J.* **101**: 103–111.

Hill, R. (1937). Oxygen evolved by isolated chloroplasts. *Nature* **139**: 881.

Hill, R. and Bendall, F. (1960). Function of the two cytochrome components in chloroplasts: a working hypothesis. *Nature* **186**: 136–137.

Hillier W. and Babcock, G.T. (2001). Photosynthetic reaction centers. *Plant Physiol.* **125**: 33–37.

Hind, G. and Jagendorf, A.T. (1963). Separation of light and dark stages in photophosphorylation. *Proc. Natl. Acad. Sci. USA* **49**: 715–722.

Hope, A.B. (1993). The chloroplast cytochrome b/f complex: a critical focus on function. *Biochem Biophys. Acta.* **1143**: 1–22.

Horton, H.R., Moran, L.A., Ochs, R.S., Rawn, J.D. and Scrimgeour, K.G. (1996). Photosynthesis, *Principles of Biochemistry*, 2nd edn, pp. 435–458. Prentice-Hall, Upper Saddle River, NJ.

Hunter, C.N., van Grondelle, R. and Olsen, J.D. (1989). Photosynthetic antenna proteins: 100 ps before photochemistry starts. *Trends Biochem. Sci.* **14**: 72–76.

Iwata, S., Lee, J.W., Okada, K., Lee, J.K., Iwata, M., Rasmussen, B., Link, T.A., Ramaswamy, S. and Jap, B.K. (1998). Complete structure of the 11-subunit bovine mitochondrial cytochrome bc_1 complex. *Science* **281**: 64–71.

Jagendorf, A.T. and Uribe, E. (1966). ATP formation caused by acid–base transition of spinach chloroplasts. *Proc. Natl. Acad. Sci. USA* **55**: 170–177.

Kortschak, H., Hartt, C.E. and Burr, G.E. (1965). CO_2 fixation in sugarcane leaves. *Plant Physiol.* **40**: 209–213.

Krauss, N., Schubert, W.D., Klukas, O., Fromme, P., Witt, H.T. and Saenger, W. (1996). Photosystem I at 4 Å resolution: a joint photosynthetic center and core antenna system. *Nat. Struct. Biol.* **3**: 965–973.

Lehninger, A.L., Nelson, D.L. and Cox, M.M. (1993). Oxidative phosphorylation and photophosphorylation, *Principles of Biochemistry*, pp. 542–597. Worth, New York.

Loomis, W. (1960). Historical introduction. In: W. Ruhland (ed.), *Encylopaedia of Plant Physiology* **5**(1): 85. Springer, Berlin.

Malkin, R. (1992). Cytochrome bc_1 and b_6f complexes of photosynthetic membranes. *Photosyn. Res.* **33**: 121–136.

Malkin, R. and Niyogi, K. (2000). Photosynthesis. In: *Biochemistry and Molecular Biology of Plants*, B. B. Buchanan, W. Gruissem, and R.L. Jones (eds), pp. 569–628. American Society of Plant Physiologists, Rockville, MD.

Malmstrom, B.G. (1989). The mechanism of proton translocation in respiration and photosynthesis. *FEBS Letters* **250**: 9–21.

Mitchell, P. (1961). Coupling of phosphorylation and hydrogen transfer by chemiosmotic type of mechanism. *Nature* (London) **191**: 144.

Prince, R.C. (1985). Redox-driven proton gradients. *BioScience* **35**: 22–26.

Ruben, S., Randall, M., Kamen, M. and Hyde, J.L. (1941). Heavy oxygen (^{18}O) as a trace in the study of photosynthesis. *J. Am. Chem. Soc.* **63**: 877.

Salisbury, F.B. and Ross, C.W. (1992). Chloroplasts and light, *Plant Physiology*, pp. 207–224. Wadsworth, Belmont, CA.

Staehlin, L. and Arntzen, C.J. (eds) (1986). Photosynthesis III. Photosynthetic membranes and light harvesting systems. In: *Encyclopedia of Plant Physiology*, New Series, Vol. 19. Springer-Verlag, Berlin.

van Niel, C.B. (1941). The bacterial photosyntheses and their importance for the general problem of photosynthesis. *Adv. Enzymol.* **1**: 263.

Walker, J.E., Lutter, R., Dupuis, A. and Runswick, M.J. (1991). Identification of the subunits of F_1F_0-ATPase from bovine heart mitochondria. *Biochem.* **30**: 5369–5378.

Wood, P.M. (1985). What is the Nernst equation? *Trends Biochem. Sci.* **10**: 106–108.

Xia, D., Yu, C.A., Kim, H., Xian, H.Z., Kachurin, A.M., Zhang, L., Yu, L. and Deisenhofer, J. (1997). Crystal structure of the cytochrome bc_1 complex from bovine heart mitochondria. *Science* **277**: 60–66.

Zhang, Z., Huang, L., Shulmeister, V.M., Chi, Y., Kim, K.K., Hung, L., Crofts, A.R., Berry, E.A. and Kim, S. (1998). Electron transfer by domain movement in cytochrome bc_1. *Nature* **392**: 677–684.

Zuber, H. (1986). Structure of light-harvesting antenna complexes of photosynthetic bacteria, cyanobacteria and red algae. *Trends Biochem. Sci.* **11**: 414.

3
CO_2 Fixation

C3 Photosynthetic Carbon Reduction (PCR) Cycle

All photosynthetic eukaryotes, from the simplest algae to the most advanced land plants, have a basic common mechanism for the reduction of CO_2 to carbohydrate. The metabolic pathway leading to carbon assimilation has several names, including the photosynthetic carbon reduction (PCR) cycle, the reductive pentose phosphate (RPP) cycle, and the C3 cycle. It is also referred to as the Calvin cycle, after Melvin Calvin, who, with his co-workers, elucidated the pathway in the early 1950s (Calvin 1989). For their efforts in mapping the complex sequence of reactions involved in the fixation of CO_2 and its conversion into carbohydrate, Calvin was awarded the Nobel Prize for Chemistry in 1961. The reactions of the pathway take place in the stroma of the chloroplast, mediated by enzymes that are not bound to the thylakoid membranes.

The solution to the problem of CO_2 fixation was made possible by the development of the technique of paper chromatography and the discovery of radioactive carbon-14 (^{14}C) in the late 1940s and early 1950s. Calvin's group cultivated the unicellular green alga *Chlorella* under conditions favourable for photosynthesis. When a steady rate of photosynthesis was reached, $^{14}CO_2$ was introduced and photosynthesis continued for various periods of time. The reaction was stopped by dropping the cells into hot methanol which denatured the enzymes and extracted the sugars for subsequent chromatographic analysis. When photosynthesis was stopped after as little as two seconds, most of the radioactivity was found in a three-carbon compound, 3-phosphoglycerate (3-PGA), which appeared to be the first stable product of photosynthesis. The next step was to determine what two-carbon molecule accepted the CO_2 to form 3-PGA. However, no two-carbon molecule was ever found. Instead, it was discovered that a five-carbon keto-sugar, ribulose-1,5-bisphosphate (RuBP), accumulated. The accumulation of RuBP was accompanied by a sharp drop in the level of radio-labelled 3-PGA, suggesting that the five-carbon RuBP may be the acceptor molecule for CO_2.

Photobiology of Higher Plants. By M. S. Mc Donald.
© 2003 John Wiley & Sons, Ltd: ISBN 0 470 85522 3; ISBN 0 470 85523 1 (PB)

Close-up on the Calvin Cycle

In order for the chloroplast to continue to take up CO$_2$, the product (3-PGA) must be continually removed and the acceptor molecule, RuBP, must be maintained. Both of these processes require the input of ATP and NADPH. Fixation of CO$_2$ by means of the Calvin cycle (Fig. 3.1) occurs in three stages:

1. *Carboxylation*: the condensation of CO$_2$ with the five-carbon acceptor, ribulose-1,5-bisphosphate (RuBP) to form two molecules of 3-phosphoglycerate (3-PGA).

2. *Reduction*: the reduction of 3-phosphoglycerate to 3-phosphoglyceraldehyde (3-PGAL), using the ATP and NADPH produced in the light reactions.

3. *Regeneration*: ribulose-1,5-bisphosphate, the initial acceptor of CO$_2$, is regenerated by means of interconversions of 3-PGAL and using ATP.

Carboxylation of ribulose-1,5-bisphosphate

Atmospheric CO$_2$ enters the PCR cycle by reacting with the five-carbon ribulose-1,5-bisphosphate to yield two molecules of 3-phosphoglycerate. The carboxylation reaction is catalysed by the enzyme ribulose-1,5-bisphosphate carboxylase (RubisCO). In this condensation reaction CO$_2$ is added to RuBP to form an intermediate six-carbon compound which is unstable and is quickly hydrolysed to two molecules of 3-phosphoglycerate (Fig. 3.2). RubisCO accounts for about 50% of the soluble protein in plant leaves and is probably one of the most abundant enzymes in nature. Its high concentration in the chloroplast stroma coupled to its

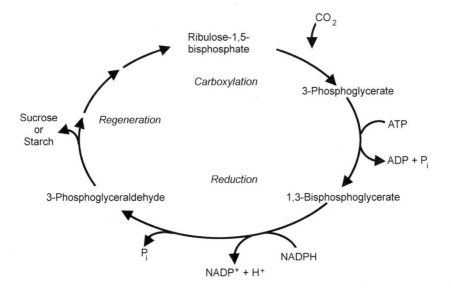

Figure 3.1   The three stages of CO$_2$ fixation in photosynthetic organisms

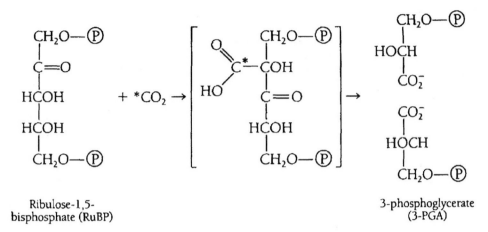

Ribulose-1,5-
bisphosphate (RuBP)

3-phosphoglycerate
(3-PGA)

Figure 3.2   The carboxylation of ribulose-1,5-bisphosphate, catalysed by the enzyme RubisCO, proceeds in two stages

high affinity for CO_2 ensures rapid fixation of the normally low (0.03%) concentration of CO_2 in the atmosphere.

Reduction of 3-phosphoglycerate

The product of CO_2 fixation, 3-phosphoglycerate (3-PGA), is removed by reduction to 3-phosphoglyceraldehyde (3-PGAL), in a two-step reaction (Fig. 3.3). The 3-phosphoglycerate is first phosphorylated to 1,3-bisphosphoglycerate (1,3-PGA) which is then reduced to 3-phosphoglyceraldehyde (3-PGAL). The chloroplast enzyme NADP-glyceraldehyde-3-phosphate dehydrogenase catalyses this reaction; the ATP and NADPH used are generated in the light reactions.

Regeneration of ribulose-1,5-bisphosphate

For the continued fixation of CO_2 it is essential to regenerate the CO_2-acceptor molecule, ribulose-1,5-bisphosphate. This is accomplished by means of inter-conversions of the carbon skeletons of 3-phosphoglyceraldehyde and dihydroxyacetone produced in the first two stages of carbon fixation. This series of enzyme-controlled reactions, in which 3-phosphoglyceraldehyde plays a central role, yields intermediates which include four-, five-, six- and seven-carbon sugars. In a reaction catalysed by aldolase, one molecule of 3-PGAL condenses with one molecule of dihydroxyacetone-3-P to form the hexose, fructose-1,6-bisphosphate; fructose-1,6-bisphosphatase removes one phosphate from this molecule to yield fructose-6-monophosphate. 3-Phosphoglyceraldehyde (3-PGAL) reacts with fructose-6-monophosphate (6C-P) in the presence of transketolase to yield xylulose-5-phosphate (5C-P) and erythrose (4C-P) (Fig. 3.4).

In another aldolase-catalysed reaction 3-PGAL combines with erythrose (4C-P) to form sedoheptulose-1,7-bisphosphate (7C-2P), which, in the presence of the enzyme

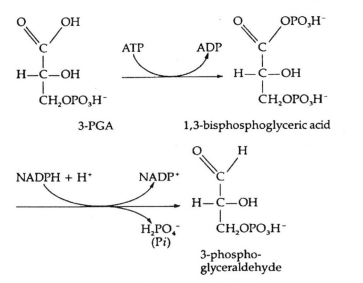

Figure 3.3 Reduction of 3-phosphoglyceric acid (3-PGA) to 3-phosphoglyceraldehyde (3-PGAL)

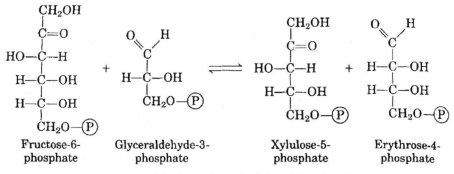

Figure 3.4 The formation of xylulose-5P and erythrose

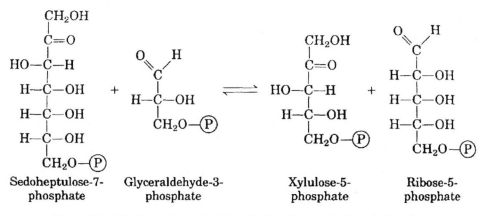

Figure 3.5 The formation of xylulose-5-phosphate and ribose-5-phosphate

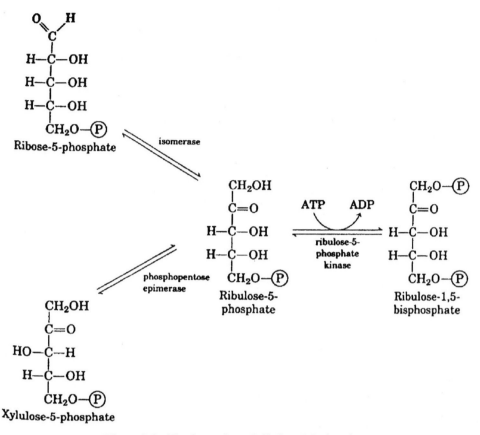

Figure 3.6   The formation of ribulose-1,5-phosphate

sedoheptulose-1,7-bisphosphatase is converted to sedoheptulose-7-monophosphate (7C-P) very quickly. 3-PGAL then combines with sedoheptulose-7-monophosphate (7C-P) in a reaction catalysed by transketolase to yield one molecule of xylulose-5-phosphate (5C-P) and one molecule of ribose-5-phosphate (5C-P), (Fig. 3.5). Both ribose-5-phosphate (phospho-riboisomerase) and xylulose-5-phosphate (phospho-pentose epimerase) can be converted to ribulose-5-phosphate (5C-P); 5C-P can be phosphorylated in the presence of ribulose-5-phosphate kinase to form ribulose-1,5-bisphosphate (RuBP) in the final step of the regeneration of the PCR cycle (Fig. 3.6).

In the generative phase, $5 \times 3C$ molecules (triose phosphates) are interconverted to yield $3 \times 5C$ molecules (pentose phosphates); a simplified schematic diagram of these interconversions is shown in Figure 3.7.

The reactions that make up the C3-photosynthetic carbon reduction cycle are catalysed by various isomerases, epimerases, aldolases, kinases, phosphatases and transketolase (Table 3.1). All of these enzymes have now been found in the stroma and all of the reactions have been demonstrated *in vitro*, at rates that would support maximal rates of photosynthesis.

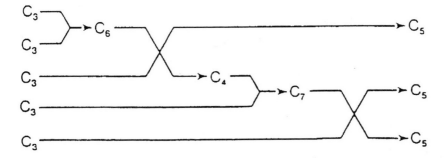

Figure 3.7 A schematic diagram showing the interconversions of triose phosphates (3C) into pentose phosphates (5C)

Table 3.1 Reactions of the photosynthetic carbon reduction cycle

Reaction	Enzyme
ribulose-1,5-bisphosphate + CO_2 + H_2O $\longrightarrow$ 3-phosphoglycerate	RubisCO
3-phosphoglycerate + ATP $\longrightarrow$ 1,3-bisphosphoglycerate	Kinase
1,3-bisphosphoglycerate + NADPH $\longrightarrow$ 3-phosphoglyceraldehyde + NADP	Dehydrogenase
3-phosphoglyceraldehyde $\longrightarrow$ dihydroxyacetone phosphate	Isomerase
dihydroxyacetone phosphate + 3-phosphoglyceraldehyde $\longrightarrow$ fructose-1,6-bisphosphate	Aldolase
fructose-1,6-bisphosphate + H_2O $\longrightarrow$ fructose-6-phosphate + P_i	Phosphatase
fructose-6-phosphate + 3-phosphoglyceraldehyde $\longrightarrow$ erythrose-4-phosphate + xylulose-5-phosphate	Transketolase
erythrose-5-phosphate + dihydroxyacetone phosphate $\longrightarrow$ sedoheptulose-1,7-bisphosphate	Aldolase
sedoheptulose-1,7-bisphosphate + H_2O $\longrightarrow$ sedoheptulose-7-phosphate	Phosphatase
sedoheptulose-7-phosphate + 3-phosphoglyceraldehyde $\longrightarrow$ xylulose-5-phosphate + ribose-5-phosphate	Transketolase
xylulose-5-phosphate $\longrightarrow$ ribulose-5-phosphate	Epimerase
ribose-5-phosphate $\longrightarrow$ ribulose-5-phosphate	Isomerase
ribulose-5-phosphate + ATP $\longrightarrow$ ribulose-1,5-bisphosphate	Kinase

Energetics of the PCR Cycle

For every three turns of the PCR cycle three molecules of ribulose-1,5-bisphosphate (15C) condense with three molecules of CO_2 (3C) to form six molecules of 3-phosphoglycerate (18C). These six molecules of 3-phosphoglycerate are reduced to six molecules of 3-phosphoglyceraldehyde using 6 ATP (in the synthesis of 1,3-bisphosphoglycerate) and 6 NADPH (in the reduction of 1,3-bisphosphoglycerate to 3-phosphoglyceraldehyde). One of these molecules of 3-PGAL (3C) is the net yield of the PCR cycle. The other five 3-PGAL molecules (15 C) are rearranged to form three molecules of RuBP (15C); this last step requires 1 ATP per RuBP formed, that is, 3 ATP. Thus, three turns of the cycle, which fix three molecules of CO_2 and yield a net one molecule of PGAL, require 6 NADPH and 9 ATP. Therefore the reduction of one molecule of CO_2 requires 2 NADPH ($2 \times 217\,kJ = 434\,kJ$) plus 3 ATP ($3 \times 29\,kJ$), that is, a total of 521 kJ. The oxidation of one mole of hexose yields about $469\,kJ\,mol^{-1}$ CO_2; this represents about 90% efficiency in energy storage for the PCR cycle. Six turns of the cycle would generate $6 \times RuBP$, producing the equivalent of one hexose; twelve turns would yield the equivalent of a sucrose molecule and so on.

Autocatalysis

In order to incorporate CO_2 the plant cell needs the acceptor molecule RuBP. To assimilate large amounts of CO_2 the cell either needs large amounts of RuBP or needs to use a small amount of RuBP frequently; the plant cell uses the latter strategy. The PCR cycle is autocatalytic, that is, the product, triose phosphate, can be recycled to generate more substrate (RuBP). If the cycle turns over five times, the amount of CO_2-acceptor, RuBP, can be doubled. In darkness, when photosynthesis is switched off, carbon is used for other metabolic processes, resulting in a sharp decrease in the levels of the intermediates, including RuBP, of the PCR cycle. When photosynthesis restarts, the rate could be severely limited by the unavailability of RuBP, the CO_2-acceptor. Normally the carbon fixed through the PCR cycle is exported from the chloroplast or accumulated as starch. The autocatalytic activity of the PCR cycle provides additional CO_2-acceptor by diverting carbon towards the generation of RuBP instead (Fig. 3.8). In export

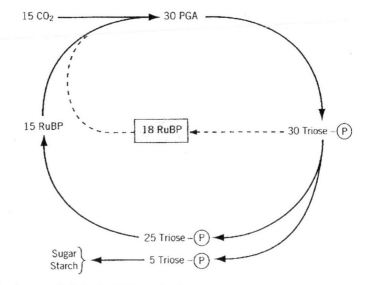

Figure 3.8 Autocatalysis in the PCR cycle. (From Hopkins 1999, reproduced with permission)

from the chloroplast this way there is always sufficient CO$_2$-acceptor in the chloroplast to support high rates of photosynthesis. This built-in autocatalytic property of the PCR cycle may help to explain why all photosynthetic organisms ultimately rely on the C3 cycle for CO$_2$ fixation. The mechanism by which autocatalysis is controlled is not known. It may occur as a consequence of competition between the enzymes favouring recycling and those leading to starch synthesis and product export from the chloroplast.

Regulation of the PCR Cycle

The high energy efficiency of the PCR cycle suggests that there must be some form of regulation to avoid wasteful cycling and to ensure adequate levels of the components of the cycle coupled to proper allocation of carbon for energy production and synthesis of sucrose and starch. The concentration of each enzyme in the stroma is controlled by the expression of both the nuclear and the chloroplast genes (Whitfield and Bottomly 1983). Short-term regulation of the enzymes of the PCR cycle is achieved mainly by light-activated mechanisms and metabolite modulation.

(a) Light-activated regulation

Regulation of RubisCO

RubisCO activity is light-regulated. RubisCO and other enzymes of the PCR cycle turn off rapidly in the dark, thereby conserving ATP that is generated in the dark for other synthetic reactions. The effect of light on RubisCO activity is apparently indirect: it involves complex interactions between Mg^{2+} and H$^+$ ion fluxes across the thylakoid membrane, pH changes in the chloroplast, CO$_2$ activation and an activating protein.

Although the reactions of the PCR cycle are called 'dark' reactions, it is light that induces the ion concentration and pH changes that promote the activity of RubisCO and other enzymes of the cycle. As previously noted (Chapter 2), light-driven electron transport is accompanied by a net movement of protons from the stroma into the lumen of the thylakoid. This movement of H$^+$ ions across the thylakoid membrane generates a proton gradient which results in an increase in the pH of the stroma from around pH 7 to around pH 8. *In vitro* pH 8 favours RubisCO activity. Light also promotes the movement of Mg^{2+} ions from the lumen outwards to the stroma; this balances the H$^+$ ion movement in the opposite direction. The light-induced increase in pH and Mg^{2+} ion level in the stroma provides an environment close to the optimum for the activation of the enzymes of the PCR cycle. For example, the darkened stroma has a pH of about 7 and an Mg^{2+} ion concentration of 1–3 mM, whereas the illuminated stroma has a pH of about 8 and an Mg^{2+} ion concentration of 3–6 mM (Leegood *et al.* 1985).

In the 1970s it was shown that, *in vitro*, RubisCO was activated by pre-incubation with CO$_2$ and Mg^{2+} ions to yield an active carbamylated form of the enzyme: RubisCO uses CO$_2$ not only as a substrate but also as an activator. The activating

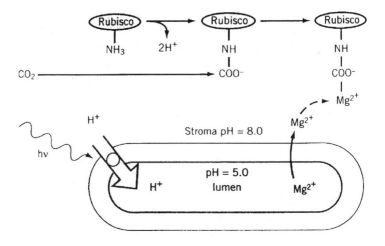

Figure 3.9   Activation of RubisCO is facilitated by the increase in stromal pH and Mg^{2+} concentration that accompanies light-driven electron transport. (From Lorimer and Moziorko 1980)

CO_2 must bind to an activating site that is separate and distinct from the substrate-binding site. Lorimer and Moziorko (1980) proposed a model for RubisCO activation which takes into account CO_2, Mg^{2+} and pH (Fig. 3.9). According to this model the enzyme is converted into a catalytically active form by the addition of the activator-CO_2 to the uncharged α-amino group of a specific lysine residue (lys-201) within the active site of the enzyme. The resulting carbamate (lysine-NH-CO_2^-) then binds Mg^{2+} to the carboxyl group to form the activated complex, the active enzyme. The Mg^{2+} in turn, forms part of the binding site for a second molecule of CO_2, which acts as the substrate in the carboxylase-catalysed reaction. The reactive molecule of CO_2 probably adds to the enolate of ribulose-1,5-bisphosphate to form 2-carboxy-3-ketoarabinitol-1,5-bisphosphate as an intermediate (Fig. 3.10). Two protons are released during these reactions so that activation is promoted by an increase in pH as well as by increases in Mg^{2+} ion concentration.

However, this *in vitro* model could not fully account for the activation of RubisCO in leaves (Portis 1990). Carbamylation *in vitro* occurs only in the presence of millimolar levels of CO_2, whereas the *in vivo* concentration of CO_2 in the leaf cell would only be about 10 μM; measured concentrations of *in vivo* Mg^{2+} and CO_2 levels and pH differences were too low. How then could RubisCO be carbamylated *in vivo*? The problem was solved when a mutant of *Arabidopsis thaliana* was found, in the early 1980s (Portis 1992), that failed to activate RubisCO in the light, even though the enzyme isolated from the mutant was apparently identical to that isolated from the wild type. Electrophoretic analysis showed that the *rca*-mutant was lacking a soluble chloroplast protein. RubisCO could be activated *in vitro* by adding this missing protein to a reaction mixture containing RubisCO, RuBP and physiological concentrations of CO_2. The protein was named RubisCO-activase because of its light-dependent enhancement of RubisCO. The mode of action of RubisCO-activase

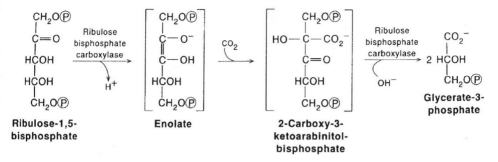

Figure 3.10   The reaction catalysed by ribulose bisphosphate carboxylase

is not clear but it is known to require ATP. Binding of RuBP to the active site of RubisCO prevents carbamylation; the activase may be involved in removing bound RuBP from the active site to allow carbamylation.

RubisCO is also subject to negative control. It was found, in the early 1980s, that in some plants, such as soybean, RubisCO in extracts from dark-grown leaves could not be activated by CO$_2$ and Mg^{2+} ions, whereas in leaf extracts of light-grown leaves RubisCO was readily activated. It has been shown that this is due to the presence of a RubisCO inhibitor, 2-carboxy-arabinitol-1,5-bisphosphate (CA1P) in dark-grown plants. CA1P is a transition state analogue of 2-carboxy-3-keto-arabinitol-1,5-bisphosphate, a β-keto acid intermediate of the RubisCO carboxylation of RuBP reaction. Transition state analogues bind to an enzyme more tightly than a substrate binds to the enzyme in the enzyme–substrate complex (because they fit the active site better); they are thus often referred to as *tight-binding inhibitors*. Due to the fact that this compound is synthesized in the dark it is often misleadingly called 'night' or even 'nocturnal' inhibitor. It is gradually broken down in the presence of light, permitting reactivation of RubisCO.

Regulation of other PCR enzymes

At least four other chloroplast enzymes essential for the operation of the PCR cycle are subject to another kind of light regulation. The activities of glyceraldehyde-3-phosphate dehydrogenase, fructose-1,6-bisphosphatase, sedoheptulose-1,7-bisphosphatase and ribulose-5-kinase are controlled via a light-regulated, thiol-based oxidation–reduction system. The activity of these enzymes is associated with the photosynthetically produced reductant, thioredoxin (Buchanan 1980).

Like ferredoxin, thioredoxin is a small, iron–sulphur protein. It contains two adjacent cysteinyl residues that undergo reversible oxidation–reduction from the reduced sulphydryl (—SH—HS—) state to the oxidized disulphide (—S—S—) state. In the illuminated chloroplast, PSI donates electrons for the reduction of ferredoxin, which, in turn, reduces thioredoxin in a reaction catalysed by ferredoxin–thioredoxin reductase. The reduced thioredoxin reduces the disulphide bonds of the cysteinyl residues of the target PCR enzyme resulting in its activation (Fig. 3.11).

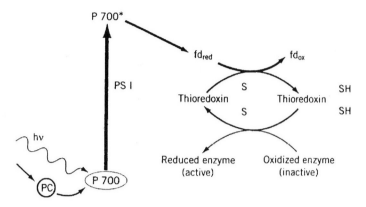

Figure 3.11 The ferredoxin–thioredoxin system for light-driven enzyme activation. (From Hopkins 1999)

In the dark, the cysteinyl groups of the target enzyme are oxidized and the enzyme remains inactive. The mechanism of thioredoxin-modulated enzyme activation is clearly seen in the regulation of fructose-1,6-bisphophatase (FBPase). Chloroplastic FBPase, like the cytosolic enzyme, is regulated by light; it differs from its cytosolic counterpart, however, in that it is also regulated by thioredoxin. The activity of the enzyme is sensitive to pH. FBPase is inactive in the dark in its oxidized form and at its optimum pH 8.8; in the light, the active form of the enzyme has an optimum pH of 7.5–8.5. In general, the stromal pH and Mg^{2+} ion concentration, both of which increase due to proton-pumping into the lumen, act as regulators of the PCR cycle. The enzyme phosphoribulokinase, which catalyses the conversion of ribulose-5-phosphate to ribulose-1,5-bisphosphate (Fig. 3.6) and which is also regulated by the thioredoxin system, is subject to another type of pH regulation, through its inhibition by 3-phosphoglycerate (3-PGA), the RubisCO reaction product. The enzyme is inhibited only by the 3-PGA^{2-} form of the 3-phosphoglycerate. In light, active photosynthesis results in the stroma becoming alkaline (pH about 8) and 3-phosphoglycerate is predominantly in the 3-PGA^{3-} form. However, in darkness, the stromal pH drops, 3-phospho-glycerate exists mainly as 3-PGA^{2-} and the enzyme phosphoribulokinase is increasingly inhibited. This inhibition prevents ATP uptake by the enzyme and so the assimilation of CO_2 is retarded.

(b) Metabolite regulation

Most of the proteins required for the metabolic pathways in the chloroplast are encoded in the nucleus. The outer membrane of the chloroplast is freely permeable to even large molecules; the inner membrane presents a more selective barrier to metabolite movement. Regulation of the translocation of intermediates of the PCR cycle across the chloroplast envelope is essential since most of them are also involved in sucrose synthesis, which occurs exclusively in the cytosol.

While the inner chloroplast membrane is impermeable to most phosphorylated compounds, including fructose-6-phosphate, glucose-6-phosphate and fructose-1,6-bisphosphate, it does possess a specific transporter that catalyses the one-for-one exchange of P_i with triose phosphates, either phosphoglycerate or dihydroxyacetone phosphate (Fig. 3.12).

Only triose phosphates in which the phosphate is attached to the end of the chain, e.g. glycerate-3P, glyceraldehyde-3P or dihydroxyacetone-3P, are translocated. In addition, compounds are translocated out of the chloroplast in the form carrying a double negative charge, such as 3-PGA^{2-}; in light, phosphoglycerate occurs as 3-PGA^{3-} and is retained in the stroma.

The rate at which carbon is removed from the PCR cycle is controlled by the balance between starch synthesis in the chloroplast and sucrose synthesis in the cytosol. The phosphate transporter provides this control by coordinating the allocation of triose phosphates between these two anabolic pathways. If sucrose synthesis is too rapid, the transport of excess P_i into the chloroplast will result in the removal of too much triose phosphate. This will have a deleterious effect on carbon fixation by the PCR cycle because five out of every six triose phosphate molecules

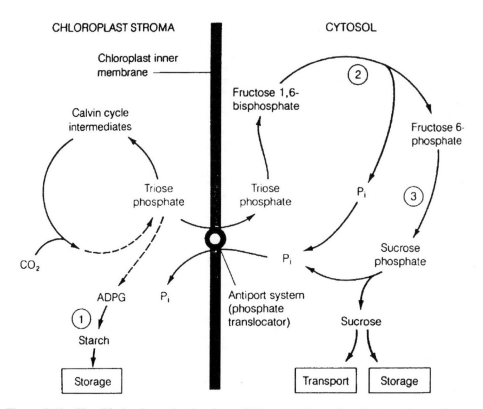

Figure 3.12 Simplified schematic drawing of P_i and triose phosphate exchange between chloroplast stroma and cytosol. (From Preiss 1982). With permission, from the *Annual Review of Plant Physiology* Volume 33 © 1982 by Annual Reviews www.annual reviews.org

are needed to regenerate RuBP. If sucrose synthesis is too slow, insufficient P_i will be available to the chloroplast for triose phosphate synthesis.

Photorespiration – the Carbon Oxidative Cycle (PCO)

In the 1920s, Otto Warburg, a German biochemist, observed that photosynthesis in the alga *Chlorella* was inhibited by O_2. This inhibition occurs in all C3 plants studied since and is termed the *Warburg effect* (Fig. 3.13). Historically it was assumed that respiration in the mitichondria and photosynthesis in the chloroplasts were effectively independent. It was postulated that photosynthesis could supply all the energy needed by the leaf and that mitochondria 'shut down' in the light. Apparent or net CO_2 fixation is the amount by which photosynthesis exceeds respiration; because respiration releases CO_2 continuously, net photosynthesis is represented by photosynthetic CO_2 uptake minus the CO_2 evolved during respiration. Respiration in green plants is not, as was believed for a long time, independent of light. In fact, respiration (CO_2 release and O_2 uptake) in higher plants is much greater in light than in darkness. This is substantiated by the observation (Decker 1955) that increased levels of CO_2 can be measured for several minutes after illuminated leaves are switched to darkness (the so-called post-illumination burst of CO_2). This light-dependent CO_2 evolution is termed *photorespiration*. The process is restricted to photosynthetically active cells and differs from normal respiration in many fundamental ways. Photorespiration has, for example, a very low affinity for O_2: in most tissues mitochondrial O_2-uptake is saturated at an atmospheric content of 10–20 ml O_2 per litre whereas photorespiration is not saturated even in a pure O_2 atmosphere. Respiration in illuminated photosynthetic tissues occurs by two processes – the normal process of respiration that occurs in all plant tissues even during darkness and the more rapid, light-dependent photorespiration. The two processes are spatially separated within the cell: normal respiration occurs within the mitochondria and the cytosol whereas photorespiration involves the cooperation of chloroplasts, peroxisomes and mitochondria (Ogren 1984; Husic *et al.* 1987). Early reports (Decker 1955) of higher respiration rates for C3 plants in light than in darkness were not readily accepted in the beginning; it is now known that total respiration in leaves of C3 species is often two or three times as rapid in light as in darkness.

Photorespiration is not coupled to oxidative phosphorylation. It appears to constitute a severe drain on chloroplast metabolism: it does not generate NADPH, it consumes ATP and O_2 and releases CO_2. Photorespiration is a wasteful process and appears to be a net liability on the cells in which it occurs since it decreases photosynthetic output by siphoning organic material from the Calvin cycle. It results in the reoxidation of the products just previously assimilated in photosynthesis (Bidwell 1983); in some plants, as much as one-third of the CO_2 that is fixed is released again immediately by photorespiration (Gerbaud and André 1987; Sharkey 1988).

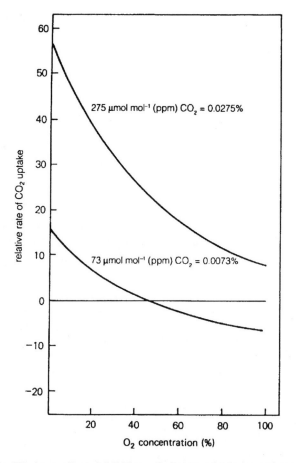

275 μmol mol⁻¹ (ppm) CO₂ = 0.0275%

73 μmol mol⁻¹ (ppm) CO₂ = 0.0073%

Figure 3.13 The Warburg effect: inhibition of photosynthesis in soybean (C3) plants by O_2. (From Forrester *et al.* 1966)

RubisCO shows oxygenase activity

Work on photorespiration during the 1960s was paralleled by interest in the synthesis and metabolism of the two-carbon compound glycolate. Glycolate was found to be extensively synthesized in green leaves in high concentrations of O_2 or low concentrations of CO_2. It gradually became apparent that glycolate metabolism was linked to photorespiration and that the enzymes involved were located in chloroplasts, peroxisomes and mitochondria. The enzymatic pathway of glycolate synthesis (Fig. 3.15), called the photosynthetic carbon pathway (PCO) was elucidated. Photosynthesis is always accompanied by photorespiration; the latter process takes place in the light, consumes O_2 and converts ribulose-1,5-bisphosphate to glycolate and CO_2. The key to photorespiratory CO_2 evolution and glycolate metabolism was established by the discovery (Bowes *et al.* 1971) of the dual nature of RubisCO. In addition to its carboxylase activity, RubisCO was

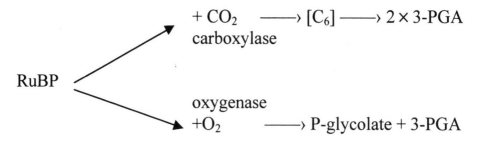

$$+ CO_2 \longrightarrow [C_6] \longrightarrow 2 \times 3\text{-PGA}$$
carboxylase

RuBP

oxygenase
$$+O_2 \longrightarrow P\text{-glycolate} + 3\text{-PGA}$$

Figure 3.14   The dual nature of RubisCO

found to have an oxygenase activity and hence the full name ribulose-1,5-bisphosphate carboxylase–oxygenase. The carboxylase activity of RubisCO adds CO_2 to RuBP to yield two molecules of 3-phosphoglyceric acid (3-PGA), whereas the oxygenase activity of the enzyme adds O_2 to RuBP to yield one molecule of 3-PGA and one molecule of phosphoglycolate (Fig. 3.14).

The phosphoglycolate is subsequently metabolized in a series of reactions that result in the release of CO_2 and the recovery of the remaining carbon by the PCR cycle. The two reactions are competitive, that is, CO_2 and O_2 compete for the same active site on the RubisCO enzyme; the outcome of the competition depends on the relative concentrations of CO_2 and O_2. The rate of the carboxylase reaction is four times that of the oxygenase reaction under normal atmospheric conditions at 25 °C. Although the concentration of O_2 (21%) in the present-day atmosphere is much higher than that of CO_2 (0.03%), CO_2 is more soluble in the stroma than O_2 and the affinity of RubisCO for CO_2 ($K_m = 12\,\mu M$) is much greater than for O_2 ($K_m = 200\,\mu M$). The environmental condition that fosters photorespiration is that of hot, dry, bright days. On such days plants close their stomata in order to prevent water loss, and photosynthesis soon depletes CO_2 and increases the concentration of O_2 in illuminated chloroplasts as a result of the photolysis of H_2O. In addition, the affinity of RubisCO for CO_2 decreases with increasing temperature, a factor which favours the oxygenase-catalysed reaction. Thus O_2 'fixation' by RubisCO contributes considerably to the catabolism of RuBP and also constitutes a significant waste of energy. Carbon dioxide is the preferred substrate for RubisCO; the active site of the enzyme probably evolved at a time when O_2 was not an important component of the atmosphere. As O_2 accumulated in the atmosphere due to photosynthesis in algae and early land plants, RubisCO necessarily began to fix O_2 because its active site could not discriminate between CO_2 and O_2.

C2 glycolate pathway

The phosphoglycolate produced by the oxygenase action of RubisCO on RuBP cannot be metabolized within the chloroplast by means of the Calvin cycle. Salvaging the carbons from the phosphoglycolate is achieved by the C2 glycolate pathway (Fig. 3.15). This pathway requires the cooperation of three cell

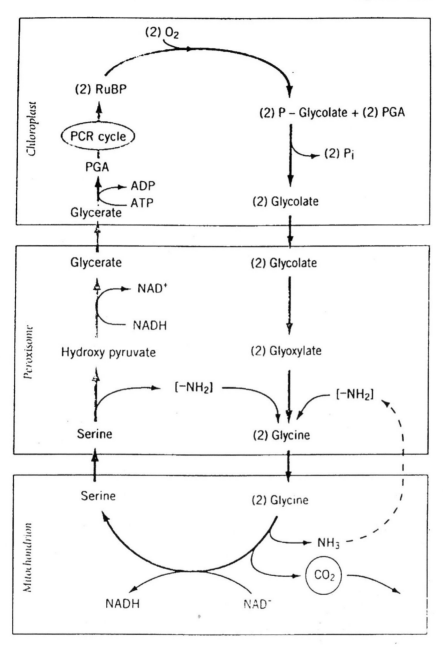

Figure 3.15 The photorespiratory C2 glycolate pathway

organelles – chloroplasts, peroxisomes and mitochondria. The key features of the C2 cycle are (a) the formation of the 2C glycolate in the chloroplast, (b) the conversion of glycolate to glycine in the peroxisome, (c) the decarboxylation of glycine in the mitochondrion to form serine, CO_2 and NH_3, (d) the conversion of

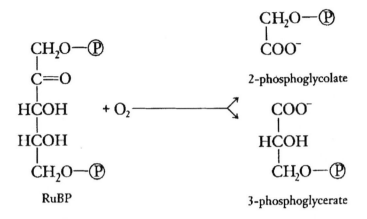

Figure 3.16   The RubisCO oxygenase reaction

serine to phosphoglycerate in the peroxisome and (e) the re-entry of the phosphoglycerate into the Calvin cycle in the chloroplast.

The C2 glycolate cycle thus results in the release of a molecule of CO_2 and the salvaging of the other carbon by the PCR cycle. The C2 glycolate pathway is initiated by the oxidation of RuBP by RubisCO to form 3-PGA and P-glycolate (Fig. 3.16). The 3-PGA can be further metabolized in the PCR cycle. The P-glycolate is rapidly dephosphorylated to form glycolate and P_i in the chloroplast. The glycolate is exported from the chloroplast into the adjacent peroxisome. Peroxisomes are small organelles that contain several oxidative enzymes. They exist almost exclusively in photosynthetic tissues and in electron micrographs appear to be associated directly with chloroplasts (Fig. 3.17).

In the peroxisome glycolate is oxidized to glyoxylate and hydrogen peroxide in a reaction catalysed by glycolate oxidase. The H_2O_2 generated is then decomposed to H_2O and O_2 by catalase, an enzyme that is abundant in peroxisomes. The glycolate is converted to glycine (a 2C amino acid) in a transamination reaction that probably involves glutamate as the amino donor. Glycine then enters the mitochondrion. Here two molecules of glycine (4C) are converted into one molecule of serine (3C) plus one molecule of CO_2 and one NH_4^+ ion (Walker and Oliver 1986). Glycine is thus the immediate source of the CO_2 released in photorespiration. The NH_4^+ released is not lost but is assimilated into amino acids in the chloroplast. The serine then leaves the mitochondrion and enters the peroxisome where it is transaminated to form hydroxy-pyruvate which is subsequently reduced to glycerate. Glycerate then enters the chloroplast and is phosphorylated to 3-PGA.

The role of photorespiration

For every two molecules of phosphoglycolate (4C) formed by photorespiration, one molecule of 3-PGA (3C) is ultimately recycled and one molecule of CO_2 is lost, i.e.

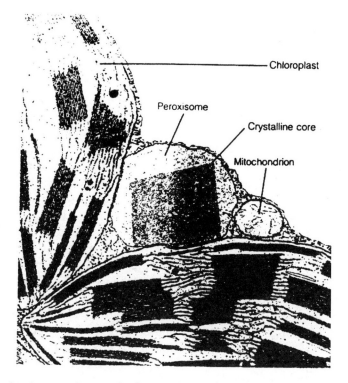

Figure 3.17   An electron micrograph of a peroxisome seen in close association with a chloroplast and a mitochondrion. The crystalline structure in the peroxisome is catalase. (From Taiz and Zeiger 1991)

one C lost for every two O$_2$ fixed. This loss of 25% of the C from the PCR cycle due to photorespiration represents a serious decrease in the efficiency of photosynthesis. There is also an energy cost associated with photorespiration. The amount of ATP and NADPH consumed in the glycolate pathway is roughly equal to that used for the reduction of CO$_2$ in the PCR cycle, yet there is a net loss of carbon in the PCO cycle. The rate of photorespiration is promoted by increases in temperature, light and the O$_2$/CO$_2$ ratio. As we have seen, changes in the rate of photorespiration with changing O$_2$/CO$_2$ ratios is due to competition between these substrates for the active site of RubisCO. The influence of temperature on photorespiration is due to the effects of temperature on the relative solubilities of O$_2$ and CO$_2$ in the chloroplast stroma. Increase in temperature increases the solubility of O$_2$ relative to CO$_2$ and so oxygenation is favoured at higher temperatures. Increase in light intensity also increases photorespiratory rates since increase in the rates of photosynthesis due to higher light intensity raises the level of O$_2$ and depletes the CO$_2$ content of the leaf.

The physiological justification of this wasteful photorespiratory C2 cycle remains unexplained. Generally C3 plants do not need photorespiration. In fact, CO$_2$-enriched air is commonly used to increase yields in glasshouse crops such as tomatoes. Mutants of *Arabidopsis thaliana* which are deficient in some of the enzymes of the photorespiratory pathway also grow normally in CO_2-enriched air. Such mutants survive if the oxygenase activity of RubisCO is suppressed. However, in normal air these plants die because of the accumulation of toxic amounts of glycolate. Photorespiration could, in this case, be regarded as a mechanism for glycolate detoxification. Some experts hold that photorespiration is a means of removing excess ADP and NADPH produced during high light intensities (Ogren 1984). When leaves are exposed to high irradiance in the absence of CO_2 they become photoinhibited because the absorbed light energy cannot be used efficiently. At low concentrations of CO_2 this energy is used to reduce oxygen, generating toxic oxygen radicals which cause photodestruction of pigments. Photorespiration may thus allow the harmless dissipation of light energy when CO_2 levels are low.

In addition to the loss of CO_2, NH_4^+ is also released at a high rate during photorespiration, as glycine is converted to serine (Fig. 3.15); the rate of NH_4^+ production in C3 plants is 10–20 times the rate of primary nitrate assimilation. Since NH_4^+ is an important reservoir of nitrogen and is also toxic to plant tissues, it must be efficiently recovered. The NH_4^+ released is reassimilated in the chloroplasts by the glutamine synthetase/glutamate synthase pathway. The amino (NH_2^+) group liberated in the deamination of serine re-enters the PCO cycle as α-ketoglutarate which is transaminated to glutamic acid. The rates of assimilation of nitrogen as NH_4^+ and NH_2^+ are high, so that the photorespiratory nitrogen cycle can be regarded as a major pathway of nitrogen metabolism in C3 plants.

C4 Photosynthetic Carbon Assimilation (PCA) Cycle

The C4 pathway of CO_2 fixation

Plants that grow in high light intensity and in hot, dry environments, must keep their stomata closed for most of the time in order to prevent excessive water loss. This causes the CO_2 level inside the leaf to fall sharply, a condition that depresses RubisCO's carboxylase activity and favours photorespiration. Many plants that can grow in hot, dry environments, including commercially important crops such as maize and sugarcane, have evolved a mechanism to avoid this wasteful process. These plants preface the Calvin cycle with a process that fixes CO_2 first into four-carbon organic acids (Fig. 3.18). Such plants are called C4 plants and their mode of CO_2 fixation is termed the C4 pathway because $^{14}CO_2$ labelling showed that the first product formed during photosynthesis was a four-carbon compound such as oxaloacetate or malate; in C3 plants the three-carbon PGA initiates the C3 pathway of the Calvin cycle.

Kranz anatomy of leaves

Most C4 species are monocots (especially grasses and sedges), although more than 300 are dicots. Several thousand species in at least 19 plant families use the C4 pathway of CO_2 fixation. C4 plants are generally tropical or subtropical and include temperate-zone crop plants (such as maize and sugarcane) that are native to the tropics. No one family has been found to consist exclusively of C4 plants. All 19 families contain both

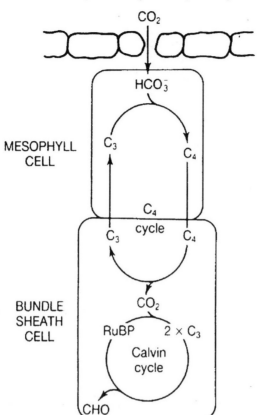

C_4 dicarboxylic acid pathway

Figure 3.18 Spatial separation of the components of the C4 pathway. (From Leegood *et al.* 1997)

C3 and C4 plants, which suggests that C4 metabolism may have evolved later in time than C3 metabolism. C4 plants typically have high photosynthetic rates, high growth rates, low photorespiration rates and low rates of water loss.

Leaves of C4 plants have a unique anatomy. C4 photosynthesis involves two distinct types of photosynthetic cells. One (sometimes two) layer of tightly packed, large, thick-walled bundle sheath cells, containing abundant large chloroplasts surrounds the vascular bundle (Fig. 3.19). This concentric arrangement of cells is referred to as *kranz* (German for 'halo' or 'wreath') *anatomy*. The bundle sheath cells, in turn, are surrounded by thin-walled, chloroplast-containing mesophyll cells which are loosely arranged and are adjacent to the air spaces of the leaf. The cell types are so arranged that a mesophyll cell is never more than one cell away from a bundle sheath cell. The thickened walls of the bundle sheath cells are often suberized and contain numerous plasmodesmata connecting them to the mesophyll cells. This unique leaf anatomy neatly accommodates the CO_2-fixing function of the

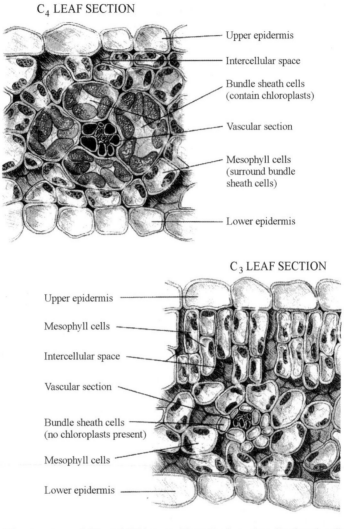

Figure 3.19   The structure of C3 and C4 leaves. Note the large bundle sheath cells of the C4 leaf containing chloroplasts and surrounded by mesophyll cells. (From Horton *et al.* 1993)

C4 plant. C4 plants possess a CO_2 pump which is divided between the two cell types; based on a cycle of carboxylation (mesophyll) and decarboxylation (bundle sheath) it generates a high level of CO_2 in the vicinity of RubisCO. The CO_2 may be concentrated within the bundle sheath, whose suberized, thickened walls are practically impermeable to gases. In addition, the arrangement of the mesophyll cells and the numerous plasmodesmata connections allows division of labour and close cooperation between the two cell types – the trapping of CO_2 by the mesophyll followed by its transfer to and fixation in the bundle sheath.

All C4 plants share the common initial step of PCR carboxylation, and, until recently, it was thought that kranz anatomy was essential for C4 photosynthesis in terrestrial plants. A number of recent studies challenge the view that kranz anatomy is essential for C4 photosynthesis by demonstrating that C4 photosynthesis functions in single cells of the central Asian chenopods *Borszczowia aralocaspica* and *Bienertia cyclopetra* (Voznesenkaya *et al.* 2001; Freitag and Stichler 2002; Sage 2002). These species have the photosynthetic features of C4 plants, yet lack kranz anatomy. They accomplish C4 photosynthesis through spatial compartmentation of photosynthetic enzymes within the same chlorenchyma cell in which no periclinal walls separate PCA and PCR regions. In other words, instead of occurring in distinct cells, PCA and PCR metabolism occur at opposite ends of the same elongated cell. Phosphoenolpyruvate (PEP) regeneration (below) occurs in chloroplasts located at the distal end of the cell away from the vascular bundles, whereas RubisCO and PCR functions are localized at the proximal end of the same cell near the vascular bundles.

CO₂ fixation in C4 plants

In the mid-1960s, H.P. Kortschack and his co-workers (1965), using $^{14}CO_2$, demonstrated that certain C4 acids were the early products of photosynthesis in sugarcane and maize; sugar phosphates were not formed until later. These findings were confirmed and extended by Hatch and Slack (1966).

When $^{14}CO_2$ was fed to leaves which were undergoing steady-state photosynthesis (similar to Calvin's work with algae) the four-carbon acids malic and aspartic were found to be the first stable products of CO_2 fixation and hence the term C4 photosynthesis was introduced. Hatch and Slack proposed a cyclic mechanism for the process in which the carbon is first incorporated into a C4 acid and subsequently transferred as CO_2 into the PCR cycle (Fig. 3.20).

PEP-carboxylase fixes CO₂ into C4 acids

Enzymatic investigations of maize and other C4 plants in the 1970s established that the mesophyll cells contain all the phosphoenolpyruvate carboxylase (PEPCO) in the leaf whereas RubisCO and the other enzymes of the Calvin cycle are exclusively located in the chloroplasts of the bundle sheath cells. Carbon dioxide diffuses into the leaf via the stomata and is hydrated to bicarbonate (HCO_3^-). Bicarbonate (and not CO_2) and phosphoenolpyruvate (PEP) in the mesophyll cells are the substrates of a carboxylation reaction catalysed by PEPCO, which produces the 4C oxaloacetate. PEPCO in the mesophyll cells has a high affinity for HCO_3^- and can fix CO_2 more efficiently than can RubisCO. Unlike RubisCO, PEPCO has no site for O_2 as an alternative substrate and so there is no competition between CO_2 and O_2 for this enzyme. The reaction serves to fix and concentrate CO_2 in the form of C4 acids. Depending on the C4 species, the oxaloacetate formed is either reduced to malate (4C) or transaminated to aspartate (4C). These four-carbon acids, which contain the fixed CO_2, are transported via plasmodesmata from the mesophyll into the adjacent bundle sheath cells. The decarboxylating malate dehydrogenase ('malic enzyme') in the bundle sheath cells, splits the imported malate into CO_2 and

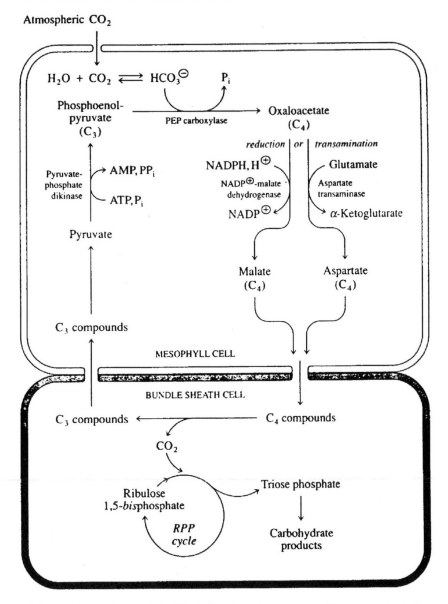

Figure 3.20   Scheme of the Hatch–Slack pathway. (From Horton *et al.* 1993)

pyruvate. The pyruvate moves back into the mesophyll cells and thus completes the cycle. The decarboxylation reaction creates a high CO_2/O_2 ratio in the bundle sheath cells, thereby enhancing the carboxylase activity of RubisCO and minimizing its oxygenase action. Thus, there is essentially no photorespiration in C4 plants. The CO_2 arising from the decarboxylation of malate is fixed again, this time by RubisCO, in exactly the same reaction that occurs in C3 plants, resulting in the incorporation of CO_2 into 3-phosphoglycerate. The pyruvate formed by the decarboxylation reaction and returned to the mesophyll cells is converted to PEP by an unusual enzymatic reaction catalysed by pyruvate phosphate dikinase. This enzyme is called dikinase because two different molecules are phosphorylated simultaneously by one molecule of ATP: pyruvate is phosphorylated to PEP, phosphate is phosphorylated to pyrophosphate:

$$\text{Pyruvate} + \text{ATP} + \text{P}_i \xrightarrow{\text{Pyruvate-phosphate dikinase}} \text{PEP} + \text{AMP} + \text{PP}_i$$

Removal of the products AMP and PP$_i$ favours the synthesis of PEP:

$$\text{PP}_i \xrightarrow{\text{pyrophosphatase}} 2\,\text{P}_i$$

$$\text{AMP} + \text{ATP} \xrightarrow{\text{adenylate kinase}} 2\,\text{ADP}$$

Since AMP (and not ADP) is the product of the dikanase-catalysed reaction, the effective cost of converting each molecule of pyruvate to ATP is 2 ATP. Within the basic pattern of the C4 pathway (Fig. 3.20) there are three variations. The initial carboxylation reaction occurs in the cytosol of the mesophyll cells and is the same for all three types. The variants are classified according to the particular C4 acid (malate or aspartate) transported into the bundle sheath cells and on the nature and location of the different enzymes that decarboxylate the C4 acids (Table 3.2).

While there is no great taxonomic distinction between plant groups, particular species favour one variant over the other. In the NAD-malic enzyme (NAD-ME) variant (bundle sheath chloroplast), oxaloacetate is first converted to malate which is exported to the chloroplast in the bundle sheath cell where it is decarboxylated to CO$_2$ and pyruvate:

$$\text{malate} + \text{NADP}^+ \longrightarrow \text{pyruvate} + \text{CO}_2 + \text{NADPH}$$

Table 3.2 Three variants of the C4 pathway of carbon assimilation

C4 acid returned to bundle sheath	Decarboxylating enzyme	Variant name	C3 acid returned to mesophyll	Example
Malate	NADP-dependent malic enzyme (chloroplast)	NADP-ME	Pyruvate	Maize, crabgrass, sugarcane, sorghum
Aspartate	NAD-dependent malic enzyme (mitochondrion)	NAD-ME	Alanine	Millet, pigweed
Aspartate	PEPCO-carboxy-kinase (cytoplasm)	PEPCK	Alkanine/ pyruvate	Guinea grass

Plant species which use malate as the main vehicle for CO_2 transport are termed 'malate formers'; species which use aspartic acid to transport CO_2 they transport from the mesophyll cells to the bundle sheath cells are termed 'aspartate formers'. In the other two variants NAD-malic enzyme (NAD-ME) and PEP carboxykinase (PCK), oxaloacetate is transaminated to aspartate which is transported to the bundle sheath cells; both of these variants are termed 'aspartate formers' (Flugge and Heldt 1991). The NAD-ME variant is located in the mitochondrion and decarboxylates malate to pyruvate and CO_2:

$$\text{malate} + NAD^+ \longrightarrow \text{pyruvate} + CO_2 + NADH$$

The third decarboxylation system, PEP-carboxykinase (PCK), located in the cytosol, operates mainly in 'aspartate formers'. PCK catalyses the decarboxylation of oxaloacetate (formed from aspartate) to phosphoenol pyruvate (PEP) and ADP and releases CO_2:

$$\text{oxaloacetate} + ATP \longrightarrow PEP + ADP + CO_2$$

The formation of either malate or aspartate is important from the energetics point of view of the C4 pathway. Malate formers can form reducing equivalents and, unlike aspartate, they transport 2 Hs per molecule of CO_2 from the mesophyll cells to the Calvin cycle in the bundle sheath cells. This could explain the absence of non-cyclic flow in the chloroplasts of bundle sheath cells of malate formers; aspartate must have PSII in the bundle sheath to generate the reductant required for carbon assimilation. The chloroplasts in the bundle sheath cells of the NADP-ME species lack thylakoid stacking (they are agranal). These chloroplasts are deficient in PSII; they do contain PSI which produces ATP but not the reductant NADPH, which is therefore provided by the malate from the mesophyll cells. Deficiency of PSII in the bundle sheath cells is probably due, in part, to the lack of O_2 in the gas-tight, kranz structure. Regardless of the variant employed, the principal effect of the C4 pathway is to concentrate CO_2 in the bundle sheath cells where the enzymes of the PCR cycle are located. The transport of CO_2 in the form of organic acids allows for the build-up of much higher concentrations of CO_2 than could be achieved by diffusion of CO_2 alone. Using radio-labelled $^{14}CO_2$, Edwards and Walker (1983) found the concentration of CO_2 in bundle sheath cells to be as high as 60 M – about ten times higher than in C3 plants. Under optimal conditions of light and temperature, C4 plants can assimilate CO_2 at rates two or three times that of C3 species. The CO_2-concentrating process of the C4 pathway hydrolyses 2 ATP per CO_2 transported. In addition, 3 ATP and 2 NADPH are required per molecule of CO_2 fixed in the PCR cycle. Thus the total energy for fixing CO_2 by the combined C4 PCA and the C3 PCR cycles is 5 ATP plus 2 NADPH per molecule of CO_2 fixed.

Regulation of C4 Metabolism

In addition to the regulation of the PCR cycle enzymes, C4 photosynthesis requires separate controls. The two metabolic pathways must be coordinated: the rates of

CO_2-pumping must parallel the rates at which RubisCO fixes CO_2 and the rates at which the electron transport system supplies ATP and NADPH. The C3 and C4 metabolic pathways are located in two different cell types; they are connected through the interconversion of 3-PGA and PEP in the mesophyll and so the withdrawal of carbon must be regulated. As might be expected, a number of the enzymes of the pathway are modulated by light, but regulation of the C4 pathway is, in general, not clearly understood.

Mesophyll enzymes

PEPCO, which is probably present in all plants, catalyses the addition of CO_2 to PEP to form the four-carbon oxaloacetete:

$$PEP + HCO_3^- \longrightarrow oxaloacetate + P_i$$

PEPCO is regulated by light–dark transitions; its activity is high in C4 plants in light in order to maximize CO_2 availability for the PCR cycle in the bundle sheath cells. The phosphoenolpyruvate, which is the substrate for the enzyme in C4 plants, is formed in the bundle sheath chloroplasts but it is also an important intermediate in glycolysis. Continued high PEPCO activity at night could deplete the PEP and thus seriously impair respiration. A wide range of *effectors* (phosphorylated intermediates such as triose-P and hexose-P, amino acids and organic acids) modulate PEPCO activity. The enzyme is activated by triose-P and hexose-P and is strongly inhibited by malate. Thus, as carbon release in the bundle sheath cells increases, triose-P and hexose-P levels will rise in the mesophyll and enhance PEPCO activity. Conversely, when carbon utilization in the bundle sheath cells falls, malate will accumulate and inhibit PEPCO. Malate blocks PEPCO activity by means of feedback inhibition whereby the product inhibits the activity of the enzyme (Jiao and Chollet 1991). PEPCO extracted from leaves of illuminated plants is less sensitive to inhibition by malate than is the enzyme extracted from darkened leaves (Doncaster and Leegood 1987). The lowered sensitivity in light to malate inhibition is important at a time when CO_2 assimilation and malate production are required to boost CO_2 levels in the bundle sheath cells. This change in sensitivity to malate has been shown by Nimmo *et al.* (1987) and Bakrim *et al.* (1992) to be associated with the phosphorylation state of the enzyme: the phosphorylated form is the biochemically active form and is less sensitive to malate than the non-phosphorylated form. These workers have identified two kinases that phosphorylate sorghum leaf PEPCO and showed that extracts from illuminated leaves have a greater capacity to phosphorylate the enzyme. PEPCO may thus be regulated by light-sensitive protein kinases. In addition, they reported that inhibitors of photosynthesis such as diuron gramicidin and DL-glyceraldehyde prevented phosphorylation of PEPCO in leaves. This may indicate the involvement of both electron transport and ATP in the signal transduction chain. Pierre *et al.* (1992) have suggested that changes in cytosolic pH and Ca^{2+} may also play a role in the light-induced signal pathway.

Pyruvate-P_i dikinase (PPDK) is another enzyme of the C4 pathway that is regulated by phosphorylation; it catalyses the reaction

$$\text{Pyruvate} + P_i + \text{ATP} \longrightarrow \text{PEP} + \text{AMP} + PP_i$$

The enzyme is activated by increased illumination and rapidly inactivated by low light intensity (Burnell and Hatch 1985). Inactivation results from ADP-dependent phosphorylation of a threonine residue on the enzyme; activation is accomplished by the phosphorolytic cleavage of this threonyl phosphate group. Phosphorylation and dephosphorylation of the threonine residue is catalysed by a single protein, pyruvate orthophosphate dikinase-regulatory protein. This a rare example of a protein which is bifunctional, that is, it has two active sites. These reactions are accompanied by the phosphorylation and dephosphorylation of a histidine residue on PPDK in a series of reactions that are not well understood. In the light the histidinyl phosphate group is normally transferred to pyruvate to generate PEP; in the dark there is probably insufficient pyruvate available. NADP-malate dehydrogenase, which catalyses the reaction

$$\text{Oxaloacetate} + \text{NADPH} + H^+ \longrightarrow \text{malate} + \text{NADP}^+$$

is regulated through the thioredoxin system; it is activated (reduced) in light and inactivated (oxidized) in the dark.

Bundle sheath enzymes

All three carboxylases (Table 3.2) which generate CO_2 for the PCR cycle in the bundle sheath cells are subject to regulation. $NADP^+$ malic enzyme (NADP-ME) is activated by illumination of leaves. When leaves of sugarcane were placed in darkness, transfer of ^{14}C from the carboxyl group of 4C acids ceased (Hatch and Slack 1966), indicating efficient modulation of decarboxylation by the enzyme. Inhibition of NAD^+-malic enzyme (NAD-ME) activity shows a sigmoidal response to malate concentration (Hatch et al. 1974). It is also inhibited by NADH so that the enzyme could be modulated by the $NADH/NAD^+$ ratio. Phosphoenolpyruvate carboxykinase (PEPCK) is inhibited by 3-PGA, fructose-6P, fructose-1,6-bisP and by dihydroxyacetone-P (Burnell et al. 1986) and would be expected to be switched off in the dark (Carnal et al. 1993). The enzyme has been shown to be subject to proteolytic cleavage in all plants (Walker et al. 1995) which may result in loss of regulatory properties.

Crassulacean Acid Metabolism (CAM)

Crassulacean acid metabolism (CAM) is the term applied to a second photosynthetic adaptation to arid conditions. This unusual photosynthesis was first discovered in members of the Crassulaceae where it had long been noted that the leaf cell sap became acidic at night and progressively basic during the day. These plants open their stomata in darkness and close them in light – the reverse of how other plants work. Closing the stomata during the day helps the plant to conserve water but also prevents CO$_2$ entering the leaves. During the night, when stomata are open, these plants take in CO$_2$ and incorporate it into organic acids which are stored in the cell vacuole. During the day, when stomata are closed, these organic acids are the source of CO$_2$ for the reactions of the Calvin cycle. As with C4 metabolism, CAM plants preface the PCR cycle with a process that first fixes CO$_2$ into organic acids: they possess a mechanism for concentrating CO$_2$ at the site of RubisCO.

Although originally found in the Crassulaceae, CAM is not restricted to that family: about 10% of vascular plants have developed this photosynthetic adaptation. CAM is more widespread than C4 photosynthesis and occurs in more than 23 families, including monocots, dicots and primitive plants. CAM occurs in succulent plants (fleshy plants having a low surface-to-volume ratio) such as members of the Crassulaceae, Cactaceae and Euphorbiaceae. Not all succulents have CAM. Halophytes for example, which grow in salty soils, are succulents but not CAM. Plants that grow in arid regions, such as Aizoaceae (including ice-plants), Liliaceae and Agavaceae; epiphytes such as members of the Bromeliaceae and the Orchidaceae of the rainforests; a few species of epiphytic ferns; and the primitive *Welwitschia* have CAM. This wide taxonomic distribution of CAM suggests that, like C4 photosynthesis, this adaptation has arisen independently many times during the course of evolution. CAM plants are especially adapted to arid and drought conditions and the mechanism of metabolism enables them to conserve water. CAM plants may lose 50–100 g water per gram of dry matter produced, compared with up to 300 g in C4 plants and up to 500 g in C3 plants. About 50% of all known CAM plants are epiphytes – a condition that has to cope with erratic supplies of water.

PEP-carboxylase fixes CO$_2$ in darkness

The principal metabolic feature of CAM plants is the assimilation of CO$_2$ at night into organic acids (Fig. 3.21). In the cytosol, HCO$_3^-$ (not CO$_2$) reacts with PEP to form oxaloacetate, in a reaction catalysed by PEPCO. The PEP for this reaction is made available by the breakdown of starch. The oxaloacetate is rapidly reduced by NAD-dependent malate dehydrogenase to form malate, which is stored in the vacuole. Transport of the malate into the vacuole, which, in CAM plants generally occupies about 90% of the total cell volume, is necessary in order to maintain a neutral pH in the cytosol. In the vacuole, the concentration of malate can be as high as 0.2 M at the end of the night when the vacuolar sap may have a pH 3–3.5. During the day malate is released from the vacuole by decarboxylation to provide CO$_2$ for fixation in the Calvin cycle. Thus, the large pool of malate accumulated in the vacuole overnight is the source of CO$_2$ for carbon assimilation during the day; the high level of CO$_2$ greatly reduces the possibility of photorespiration.

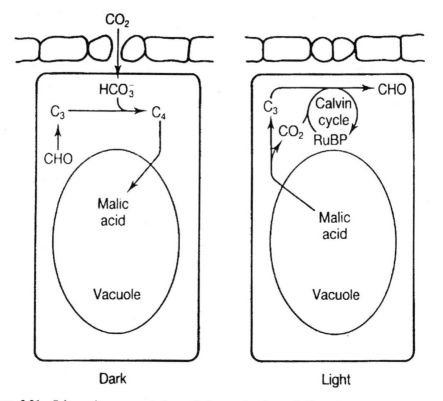

Figure 3.21   Schematic representation of the mechanism of Crassulacean acid metabolism, showing the temporal separation of CO_2 fixation. (From Leegood *et al.* 1997)

Comparison of C4 and CAM

The CAM pathway is similar to the C4 pathway in that CO_2 is first incorporated into organic acids before it enters the Calvin cycle (Fig. 3.22). In both pathways, PEPCO catalyses the reaction of bicarbonate and phosphoenolpyruvate to yield oxaloacetate which is rapidly reduced to malate.

The three alternative decarboxylation pathways (Table 3.3) are the same in CAM as in C4 metabolism. One significant difference between C4 and CAM is, however, that the C4 pathway involves the spatial separation of the carboxylation (mesophyll) and decarboxylation (bundle sheath) phases (Fig. 3.18), whereas CAM, which occurs in the mesophyll cells, involves a temporal separation – carboxylation at night (a quick fix!) and decarboxylation in the daytime (Fig. 3.21).

A second important difference is that in CAM plants the PEP required for carboxylation to malate is derived from starch or soluble sugars and not from a cycle of carbon intermediates as in C4 plants. During the day, the PEP formed from malate decarboxylation is converted to starch by gluconeogenesis and is stored in the chloroplasts. Thus, in CAM plants, not only are there large diurnal fluctuations in the level of malate but also of that of starch (Fig. 3.23). In addition, C4 and

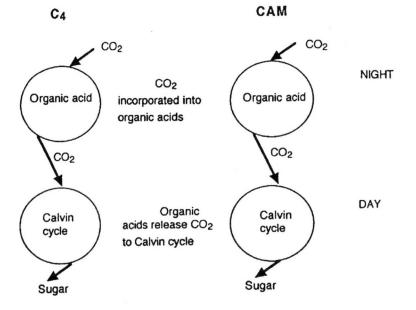

Figure 3.22   Comparison of C4 and CAM photosynthetic adaptations

Table 3.3   The three pathways of CO_2 fixation

Feature	C3	C4	CAM
Leaf structure	Bundle sheath cells lack chloroplasts	Bundle sheath cells have chloroplasts	Large vacuoles in mesophyll cells
Enzyme	RubisCO	PEPCO	PEPCO
Optimum temperature	15–25 °C	30–40 °C	35 °C
Productivity tonnes/ hectare/year	20	35	low (variable)

CAM plants may be separated on an anatomical basis. C4 plants have very distinct mesophyll and bundle sheath cells (Fig. 3.19), whereas CAM plants have large, thin-walled cells with large vacuoles. CAM plants generally have succulent stems and leaves, with fewer stomata, which are usually sunken, and a thick cuticle to restrict gas exchange.

Regulation of CAM

In addition to the regulation of the PCR cycle enzymes, successful operation of CAM also requires modulation. In CAM plants, PEPCO and the decarboxylating enzymes (NADP-ME, NAD-ME and PEPCK) function at different times in the same cell. To avoid futile carboxylation–decarboxylation cycling, PEPCO must be

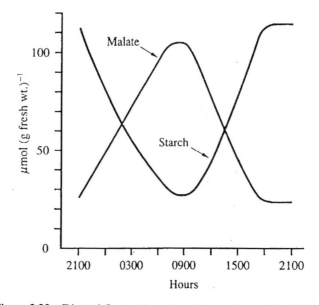

Figure 3.23   Diurnal fluctuations in malate and starch in CAM plants

switched on at night when the CO_2 concentration is high and switched off during daylight when the CO_2 is being used up in the PCR cycle. Competition for CO_2 at night is not a problem since RubisCO and the PCR cycle are switched off at night. An important regulatory feature of CAM is the inhibition of PEPCO by malate and low pH. In the daytime, high cytosolic concentrations of malate and low pH inhibit the enzyme. Sensitivity of PEPCO to inhibition by malate in CAM may be under the control of an endogenous circadian rhythm (Nimmo *et al.* 1987). PEPCO extracted from dark-grown plants has a high affinity for PEP and low sensitivity to malate inhibition; PEPCO from light-grown plants is inhibited by even low levels of malate. This short-term (day–night) regulation of PEPCO is due to the existence of two forms of the enzyme – the biochemically active form which is phosphorylated in darkness and an inactive form which is dephosphorylated in light (Jiao and Chollet 1991). It thus appears that PEPCO activity is modulated by a light-sensitive protein kinase; the light-induced transduction chain has not yet been worked out. It is interesting to note that light appears to have opposite effects in CAM and C4 systems – the protein kinase is inactive in light in CAM plants but active in the daytime in C4 plants. Crassulacean acid metabolism is usually not as efficient as either C3 or C4 photosynthesis, but it does allow plants to live under stressful conditions. Each of the three forms of photosynthesis has its advantages and disadvantages (Table 3.3).

CAM plants are not efficient photosynthesizers but they can thrive under stressful, arid conditions. C4 plants, on the other hand, are efficient photo-synthesizers but do not compete well with C3 plants at temperatures below 25 °C; they appear to be sensitive to cold.

Fate of photoassimilates

All higher plants generally convert their recently fixed carbon into either starch or sucrose. Starch is synthesized in the chloroplast; an activated nucleotide sugar ADP-glucose is the substrate for starch synthesis in plants. Starch is a storage form of carbohydrate and accumulates in the chloroplast during daylight. At night, when there is no photosynthesis, starch is broken down and converted to sucrose for translocation throughout the plant. In rapidly growing plants, sucrose, rather than starch, is the major end-product of photosynthesis. Sucrose is synthesized in the cytosol of photosynthetic cells; the substrate for sucrose synthesis is the activated nucleotide sugar UDP-glucose. Sucrose is the form in which carbohydrate is translocated throughout the plant.

Starch synthesis

The main storage carbohydrate in higher plants is the polysaccharide starch. The starch molecule is composed of two polymers – amylose and amylopectin. Amylose is a straight-chain polymer of glucose units joined together by an $\alpha(1-4)$ linkage; amylopectin is a branched polymer containing additional $\alpha(1-6)$ linkages which occur about every 24 to 30 glucose residues (Fig. 3.24). Amylopectin is similar to glycogen, the principal storage carbohydrate in animal tissues; glycogen is more highly branched, with the $\alpha(1-6)$ linkages occurring about every 10 glucose residues. The site of starch synthesis in leaves is the chloroplast. Large deposits of starch are clearly seen as granules in electron micrographs of chloroplasts from leaves of C3 plants (Fig. 3.25) and bundle sheath chloroplasts of C4 plants.

In addition, ADP-glucose phosphorylase and starch synthase, the two main enzymes involved in starch synthesis, have been located in chloroplasts (Preiss 1982). Excess triose phosphate from the Calvin cycle can be converted to starch. The triose phosphates glyceraldehyde-3-phosphate and dihydroxyacetone phosphate condense to form fructose-1,6-bisphosphate in an aldolase-catalysed reaction; this compound is then converted by fructose-1,6-bisphosphatase into fructose-6-phosphate (Fig. 3.26). Fructose-6-phosphate is then isomerized to glucose-6-phosphate by glucose-6-phosphate isomerase and converted to glucose-1-phosphate by phosphoglucomutase. Glucose-1-phosphate is converted to ADP-glucose; in this reaction, which is catalysed by ADP-glucose pyrophosphorylase, one molecule of ATP is consumed and pyrophosphate is released. ADP-glucose thus formed is an activated form of glucose and serves as an immediate glucose donor for starch synthesis. Starch synthase transfers glucose from ADP-glucose to the non-reducing end of pre-existing starch molecules that act as primers. The glucose molecule is added to the 4′-hydroxyl of the primer to form the characteristic $\alpha(1-4)$ linkage of amylose.

A branching enzyme known as Q-enzyme forges the formation of $\alpha(1-6)$ linkages between the short chains of glucose units (optimum 40 units) already joined in amylose fashion, to form the branched polymer (Fig. 3.27).

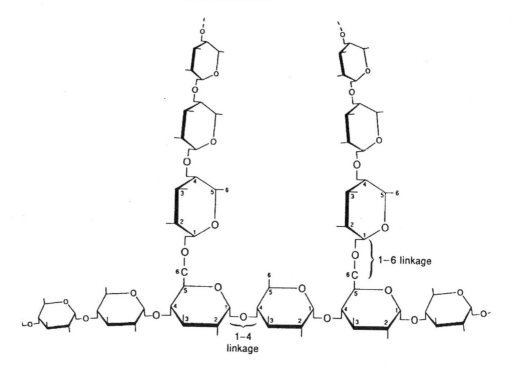

Figure 3.24   Structure of starch

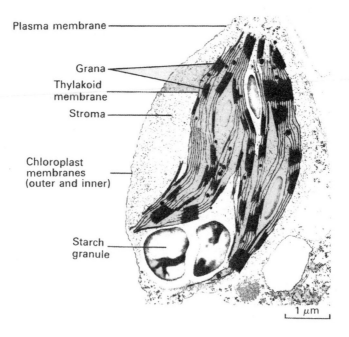

Figure 3.25   Starch granules in chloroplasts of C3 plants. (From Lodish *et al.* 1995)

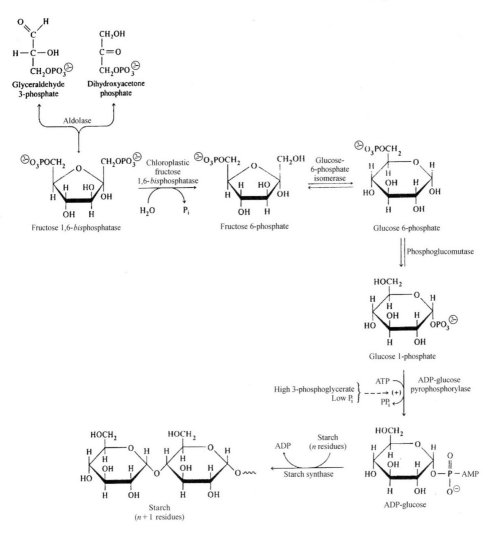

Figure 3.26 Pathway of starch biosynthesis in chloroplasts

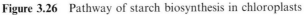

donor (optimum 40 units)   acceptor

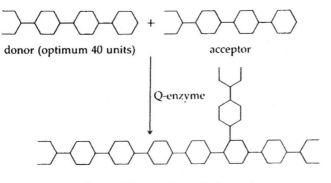

Q-enzyme

Figure 3.27 Amylopectin formation

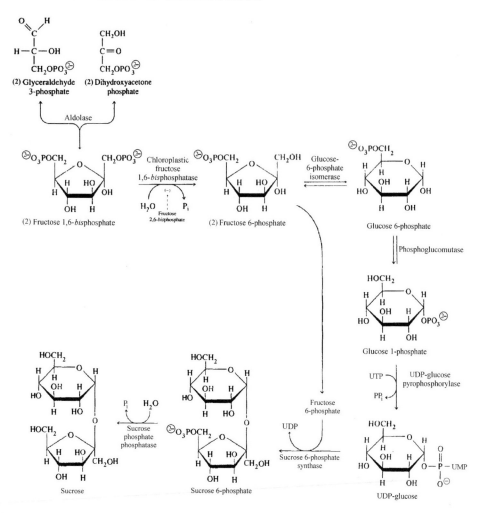

Figure 3.28   Pathway of sucrose biosynthesis

Sucrose synthesis

Sucrose may have evolved as the transport form of carbon because its unusual linkage which joins C-1 of glucose with C-2 of fructose is not hydrolysed by amylases or other common carbohydrate-degrading enzymes. The triose phosphates dihydroxyacetone phosphate and glyceraldehyde-3-phosphate from the Calvin cycle can be exported from the chloroplast and converted to sucrose in the cytosol (Fig. 3.28). In sucrose synthesis, as in starch synthesis, triose phosphate is condensed to fructose-1,6-bisphosphate (aldolase) which is then hydrolysed to fructose-6-phosphate (fructose-1,6-bisphosphatase). The fructose-6-phosphate may be isomerized to glucose-6-phosphate (glucose-6-phosphate isomerase) and converted to glucose-1-phosphate (phosphoglucomutase) which is then activated by UTP to form UDP-glucose (Fig. 3.28). The enzyme sucrose-6-phosphate synthase catalyses the reaction of fructose-6-phosphate with UDP-glucose to form sucrose-6-phosphate. Finally, sucrose-6-phosphatase removes the phosphate group to release the sucrose for transport out of the photosynthetic cells.

Starch–sucrose balance

Continuous synthesis of sucrose is required to provide a source of carbon for respiration and growth of the whole plant. During the light period, regulation of the transport of carbon compounds towards sucrose synthesis in the cytosol or starch synthesis in the chloroplast is crucially important for the control of their metabolism. During the dark period, degradation of starch in the chloroplast by phosphorylase produces glucose-1-phosphate which is the source of triose phosphate for sucrose synthesis in the cytosol.

When photosynthesis slows down, triose phosphates are less available for export to the cytosol for sucrose synthesis. When sucrose synthesis slows down, cytosolic P$_i$ is less available and more triose phosphate is retained in the chloroplast for conversion to starch. On the other hand, if the rate of sucrose formation should exceed the rate of carbon assimilation, demand for triose phosphate in the cytosol could deplete the pool of PCR-cycle intermediates and hence decrease the capacity of the Calvin-cycle enzymes to regenerate RuBP; this would seriously inhibit photosynthesis.

Summary

All photosynthetic eukaryotes use the PCR cycle to reduce CO$_2$. The PCR cycle reactions occur in the stroma of the chloroplast. The enzyme ribulose bisphosphate carboxylase (RubisCO) combines CO$_2$ with the five-carbon sugar ribulose bisphosphate (RuBP). Using the ATP and NADPH$_2$ from the light reactions, the PCR cycle produces glyceraldehyde-3-phosphate (3-PGAL), a three-carbon sugar which is the starting point for a number of metabolic pathways. Most of the 3-PGAL is used in the regeneration of RuBP; some exits the cycle and is the substrate for conversion to other essential organic compounds. Plants that use only the PCR cycle are called C3 plants. The high energy efficiency of the PCR cycle is achieved mainly through light-activated mechanisms and metabolite modulation of the enzymes involved.

On dry, hot days, C3 plants close their stomata in order to conserve water; this closure blocks gas exchange between the leaf and the atmosphere. Photosynthesis depletes the CO$_2$ in the leaf and allows the O$_2$ from the light reactions to build up. Under these conditions RubisCO displays oxygenase activity: O$_2$ rather than CO$_2$ is fixed. This process is called photorespiration and yields the two-carbon phosphoglycolate which is oxidized to CO$_2$ and H$_2$O in the peroxisomes and mitochondria. Photorespiration is wasteful, consuming carbon without producing ATP.

C4 plants, such as tropical grasses, have evolved an adaptation to overcome the deleterious effect of photorespiration: they preface the PCR cycle with the formation of 4C organic acids. Phosphoenolpyruvate carboxylase (PEPCO) catalyses the addition of CO_2 from bicarbonate to phosphoenolpyruvate to form the 4C oxalocetate in the mesophyll cells. These 4C organic acids move to the distinctive bundle sheath cells where they release the CO_2 for fixation in the PCR cycle. The initial concentration of the CO_2 eliminates photorespiration and makes C4 plants efficient photosynthesizers under hot, dry climatic conditions.

Crassulacean acid metabolism (CAM) is another adaptation by plants growing in hot, dry habitats, to by-pass photorespiration. Like the C4 plants, they preface the PCR cycle by incorporating CO_2 into organic acids. CAM plants open their stomata at night when they use PEPCO to fix CO_2 into organic acids which are stored in large vacuoles until daylight. During daytime, when stomata are closed to conserve water, the CO_2 is released from the vacuoles for use in the PCR cycle. Whereas the C4 pathway involves a spatial separation of carboxylation and decarboxylation, CAM uses a temporal separation of the two processes.

All higher plants generally convert their recently fixed carbon into either sucrose or starch. Sucrose is synthesized in the cytosol using UDP-glucose as substrate; starch is formed in the chloroplast using ADP-glucose as substrate. The sucrose/starch balance in plants is controlled mainly by a phosphate translocator on the inner membrane of the chloroplast.

References

Bakrim, N., Echevarria, C., Cretin, C., Arrio-Dupont, M., Pierre, J.N., Vidal, J., Chollet, R. and Gadal, P. (1992). Regulatory phosphorylation of Sorghum leaf phosphoenolpyruvate carboxylase. *Eur. J. Biochem.* **204**: 821–830.

Bidwell, R.G.S. (1983). Carbon metabolism of plants: photosynthesis and photorespiration. In: *Plant Physiology. A Treatise. Vol VII: Energy and Carbon Metabolism*, F.C. Steward (ed.), pp. 287–458. Academic Press, New York.

Bowes, G., Ogren, W.L. and Hageman, R.H. (1971). Phosphoglycolate production catalyzed by ribulose diphosphate carboxylase. *Biochemical Biophysical Research Communications* **45**: 716–722.

Buchanan, B.B. (1980). Role of light in the regulation of chloroplast enzymes. *Ann. Rev. Plant Physiol.* **31**: 341–374.

Burnell, J.N. and Hatch, M.D. (1985). Light–dark modulation of leaf pyruvate, P_i dikinase. *Trends in Biochemical Sciences* **10**: 288–290.

Burnell, J.N., Jenkins, C.L.D. and Hatch, M.D. (1986). Regulation of C_4 photosynthesis: a role for pyruvate in regulating pyruvate, P_i dikinase activity in vivo. *Aust. J. Plant Physiol.* **13**: 203–210.

Calvin, M. (1989). Forty years of photosynthesis and related activities. *Photosynthesis Research* **21**: 3016.

Carnal, N.W., Agostino, A. and Hatch, M.D. (1993). Photosynthesis in phosphoenolpyruvate carboxykinase type C4 plants: mechanism and regulation of C_4 acid decarboxylation in bundle sheath cells. *Arch. Biochem. Biophys.* **306**: 360–367.

Decker, J.P. (1955). A rapid, post-illumination deceleration of respiration in green leaves. *Plant Physiology* **30**: 82–84.

Doncaster, H.D. and Leegood, R.C. (1987). Regulation of phosphoenolpyruvate carboxylase in maize leaves. *Plant Physiology* **84**: 82–87.

Edwards, G.E. and Walker, D.A. (1983). *C3, C4: Mechanisms and Environmental Regulation of Photosynthesis*. Blackwell, Oxford.

Flugge, U.I. and Heldt, H.W. (1991). Metabolite translocators of the chloroplast envelope. *Ann. Rev. Plant Physiol. Plant Mol. Biol.* **42**: 129–144.

Forrester, M.L., Krotkov, G. and Nelson, C.D. (1966). Effect of oxygen on photosynthesis, photorespiration and respiration in detached leaves. 1. *Soybean Plant Physiol.* **41**: 422–427.

Freitag, H. and Stichler, W. (2002). *Bienertia cyclopetra* Bunge ex Boiss., *Chenopodiaceae*, another C4 plant without Kranz tissues. *Plant Biol.* **4**: 121–132.

Gerbaud, A. and André, M. (1987). An evaluation of the recycling in measurements of photorespiration. *Plant Physiol.* **83**: 933–937.

Hatch, M.D. and Slack, C.R. (1966). Photosynthesis by sugarcane leaves. A new carboxylation reaction and the pathway of sugar formation. *Biochemical Journal* **101**: 103–111.

Hatch, M.D., May, S.-L. and Kagawa, T. (1974). Properties of leaf NAD malic enzyme from plants with C$_4$ pathway photosynthesis. *Arch. Biochem. Biophys.* **165**: 188–200.

Hopkins, W.G. (1999). Photosynthesis: carbon metabolism, *Introduction to Plant Physiology*, pp. 189–214. Wiley, Toronto.

Horton, H.R., Moran, L.A., Ochs, R.S., Rawn, J.D. and Scrimgeour, K.G. (1993). Photosynthesis, *Principles of Biochemistry*, pp. 16.1–16.32. Neil Patterson Publishers/Prentice-Hall, Englewood Cliffs, NJ.

Husic, D.W., Husic, H.D. and Tolbert, N.E. (1987). The oxidative photosynthetic carbon cycle or C$_2$ cycle. *CRC Critical Reviews in Plant Sciences* **5**: 45–100.

Jiao, J. and Chollet, R. (1991). Post-translational regulation of phosphoenolpyruvate carboxylase in C4 and crassulacean acid metabolism in plants. *Plant Physiology* **95**: 981–985.

Kortschack, H.P., Hart, C.C.E. and Burr, G.O. (1965). Carbon dioxide fixation in sugarcane leaves. *Plant Physiology* **40**: 209–213.

Leegood, R.C., Walker, D.A. and Foyer, C.H. (1985). Regulation of the Benson–Calvin cycle. In: *Photosynthetic Mechanisms and the Environment*, J. Barber and N.R. Baker (eds), pp. 189–258. Elsevier, Amsterdam.

Leegood, R.C., von Caemerer, S. and Osmond, C.B. (1997). Metabolic transport and photosynthetic regulation in C4 and CAM plants. In: *Plant Metabolism*, D.T. Dennis, D.H. Turpin, D.D. Lefebvre and D.B. Layzell (eds), pp. 341–369. Longman, Harlow.

Lodish, H., Baltomore, D., Berk, A., Zipursky, S.L., Matsudaira, P. and Darnell, J. (1995). Photosynthesis, *Molecular Cell Biology*, 3rd edn, pp. 779–808, Scientific American Books, New York.

Lorimer, G.H. and Moziorko, H.M. (1980). Carbamate formation in the e-amino group of a lysyl residue as the basis for the activation of ribulosebisphosphate carboxylase by CO$_2$ and Mg^{2+}. *Biochemistry* **19**: 5321–5328.

Nimmo, G.A., Wilkins, M.B., Fewson, C.A. and Nimmo, H.G. (1987). Persistent circadian rhythms in the phosphorylation state of phosphoenolpyruvate carboxylase from *Bryophyllum fedtschenkoi* leaves and its sensitivity to inhibition by malate. *Planta* **170**: 408–415.

Ogren, W.L. (1984). Photorespiration: pathways, regulation and modification. *Ann. Rev. Plant Physiol.* **35**: 415–442.

Pierre, J.N., Pacquit, V., Vidal, J. and Gadal, P. (1992). Regulatory phosphorylation of phosphoenolpyruvate carboxylase in protoplasts for Sorghum mesophyll cells and the role of pH and Ca^{2+} as possible components of the light-transduction pathway. *J. Biochem.* **210**: 531–537.

Portis, A.R. (1990). Rubisco activase. *Biochemica et Biophysica Acta* **1015**: 15–28.

Portis, A.R. (1992). Regulation of ribulose 1,5-bisphosphate carboxylase/oxygenase activity. *Ann. Rev. Plant Physiol. Plant Mol. Biol.* **43**: 415–437.

Preiss, J. (1982). Regulation of the biosynthesis and degradation of starch. *Ann. Rev. Plant Physiol.* **33**: 431–454.

Sage, R.F. (2002). C4 photosynthesis in terrestrial plants does not require Kranz anatomy. *Trend Plant Sci.* **7**: 283–285.

Sharkey, T.D. (1988). Estimating the rate of photorespiration in leaves. *Physiologia Plantarum* **73**: 147–152.

Taiz, L. and Zeiger, E. (1991). Plant and cell architecture, *Plant Physiology*, pp. 3–27. Benjamin/Cummings, Redwood City, CA.

Voznesenskaya, E.V., Franceschi, V.R., Kiirats, O., Freitag, H. and Edwards, G.E. (2001). Kranz anatomy is not essential for terrestrial C4 plant photosynthesis. *Nature* **414**: 543–546.

Walker, J.L. and Oliver, D.J. (1986). Glycine carboxylase multienzyme complex. Purification and partial characterization from pea leaf mitochondria. *Journal Biol. Chem.* **261**: 2214–2221.

Walker, R.P., Trevanion, S.J. and Leegood, R.C. (1995). Phosphoenolpuruvate carboxykinase from higher plants: purification from cucumber and evidence of rapid proteolytic cleavage in extracts from a range of plant tissues. *Planta* **196**: 58–63.

Whitfield, P.R. and Bottomly, W (1983). Organization and structure of chloroplast genes. *Ann. Rev. Plant Physiol.* **34**: 279–310.

4

Photosynthesis – Physiological and Environmental Factors

Carbon Economy

The concentration of CO_2 in the air is small, about 0.03% by volume. This amount is relatively constant and provides an adequate and constant supply of CO_2 for the plant world. The balance in CO_2 removal and release into the atmosphere, over geological time, has been kept constant by a number of mechanisms. The overall cycle of CO_2 in nature involves a complex maze of reactions, termed the carbon cycle (Fig. 4.1), which shows the various ways in which CO_2 is added to or removed from the atmosphere.

Photosynthesis is one important mechanism whereby CO_2 is removed from the atmosphere. During the Carboniferous Age (about 300 million years ago) conditions for plant growth were at their best ever in the earth's history. The earth resembled a large greenhouse with high humidity and CO_2 concentrations, much higher than they are today. The vast amount of photosynthesis that occurred during that period produced millions of tonnes of carbon that accumulated and was stored within plant tissues. A large proportion of this plant material became trapped under the mud and water, in conditions that prevented decay and eventually formed the great reservoirs of coal and oil of today.

By far the most important mechanism in the stabilization of atmospheric CO_2 is the dissolving of CO_2 in the water of the oceans. Oceanic water thus represents a vast store of CO_2, in many forms, and a reservoir of CO_2 for photosynthesis. Today nearly three-quarters of the earth's surface is covered with ocean water and it is estimated that it contains eighty times as much carbon as the atmosphere in forms available for photosynthesis. Other forces help to reduce the CO_2 concentration of the atmosphere. Atmospheric CO_2 dissolved in water forms carbonic acid (H_2CO_3) which dissolves the calcium and magnesium carbonates of rocks and soil to form bicarbonate ions (HCO_3^-). Carbonic acid also weathers and decomposes silicate minerals such as the feldspars found in granites and basalts to produce bicarbonate

Photobiology of Higher Plants. By M. S. Mc Donald.
© 2003 John Wiley & Sons, Ltd: ISBN 0 470 85522 3; ISBN 0 470 85523 1 (PB)

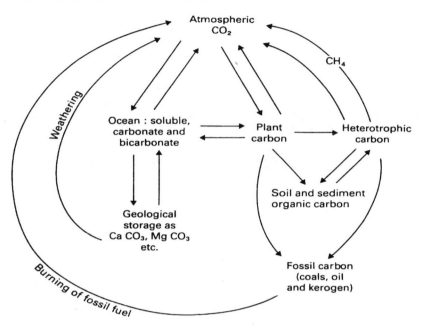

Figure 4.1   The Carbon Cycle

ions. Because the silicate contains no carbon, when these bicarbonate ions are converted by marine organisms to dolomite and limestone, carbon is removed from the atmosphere and tied up in an unusable, chemical form in sedimentary rocks (Table 4.1). The chemical decomposition of orthoclase, one mineral of the feldspar class, occurs as follows:

$$\underset{\text{orthoclase}}{KAlSiO_8} + H_2O + CO_2 \longrightarrow clay + \underset{\text{silica}}{SiO_2} + K_2CO_3$$

Table 4.1   Relative amounts of carbon, in various forms, found on earth. (Modified from Berner and Lasaga 1989)

Form	Carbon mass $(10^{15}\,kg)$	% of total
Calcium carbonate (limestone) and Ca–Mg carbonate (dolomites); mostly in sedimentary rocks	60,000	80.0
Sedimentary organic matter (kerogen)	15,000	20.0
Oceanic dissolved bicarbonate and carbonate	42.0	0.05
Recoverable fossil fuels (coal and oil)	4.0	0.005
Dead surficial carbon (humus, caliche, etc.)	3.0	0.004
Atmospheric carbon dioxide	0.75	0.001
All life (plants and animals)	0.56	0.00075

The carbon locked up in sedimentary rocks is eventually released by volcanic eruptions. In the short term, the largest contribution of CO_2 to the atmosphere is due to the respiration of plants, animals and bacteria and to the burning of fossil fuels. In the long term, over geological time, volcanic emissions and soda springs add CO_2 to the atmosphere. The largest single contribution of CO_2 is made by bacteria found in the soil, in fresh water and in the oceans. These bacteria cause the decay (oxidation) of organic material and thus release carbon as CO_2. Much of the CO_2 trapped by marine plants in photosynthesis is released either by their own respiration or by the respiration of the organisms that consume them or when the organisms die and decay. Also, the lime for the shells of many different marine animals is obtained from the conversion of calcium bicarbonate, $Ca(HCO_3)_2$ to calcium carbonate, $CaCO_3$. Half of the CO_2 tied up in the calcium bicarbonate is released in this reaction. Certain marine animals whose shells contain calcium phosphate carry this reaction still further and release all of the CO_2 held by the calcium.

The CO_2 concentration in the atmosphere is further added to by the burning of fossil fuels and by large-scale deforestation. High levels of CO_2 in the atmosphere (over and above what can be fixed by photosynthesis) and of other greenhouse gases such as methane are responsible for the overheating near the earth's surface. These gases absorb long wavelengths of radiant energy more than short wavelengths. The shorter wavelengths predominate in sunlight and penetrate the atmosphere to warm the earth. Since the earth is much cooler than the sun, it radiates the longer wavelengths that are absorbed by the greenhouse gases, which, in turn, radiate some of this energy back to the earth, warming it further. Plants in a greenhouse are warmed in the same way: greenhouse glass transmits short wavelengths and absorbs and then radiates long wavelengths.

Factors Affecting Photosynthesis

Photosynthesis, like any other physiological process, is affected by the environment in which it occurs. According to a theory of the 'three cardinal points' proposed by Julius von Sachs, in 1880, there is a minimum, optimum and maximum for each factor that relates to photosynthesis (Fig. 4.2). For example, for a given species, there is a minimum temperature below which no photosynthesis takes place, an optimum temperature at which a high rate occurs and a maximum temperature above which no photosynthesis will take place. However, the optimum temperature for one experiment changes in another experiment where different light and CO_2 levels prevail. Obviously external factors affecting photosynthesis cannot be treated in isolation but must be considered in relation to one another.

In 1905, the British plant physiologist F.F. Blackman postulated a principle of 'limiting factors' – under any given set of conditions the rate of photosynthesis will be limited by the slowest step. For example, at any given time, photosynthesis can be limited by either light or CO_2, but not by both. Blackman found that the effect of external factors on the rate of photosynthesis can be measured individually, within

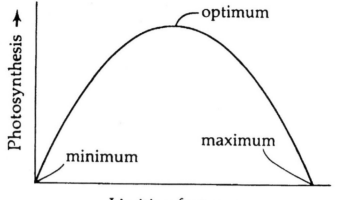

Figure 4.2   Concept of three cardinal points

certain limits; in other words, an approximation of the effects of these factors can be obtained.

Photosynthesis may be limited by a wide range of variables, including light, CO_2 concentration, O_2 concentration, temperature, water, supply of mineral ions, morphology of the leaf and chlorophyll content. These factors show not only a great variation with time but are also frequently involved in complex mutual interactions. It is almost impossible to investigate quantitatively this system as a whole. Finding a general equation to describe the rate of photosynthesis as a function of only the most important external and internal factors is a daunting task at this time. Dose–response curves for photosynthesis always show a gradual curvature in the middle range and the principle of limiting factors thus only applies at extreme situations (Fig. 4.4). It can be seen from these dose–response curves that light is the sole limiting factor only near zero and after light saturation has been reached. In the middle range photosynthesis also depends on CO_2 concentration, which becomes more important with increasing light flux. Thus the rate of photosynthesis depends on both factors over a wide range and they act together as limiting factors, each being dependent on the appropriate level of the other.

Light

Approximately $1.36\,kW\,m^{-2}$ (the solar constant) of radiant energy from the sun reaches the earth. Only about $0.9\,kW\,m^{-2}$ of this reaches plants since much energy is lost (depending on the time of day and year, elevation, latitude and many other factors) by absorption and scattering caused by water vapour, dust, CO_2 and ozone. Watts per square metre are units of irradiance, the flux of radiant energy per unit time; joules measure total energy. Since $1\,W$ equals $1\,J\,s^{-1}$, then $1.36\,kW\,m^{-2}$ equals $1.36\,kJ\,s^{-1}\,m^{-2}$. Another unit used in light measurements is the mole of light per unit surface per time, $mol\,m^{-2}\,s^{-1}$. One mole of light equals 6.02×10^{23} quanta (Avogadro's number of quanta). When a photobiological process, such as

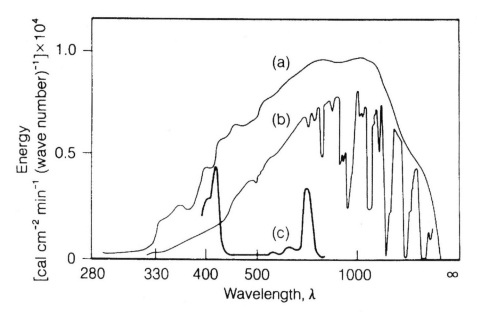

Figure 4.3   The solar spectrum in relation to the absorption spectrum of chlorophyll. Curve a is the energy output of the sun as a function of wavelength. Curve b is the energy that actually strikes the earth's surface. The sharp valleys above 700 nm (b) are due to absorption of solar energy by molecules of water vapour, dust, etc. in the atmosphere. Curve c is the absorption spectrum of chlorophyll *a*. (From Calvin 1982)

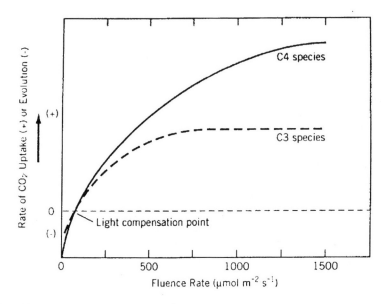

Figure 4.4   Dose–response curves of CO_2 fixation as a function of fluence rate for leaves of C3 and C4 plants

photosynthesis, depends on the number of quanta absorbed rather than on the absorbed energy, then the response to light is more accurately expressed in moles rather than in watts or joules. Energy fluxes, expressed in watts or joules can be converted to photon flux densities and expressed in moles. The term 'photon fluence rate' ($mol^{-2}s^{-1}$) is used to express the number of quanta impinging on a surface. *Photon flux densities*, in moles, are measured by instruments called quantum sensors.

Only about 5% of the solar energy reaching the earth is used in photosynthesis. This is because the bulk of the incident light is of wavelengths either too short or too long to be absorbed by the photosynthetic pigments (Fig. 4.3). In addition, of the radiation absorbed by the leaf and capable of causing photosynthesis, about 95% is lost as heat or used in the general metabolism of the leaf and only less than 5% is captured in photosynthesis. The range 400–700 nm is most active in photosynthesis. When expressed in energy units (watts or joules per second) this is called *photosynthetically active radiation* (PAR). Between 10 and 15% of the PAR reaching the leaf will not be absorbed by the photosynthetic pigments but will be either reflected at the leaf surface or transmitted through the leaf (Fig. 4.10). Since chlorophyll absorbs strongly in the blue and red, the reflected light and transmitted light is mainly green, which accounts for the green colour of the leaf.

Light intensity and photosynthetic rate

Plants differ in respect of the radiant energy required to balance photosynthesis exactly with respiration. Measurement of CO_2 fixation in intact leaves at increasing photon flux densities gives a good comparison of these two limiting factors (Fig. 4.4). In the dark, CO_2 fixation is negative; there is net CO_2 evolution due to dark respiration. As the photon flux density increases, photosynthesis also increases, resulting in a decrease in CO_2 evolution and a shift towards net CO_2 fixation. The intensity of light (photon flux density) at which the CO_2 fixed in photosynthesis is just balanced by the CO_2 evolved in respiration is called the 'light compensation point'. The light compensation point varies with species and must be appreciably exceeded by a plant for it to survive and develop. For normal C3 plants grown in the sun it is usually about 2% of full sunlight (10–40 mmol $m^{-2}s^{-1}$), which is about the light intensity in a well-lighted room. Increasing fluence rates above the compensation point increases the rate of photosynthesis; the linear portion of the dose–response curve represents the part of the process in which light is the limiting factor. At higher photon flux densities, the photosynthetic response to light begins to level off and light saturation is reached; further increases in photon flux densities no longer affect photosynthetic rates.

In most C3 plants such as potatoes, sugar beet and tomatoes, at normal atmospheric levels of CO_2, light saturation is reached at about 500–1000 mmol $m^{-2}s^{-1}$ (Fig. 4.4), that is, about one-quarter to one-half that of full sunlight. Light saturation occurs because some other factor, usually CO_2, levels off and becomes limiting. In most cases, both the saturation level and the fluence rate at which saturation occurs can be increased by increasing the ambient CO_2 level. The photosynthetic light responses of leaves of C4 plants shown in Figure 4.4 are typical of species such as maize, sugarcane, millet and sorghum, which are native to sunny habitats. Leaves of C4 plants never achieve light saturation up to or even beyond full sunlight and can have maximum rates more than twice that of most C3 species.

Adaptations to sun and shade

Optimum or saturation photon flux densities may vary considerably for different species. Some plants grow very well in shaded habitats (*shade plants*) while others require full sunlight (*sun plants*). Certain plants can adapt to a range of light regimes, growing as sun plants in sunny habitats and as shade plants in shaded habitats. Figure 4.5 illustrates the difference in CO_2 assimilation of leaves grown at high and low photon flux densities. Typical light saturation rates for shade plants are substantially lower than those for sun plants. Shade species have much lower photosynthetic rates and their photosynthetic responses are saturated at much lower irradiance levels than for crop plants. In addition, under such low irradiance levels, they usually photosynthesize at higher rates than do other species and their light compensation points are unusually low. These properties enable them to grow slowly and survive in their natural, shady habitats, where species with higher light compensation points would not get enough light and would die.

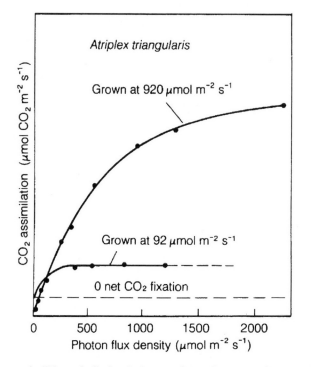

Figure 4.5 Changes in CO_2 assimilation in leaves of *Atriplex triangularis* as a function of photon flux densities. (From Björkman 1991)

In the leaf of a typical C3 plant photosynthesis saturates at a much lower photon flux density when grown at lower light intensities, indicating that the photosynthetic properties of a leaf depend on its growing conditions. In general, leaves or plants adapted to sunny or shaded environments are unable to survive when transferred to the other type of environment. While some plants, for example *Alocasia*, are obligate shade plants and others, such as *Helianthus*, are obligate sun plants, most plants are, in fact, facultative shade or sun plants. Sun plants that adapt to shade develop morphological and photosynthetic characteristics typical of those of shade plants. Thus their light compensation points decrease (mainly because they respire much more slowly); they photosynthesize more slowly and photosynthesis is saturated at lower irradiance levels. Sun plants gradually develop the ability to grow in shade. The reverse adaption, from shade to sun conditions, is less common. Shade plants usually cannot be moved into direct sunlight; high photon flux densities cause photoinhibition and bring about the death of older leaves. Some conifers are also sensitive to excess light which usually causes chlorosis and death of the plants. These symptoms are due to 'solarization', a light-dependent inhibition of photosynthesis followed by an oxygen-dependent bleaching of chloroplast pigments. Carotenoid pigments, because they absorb strongly in the blue, may function to protect against solarization by absorbing the excess radiant energy that might otherwise destroy chlorophyll (Demmig-Adams and Adams 1996).

Modification of the photosynthetic capacity due to the environment (Fig. 4.5) is a complex morphogenetic process embracing many functional and structural properties of the leaf. Bright light shifts the light compensation point to higher values. The leaves of many trees and shrubs and some herbaceous plants develop in the shade of others and during their development they acquire characteristics much like those of true shade plants (Corré 1983). These are called *shade leaves*, as opposed to *sun leaves* that develop in bright light. Shade leaves exhibit morphological and anatomical features that differ from those of plants growing in open sunlight. As would be expected, plants growing under a forest canopy have leaves that tend to be thinner, have more surface area and contain more chlorophyll than those of sun plants. In addition, shade plants generally have more chlorophyll per reaction centre and have a higher chlorophyll *b* to chlorophyll *a* ratio. Sun leaves become thicker than shade leaves because they develop longer palisade cells or an additional layer of palisade cells (Fig. 4.6).

Tomato plants have about 100 stomata per mm^2 on the lower epidermis in dim light (20 W m^{-2}); in bright light (100 W m^{-2}) additional stomata are found on the upper epidermis. This response can be observed within three days after a change in light conditions (Mohr and Schopfer 1995). In addition to their anatomical adaptations, leaves maximize light absorption by changing the spatial relationship between the leaf and the incident light. Many plants can extend the time during which the leaf blades are aligned perpendicular to sunlight by using a system of 'solar tracking' whereby the laminae follow the path of the sun during daylight hours. In some species, especially those growing in deserts, leaves turn away from the sun in order to minimize heat absorption and excess water loss. When the photon flux densities reaching a leaf are low, chloroplasts align themselves at the

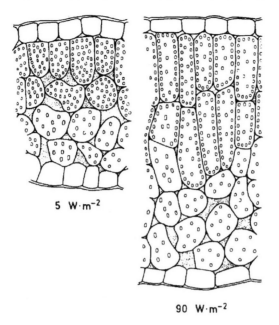

5 W·m⁻²

90 W·m⁻²

Figure 4.6 Morphogenetic modifications of the leaf to dim light (left) and bright light (right) conditions

leaf surfaces parallel to the plane of the leaf and at right angles to the direction of the incident light (Fig. 4.7) in order to maximize absorption. When photon flux densities are too high, chloroplasts become aligned along the surfaces parallel to the

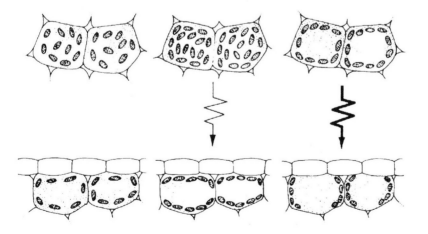

Figure 4.7 Distribution of chloroplasts in chlorenchyma cells of the duckweed *Lemna* in darkness (left), low photon flux (centre) and high photon flux (right). The upper sequence shows a surface view, the lower sequence a cross-sectional view. (From Haupt 1986)

direction of the incident light in order to avoid excess absorption and prevent chlorophyll damage.

An important environmental factor affecting photosynthesis is the canopy structure. Only the upper, outer leaves of plants in a stand are exposed to full sunlight, and irradiance reaching lower leaves may be reduced to 10%, or lower. Since leaves absorb little of the far-red light ($\lambda > 700$ nm), habitats below canopies are enriched in far-red light. While leaves of shade plants may be damaged by photoinhibition on exposure to high photon fluxes, they may also benefit from the phenomenon of *sunflecks*. These are patches of full sunlight that pass through the canopy (possibly due to disturbances caused by a sudden breezes) and travel along the shaded leaves as the sun moves. In a dense forest, sunflecks can, within seconds, increase ten-fold the photon flux density impinging on a shaded leaf. Sunflecks can contain up to 50% of the total light energy available during the day, but this is available for only a few minutes duration.

The presence of phytochrome, a red/far-red reversible photoreceptor, enables the plant to adjust to its environment. The ratio of red light (R) to far-red light (FR) varies greatly in different environments. Compared with direct daylight, there is relatively more FR in sunset light, under 5 mm of soil and especially under the canopy of other plants (such as on the forest floor). The canopy phenomenon results from the fact that green leaves absorb red light due to their high chlorophyll content, but do not absorb far-red light. Phytochrome can act as an indicator of the degree of shading of a plant by other plants. As shading increases, R/FR decreases and the ratio of far-red-absorbing phytochrome to total phytochrome (P_{fr}/P_{total}) decreases. Shade light is far-red-enriched and so tends to preferentially excite PSI. Because of this, plants tend to invest more in the chlorophyll *b*-enriched PSII antenna rather than increasing PSI and PSII equally. A 'shade-avoiding' plant applies its resources towards rapid extension growth in order to enhance its chances of growing above the canopy to receive more unfiltered, photosynthetically active light. As a result, plants growing in such environments are tall and spindly and usually have reduced branching and reduced leaf area.

CO_2 availability

The concentration of CO_2 in the atmosphere is low, at about 0.03% by volume (350 ml l^{-1}). While this amount is relatively constant and provides a steady and adequate supply of CO_2 to the plant world, it is still well below the CO_2 saturation level of most C3 plants at normal fluence rates. Enriching the atmosphere with CO_2 increases the CO_2/O_2 ratio, which causes a decrease in photorespiration and leads to greater net photosynthesis. In C3 plants, high CO_2 concentrations are needed both to enhance photosynthesis at high irradiance levels and to saturate photosynthesis at these levels (Fig. 4.8).

In contrast, for C4 plants, photosynthesis may be saturated by CO_2 levels of 400 mmol l^{-1}, which is just above the normal atmospheric concentration. The relative CO_2 requirements for C3 and C4 plants can be compared if CO_2 levels are decreased below atmospheric levels (Fig. 4.9). If irradiance levels are kept above the

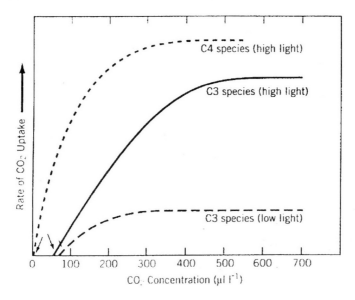

Figure 4.8 Effects of CO_2 enrichment of the atmosphere on the rate of photosynthesis at different irradiance levels. (From Gaastra 1959)

light compensation points for each, then net photosynthesis of C3 plants usually reaches zero at concentrations of CO_2 between 35 and 45 mmol l^{-1} (Bauer and Martha 1981), whereas C4 plants continue net CO_2 fixation down to levels of between zero and about 5 mmol l^{-1}. The atmospheric concentration at which photosynthesis just compensates for respiration is referred to as the *CO_2 compensation point*. That the CO_2 compensation point for C4 plants is lower than that for C3 species implies that C4 plants have high CO_2 levels in the bundle sheath chloroplasts and high pools of CO_2 in the mesophyll cells. The CO_2 absorbed by C4 leaves is fixed into organic acids, which thus maintain high levels of CO_2. The mesophyll of C3 plants has no such mechanism of fixing CO_2.

In C4 plants, the high levels of CO_2 inhibit photorespiration because, at high concentrations CO_2 competes better than O_2 for the RubisCO carboxylase site and is therefore fixed at a greater rate than O_2. If the O_2 level to which plants are exposed is decreased from the normal 21% down to about 2%, then the difference in compensation points between C3 and C4 species essentially disappears. Under such conditions, the CO_2 compensation points of C4 plants remain the same but those of C3 plants also approach zero; insufficient O_2 is available to compete with CO_2 for RubisCO and so photorespiration becomes negligible. The CO_2 from the atmosphere enters the leaf by diffusion through the stomata. Stomata are much more important in regulating CO_2 availability for photosynthesis than in controlling O_2 exchange in the leaf. Oxygen can readily penetrate the cuticle of leaves, but CO_2 cannot. Thus stomatal opening and closing has an important effect on the regulation of photosynthesis, especially in C3 plants, which incorporate CO_2 directly into phosphorylated sugar derivatives. At least two photoregulatory

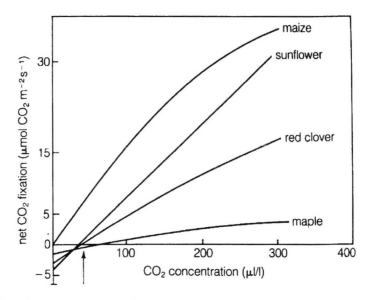

Figure 4.9 Influence of reduced CO_2 levels on photosynthesis in C3 and C4 plants. (From Hesketh 1963)

systems work together to control stomatal movement: one system is activated by the absorption of red light by chlorophyll and the second by a blue light receptor, probably a flavoprotein (see Chapter 8).

Although the concentration of CO_2 in the atmosphere is relatively constant, deviations from the average 0.03% do occur. Leaves exposed to bright sunlight during a summer growing season may show suboptimal photosynthetic rates. For example, the CO_2 concentration above a forest canopy or immediately over a field of a dense cereal crop is significantly diminished during daylight hours. In such instances, even gentle breezes, which increase convection currents and enhance stomatal evaporation, can return photosynthesis to normal rates. Glasshouse crops often lack sufficient CO_2 for maximal photosynthesis and growth. In practice, glasshouse atmosphere is sometimes enriched with added CO_2; very high levels of CO_2 may, however, cause stomata to close and create other effects which may be toxic and reduce photosynthesis.

Temperature

Absorption of radiant energy by the leaf results in heating. However, the heat absorbed is dissipated by emission of long-wave radiation and by cooling due to transpiration and to conduction and convection currents (Fig. 4.10). If there is a gradient of temperature between the leaf and the surrounding air, then conduction and convection currents lower the temperature at the leaf surface. Under conditions

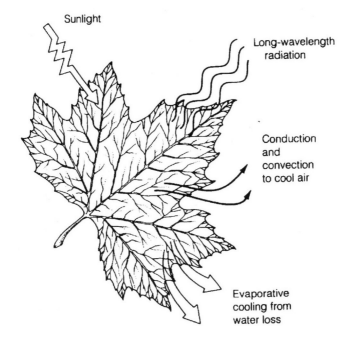

Figure 4.10   Absorption and dissipation of radiant energy by a leaf. (From Taiz and Zeiger 1991)

favourable for transpiration, the evaporation of water vapour withdraws heat energy from the leaf and thus cools it.

Photosynthesis, like most other biological processes, is temperature-dependent. Temperature affects all the biochemical reactions of photosynthesis and so it is to be expected that the responses to temperature are complex. These responses can be characterized by the three cardinal points – minimum, optimum and maximum; when photosynthetic rates are plotted as a function of temperature, a typical bell-shaped curve results (Fig. 4.2). In general, increase in temperature results in an acceleration of photosynthesis, when other factors are not limiting. The temperature response of biological reactions is frequently characterized by a *temperature coefficient*, Q_{10}, which is an empirically measured change in the rates of reaction at two temperatures $10\,^{\circ}C$ apart and given by

$$Q_{10} = \frac{kT + 10}{kT}$$

The value of Q_{10} for enzyme-catalysed reactions is about 2, meaning that the rate of reaction doubles for every $10\,^{\circ}C$ rise in temperature. The photochemical reactions are not enzyme-dependent and their Q_{10} value is close to 1; the biochemical reactions of CO_2 fixation, on the other hand, are temperature-dependent.

Temperature dependence of photosynthesis increases with increase in light saturation; there is an increase in the rate of photosynthesis as temperature is

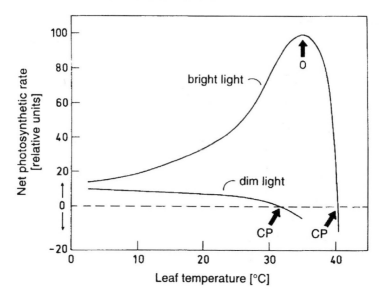

Figure 4.11   Effect of temperature on net photosynthesis in bright and dim light at saturating CO_2 concentration. In the range 0–30 °C the rate of photosynthesis increases in bright light ($Q_{10} > 10$). Above the temperature optimum (35 °C) the rate of photosynthesis decreases sharply due to photorespiration. The upper temperature compensation point (CP) is exceeded at approximately 40 °C. (From Mohr and Schopfer 1995)

increased from 0 °C to 30 °C in bright light, but not in dim light (Fig. 4.11). The effects of temperature on photosynthesis depends on the species and on the environmental conditions under which the plant is grown. High CO_2 levels ensure adequate carboxylase activity by RubisCO; under these conditions the rate of photosynthesis increases with increase in temperature up to about 30 °C. At ambient CO_2 concentrations, photosynthesis is limited by RubisCO activity and temperature changes result in two opposing responses: increase in temperature up to about 30 °C increases the rate of carboxylation, whereas higher temperatures decrease the affinity of RubisCO for CO_2. Generally C4 plants have higher temperature optima than do C3 species, due mainly to the lower rates of photorespiration in the C4 plants.

Increase in temperature increases the loss of CO_2 by respiration and this is reflected during photorespiration by increase in the ratio of dissolved O_2 to CO_2. Photorespiration thus more or less balances the increased photosynthesis resulting from temperature rises above the 15–30 °C range at which C3 plants normally grow. Since photorespiration is not significant in C4 plants, temperatures of 30–47 °C are often optimal for photosynthesis in these species (Table 4.2). Plants adapted to heat generally have a low relative photorespiration and the ratio between photosynthesis and respiration becomes an important parameter for the adaptation of photosynthesis to the environment. At high temperatures, stomata close in order to prevent water loss. This hinders CO_2 diffusion and, taken with the reduced solubility of CO_2 in water with higher temperatures, may act as a limiting factor in photosynthesis.

Optimal temperatures for photosynthesis depend largely on genetical and physiological factors. Plants of many species can adapt, within a narrow range, to changes in temperature; this helps plants to adjust to seasonal change and to adapt to different environments. For example, evergreen species, such as conifers, can photosynthesize both in summer and in winter. Plants, especially C3 species, growing at low temperatures, can maintain higher photosynthetic rates at low temperatures than can plants grown at high temperatures.

Oxygen

Under conditions of both low and high light flux, photosynthesis is more efficient when the O_2 concentration is lower than that of normal air (209 ml l^{-1}); this is especially true of C3 plants. When light flux levels are high, more light energy is absorbed than can be used for CO_2 fixation; this leads to light energy transfer to O_2 molecules and results in the wasteful process of oxidative photodestruction of organic molecules. This phenomenon, which is called the *Warburg effect*, inhibits the rate of photosynthesis. Despite the mechanisms that have evolved to enable plants to deal with wide fluctuations of light flux, oxidative photodestruction of organic molecules does occur. Short-lived, activated 'species of oxygen' are formed, which can react non-specifically with proteins, nucleic acids and chlorophyll molecules and destroy them. Biological electron transport systems are often linked to the formation of *free radicals* (molecules with one or more unpaired electrons), especially oxygen radicals, which can be toxic. The transfer of electrons to oxygen is particularly marked in illuminated chloroplasts because of the high O_2 levels occurring there. An important source of oxygen free-radicals during photosynthetic electron transport are the reduced electron acceptors of photosystem I. Normally electrons are passed from PSI to NADP$^+$, but PSI may also pass electrons directly to O_2 to form superoxide (O_2^-):

$$O_2 \xrightarrow{e^-} O_2^-$$

Superoxide formation is especially favoured when the rate of reduction of ferredoxin by PSI exceeds the rate of NADPH utilization for CO_2 fixation. An enzyme, superoxide dismutase (SOD), present in the chloroplast matrix and in the thylakoid membrane, disproportionates the superoxide to form hydrogen peroxide:

$$O_2 \xrightarrow[e-,\ 2H]{SOD} H_2O_2$$

When both the superoxide and the hydrogen peroxide are present at the same time, the formation of the especially toxic hydroxyl radical (HO*) is favoured:

$$H_2O_2 + O_2^- \longrightarrow HO^* + OH^- + O_2$$

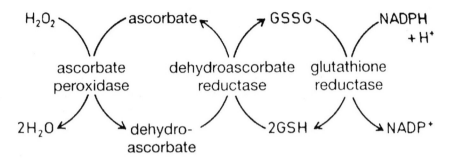

Figure 4.12 Degradation of hydrogen peroxide via the ascorbate–glutathione (reduced-form GSH; oxidized-form GSSG) redox chain in the chloroplast. (From Halliwell and Gutteridge 1985)

The H_2O_2 must be scavenged both to prevent its reaction with O_2^- to form hydroxyl radicals and because hydrogen peroxide itself directly interacts with several steps in the fixation of CO_2. In plant cells, catalase destroys H_2O_2 effectively, but this enzyme is restricted to peroxisomes. In the chloroplast, an ascorbate–glutathione redox chain scavenges the H_2O_2 in a reaction that couples the reduction of H_2O_2 to water with the oxidation of NADPH to NADP (Fig. 4.12).

Another important source of oxygen free-radicals is the electronically excited oxygen (*singlet oxygen*, $^1O_2^*$). When excess light energy is absorbed, then the excitation energy from the triplet state of chlorophyll ($^3Chl^*$) can be directly transferred to O_2, to produce excited, singlet oxygen, $^1O_2^*$:

$$^3Chl^* \longrightarrow {}^1Chl$$
$$O_2 \longrightarrow {}^1O_2^*$$

This excited oxygen must be scavenged before it stimulates peroxidation of membrane lipids. There are at least two ways to prevent photo-oxidation – either remove superoxide as it is formed or prevent its formation. Superoxide can be effectively removed by the enzyme superoxide dismutase (SOD), which is found in many cell compartments, including the chloroplast. This enzyme is able to scavenge and inactivate superoxide radicals by forming hydrogen peroxide and molecular oxygen:

$$O_2^- + O_2^- + 2H^+ \longrightarrow H_2O_2 + O_2$$

Scavenging is also accomplished by carotenoid pigments, which are efficient quenchers of $^1O_2^*$ and can deactivate superfluous excitation energy in the photosynthetic apparatus (Fig. 4.13). An important link has been established (Demmig Adams and Adams 1996) between energy dissipation and the presence of the xanthophyll zeaxanthin. Zeaxanthin is formed from the de-epoxidation (removal of two epoxy groups) of violaxanthin by a process known as the

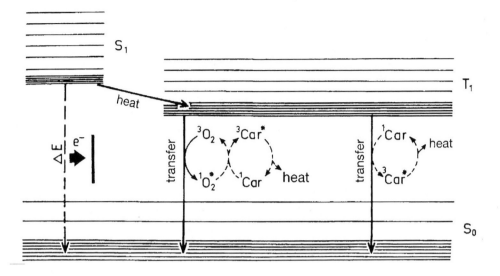

Figure 4.13   Deactivation of superfluous excitation energy in the photosynthetic apparatus by carotenoids. (From Mohr and Schopfer 1995)

xanthophyll cycle (Fig. 1.20). This reaction is induced by a low pH in the lumen, which is a normal consequence of electron transport under high light conditions; epoxidation, the reverse reaction, occurs in low light or darkness. The xanthophyll cycle operates as an effective switch generating zeaxanthin whenever dissipation of excess energy is required, but removing the zeaxanthin under conditions of low irradiance when the energy is required for photosynthesis.

As a result of oversaturation of the photosynthetic reaction centres with light energy, there are transitions in the antenna chlorophylls from the singlet to the triplet state ($S_1 \longrightarrow T_1$) (Fig. 4.13). During the deactivation of triplet chlorophyll, excited singlet oxygen ($^1O_2^*$) can be formed, which reverts to the energetic ground state ($^3O_2$) via the reversible formation of excited triplet carotenoid. Direct energy transfer between neighbouring pigment molecules can deactivate triplet chlorophyll, resulting in the formation of triplet carotenoid. This reaction is mainly responsible for the removal of superfluous light energy and its conversion into non-damaging heat energy. In addition, carotenoids act as shielding pigments in the blue/ultraviolet range.

Photosynthetic Efficiency

Less than 1% of the 1.3 kW m^{-2} radiant energy reaching the earth is absorbed by plant pigments and used in the synthesis of energy-rich molecules. It has been estimated that 3×10^{18} kJ chemical energy derived from sunlight per year are fixed globally in the form of 2×10^{11} tons fixed carbon. Despite this low yield, photosynthesis is the basic energy-supplying process on the earth. Photosynthetic rates differ greatly for species living in such diverse conditions as afforded by deserts, high mountains and tropical forests (Table 4.2).

Leaf photosynthetic capacity (which is defined as the rate of photosynthesis per unit leaf area when irradiance is saturating, CO_2 and O_2 levels are normal, temperature is optimum and relative humidity is high), varies by almost two orders of magnitude. Differences result mainly from variations in light, temperature and the availability of water and nutrients. The highest photosynthetic capacities occur among desert annuals and grasses, when water is available and not a limiting factor. Grasses with C4 metabolism generally have the highest photosynthetic rates, whereas slow-growing succulents with crassulacean acid metabolism have the lowest rates. Photosynthetic efficiency is best calculated by dividing the total PAR (photosynthetically active radiation – the light quality effective in photosynthesis) energy absorbed over a growing season into the total chemical-bond energy of the hexose produced during photosynthesis. A minimum of 12 mole photons are required to fix one mole of CO_2 in C3 plants.

Photons in the PAR region have a wavelength range of 300–800 nm corresponding to an energy range of 400–500 kmol^{-1}. The quantum energy can be calculated from Planck's equation, which relates photon energy to wavelength:

$$E = hn = hcl^{-1}$$

where $h = 6.626 \times 10^{-34}$ J s, Planck's constant; $n =$ frequency; $c = 3 \times 10^8$ m s^{-1}, speed of light; $l =$ wavelength.

One mole of photons of an average PAR wavelength of 550 nm has an energy value of 52 kcal. Twelve moles of photons would therefore have an energy value of

Table 4.2   Maximum photosynthetic rates of major plant types under natural conditions. (From Salisbury and Ross 1992)

Type of plant	Example	Max. photosynthesis (CO_2) μmol^{-2}s^{-1}
CAM	*Agave americana* (century plant)	0.6–2.4
Tropical, subtropical and Mediterranean evergreen trees and shrubs; temperate-zone evergreen conifers	*Pinus sylvestris* (Scotch pine)	3–9
Temperate-zone deciduous trees and shrubs	*Fagus sylvatica* (European beech)	3–12
Temperate-zone herbs and C3-pathway crop plants	*Glycine max* (soybean)	10–20
Twelve herbacious alpine plants	*Ligusticum mutellina*, *Taraxacum alpinum* and others	10–24
Tropical grasses, dicots, and sedges with C4 pathway	*Zea mays* (corn or maize)	20–40

624 kcal. If it takes 114 kcal of energy to reduce 1 mole (44 g) of CO_2 to hexose, then the theoretical efficiency of photosynthesis is $114/624 = 18\%$. Many crops, including forest plantations, have photosynthetic efficiencies ranging from 0.1% to only 3%. Since only $0.83 \, kW \, m^{-2}$ (64%) of the total $1.3 \, kW \, m^{-2}$ radiant energy reaching the earth is in the PAR region, the theoretical maximum photosynthetic efficiency is only about 12% (64% of 18%). At normal atmospheric CO_2 and O_2 levels and within a temperature range of 10–25 °C, photosynthetic efficiencies are about the same for C3 and C4 plants, requiring about 15 photons of light flux to fix one molecule of CO_2 (Ehleringer and Pearcy 1983). At lower temperatures or at less than normal atmospheric O_2 levels, photorespiration is essentially eliminated from C3 plants which then become more efficient than C4 plants and require only 12 photons per molecule of CO_2 fixed. Under these conditions C4 plants still require at least 14 photons, since they need three ATP molecules per CO_2 for the Calvin cycle and two more to operate the C4 pathway. The normal concentration of CO_2 (0.03%) in the atmosphere is well below the saturation level for C3 plants. Increase in CO_2 levels with concomitant decrease in O_2 levels results in a decrease in photorespiration and an increase in the rate of photosynthesis. At high irradiance, C3 plants need high levels of CO_2 to maintain increased rates of photosynthesis whereas C4 species reach saturation at CO_2 levels slightly higher than the normal 0.03%. High CO_2 levels inhibit photorespiration in C4 plants which always have a lower CO_2 compensation point than C3 species.

Crop Productivity

At rate-limiting photon flux densities, the amount of CO_2 taken up can be related to the amount of light absorbed, by the *quantum yield* (moles of CO_2 fixed per mole of photons absorbed). The quantum yield provides a direct measure of the energy required to fix carbon dioxide. Measurements of quantum yields in C3 and C4 plants have shown that in normal air between 25 and 30 °C the quantum yield for both C3 and C4 is comparable (about 0.05 mol CO_2 per mole of photons). In 2% oxygen, the quantum yield rises to 0.07–0.08 in C3 species as photorespiration is repressed, but remains constant in C4 species. Thus in the absence of photorespiration C3 plants are more efficient than C4 plants in low light. In addition, the quantum yield of C3 plants falls with rise in temperature but it remains constant in the leaves of C4 plants (Fig. 4.14). These changes in quantum yield are consistent with the known charcteristics of photorespiration which is diminished in low oxygen but increases with rise in temperature. As oxygen inhibition of photosynthesis increases, the quantum yield falls exponentially (Fig. 4.14). The evolutionary success of C4 photosynthesis is due to improved water-use efficiency and nutrient-use efficiency, as well as high photosynthetic capacity at higher temperature, all of which follow from RubisCO function in bundle sheath cells served by a CO_2-concentrating mechanism (Edwards *et al.* 2001). The productivity of C4 crops also stems from their longer growth cycles in the tropics.

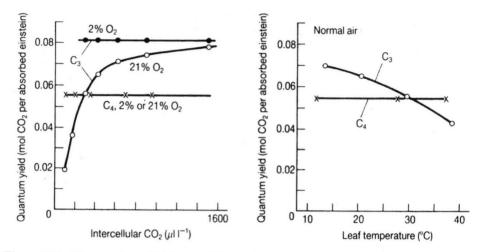

Figure 4.14   The quantum yield for net CO_2 uptake in C3 (*Encelia californica*) and C4 (*Atriplex rosea*) as a function of both CO_2 and O_2 concentration (left) and temperature in air. (From Osmond *et al.* 1980)

In CAM plants, initial night-time fixation of CO_2 by phosphoenolpyruvate carboxylase (PEPCO) (Chapter 3) occurs when stomata are open and loss of water due to transpiration is low. CO_2 release during the day promotes stomatal closure and concentrates CO_2 for RubisCO, suppressing its oxygenase activity, thereby minimizing photorespiration. The net effect of this CO_2-concentrating strategy is that CAM plants exhibit water-use efficiency rates several-fold higher than C3 and C4 plants under comparable conditions (Drennan and Nobel 2000). Thus, CAM is typically, although not exclusively, associated with plants that inhabit extremely arid environments (e.g. deserts), semi-arid regions with seasonal water availability (e.g. Mediterranean climates) or habitats with intermittent water supply (e.g. tropical epiphytic habitats) (Cushman, 2001). Most notable among these are commercially or horticulturally important plants such as pineapple (*Ananas comosus*), agave (*Agave* sp.), cacti (Cactaceae) and orchids (Orchidaceae).

CAM is also correlated with various anatomical and morphological features that minimize water loss, including thick cuticles, low surface-to-volume ratios, large cells and vacuoles with enhanced water-storage capacity (i.e. succulence) and reduced stomatal size and frequency. The selective advantage of high water-use efficiency probably accounts for the extensive diversification and speciation among CAM plants principally in water-limiting environments. Intensive ecophysiological studies over the last 20 years have documented that CAM is present in approximately 7% of vascular plant species, a much larger percentage than that of C4 species (Winter and Smith 1996).

In CAM plants, CO_2 fixation and photosynthesis are separated in time; this achieves an even greater conservation of water. A feature of CAM plants is their regulation of stomatal opening and closing. Stomata remain open during the cool and relatively humid night, allowing CO_2 entry with minimum water loss. During the heat and dryness of the day, the stomata are closed, preventing water loss. On average, a CAM plant transpires approximately 50–100 g H_2O per g dry weight, i.e. about ten times less than that transpired by a comparable C3 plant.

In C4 plants, improvement in the ratio of CO_2 fixation to transpiration is achieved by virtue of the fact that the two sections of the C4 dicarboxylate cycle (Chapter 3) occur simultaneously but spatially separated between two different types of cells. Interestingly, recent studies (Vosnesenskaya *et al.* 2001; Sage 2002) have demonstrated that C4 photosynthesis may occur in single cells of the Asian chenopods *Borszczowia aralocaspica* and *Bienertia cyclopetra* (see Chapter 3).

CAM plants in general possess a physiological and metabolic plasticity in the extent to which night-time CO_2 fixation contributes to net carbon gain. This variable CAM pathway of photosynthesis ranges from typical C3 species to those that exhibit obligate CAM. The list of plants exhibiting C3–CAM intermediate photosynthesis recorded in Table 4.3 includes 18 genera in 12 families, showing the floristic diversity of CAM plants with physiological and biochemical plasticity as an adaptive strategy (Orsenigo *et al.* 1995).

Certain CAM species which show a shift from C3 to CAM or vice versa are known to exhibit 'CAM-idling' or 'CAM-cycling' (Ting 1985). During CAM-idling, plants show low diurnal acid fluctuation with no gas exchange (Ting 1985); it is hypothesized that these CAM plants are recycling respiratory CO_2. However, some CAM plants show organic acid fluctuation but little or no night-time CO_2 fixation, a phenomenon known as CAM-cycling (Ting 1985). In addition to idling or cycling modes, facultative CAM plants may shift from a C3 to a CAM pathway during ontogeny or in response to environmental variables such as water stress, salinity or

Table 4.3   List of C3–CAM intermediate species. (From Orsenigo *et al.* 1995)

C_3–CAM Plant	Family
Cissus quadrangularis	Vitaceae
Condonanthe crassifolia	Gesneriaceae
Clusea rosea	Clusiaceae
Guzmania monostachia	Bromeliaceae
Kalanchoe uniflora	Crassulaceae
Mesembryanthemum crystallinum	Aizoaceae
Nidularium innocenti	Bromeliaceae
Peperomia camptotricha	Piperaceae
Pereskia aculeata	Cactaceae
Pereskia grandiflora	Cactaceae
Portulacaria afra	Portulaceae
Pyrrosia confluens	Polypodiaceae
Sedum acre	Crassulaceae
Sedum telephium	Crassulaceae
Sempervivum montanum	Crassulaceae
Talinum calycinum	Portulaceae
Welwitschia mirabilis	Welwitschiaceae
Yucca gloriosa	Liliaceae

photoperiod. The shift from C3 to CAM photosynthesis is usually accompanied by several changes in physiology, biochemistry and gene expression: environmental factors influence the expression of photosynthetic machinery in some plants. In facultative CAM plants, such as the common ice-plant (*Mesembryanthemum crystallinum*), high salinity, osmotic stress or dehydration causes the activity and gene expression of many enzymes involved in CAM (e.g. glycolysis, gluconeogenesis and malate metabolism) to increase (Cushman and Bohnert 1999). In *M. crystallinum* a switch from C3 to CAM mode is induced by NaCl stress (Winter and Smith 1996); in some submerged aquatic plants, such as *Hydrilla verticillata*, a change from C3 to C4 metabolism occurs if CO_2 is limited in the water (Bowes and Salvucci 1989; Reiskind *et al.* 1997). In these facultative CAM plants and aquatic plants, the photosynthetic carbon metabolism operates in a single type and no differentiation of two cell types is required. In a cultivar of the crassulacean plant *Kalanchoe blossfeldiana*, which is a short-day plant with respect to floral induction, the expression of CAM is also regulated by daylength. During long days this plant behaves as a normal C3 plant, and is converted to CAM when the day length is shorter than a critical value (Fig. 4.15). Other cultivars of this species use CAM also under long-day conditions (Zabka and Chaturvedi 1975).

While C4 photosynthesis requires structural differentiation and biochemical specialization of photosynthetic cells, transcriptional activation controls mRNA accumulation in CAM-associated genes (Cushman *et al.* 1999). In the common ice-plant the expression of a CAM-specific phosphoenolpyruvate carboxylase isoform can also be enhanced by high light intensity, light quality and long photoperiods (McElwain *et al.* 1992; Cockburn *et al.* 1996), suggesting the involvement of phytochrome in stress modulation.

Sun and Shade Plants

Plants adapt to a wide range of light environments, ranging from the deep shade of rainforest understoreys and underwater habitats to the high-radiation environments of deserts and mountain tops. Plants have evolved a number of mechanisms to optimize their use of sunlight over such a wide diversity of habitats (Table 4.4). Optimum or saturation photon flux densities may vary considerably for different species. Some plants grow very well in shaded habitats (*shade plants*), while others require full sunlight (*sun plants*).

Certain plants can adapt to a range of light regimes, growing as sun plants in sunny habitats and as shade plants in shaded habitats. Figure 4.15 illustrates the difference in CO_2 assimilation of leaves grown at high and low photon flux densities. In the leaf of a typical C3 plant photosynthesis saturates at a much lower photon flux density when grown at lower light intensities, indicating that the photosynthetic properties of a leaf depend on its growing conditions. In general, leaves or plants adapted to sunny or shaded environments are unable to survive when transferred to the other type of environment. While some plants, for example

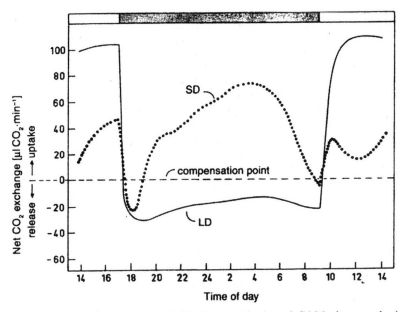

Figure 4.15   The switch between normal C3 photosynthesis and CAM photosynthesis resulting from a change in the daily photoperiod, in *Kalanchoe blossfeldiana* cv. Tom Thumb, a short-day plant. (From Zabka and Chaturvedi 1975)

Alocasia, are obligate shade plants and others, such as *Helianthus*, are obligate sun plants, most plants are, in fact, facultative shade or sun plants.

Sun plants that adapt to shade develop morphological and photosynthetic characteristics typical of those of shade plants. Thus their light compensation points decrease (mainly because they respire much more slowly); they photosynthesize more slowly and photosynthesis is saturated at lower irradiance levels. Sun plants gradually develop the ability to grow in shade. The reverse adaption, from shade to sun conditions, is less common. Shade plants usually cannot be moved into direct sunlight; high photon flux densities cause photoinhibition and bring about the death of older leaves.

The leaves of shade trees exhibit morphological and anatomical features that differ from those of plants growing in open sunlight. As would be expected, plants growing under a forest canopy have leaves that tend to be thinner, have more surface area and contain more chlorophyll than those of sun plants. Shade plants tend to have elongated stems and growth orientated towards the light. In addition, shade plants generally have more chlorophyll per reaction centre and have a higher chlorophyll *b* to chlorophyll *a* ratio. Sun leaves become thicker than shade leaves because they develop longer palisade cells or an additional layer of palisade cells (Fig. 4.16). Tomato plants have about 100 stomata per mm^2 on the lower epidermis in dim light (20 W m^{-2}); in bright light (100 W m^{-2}) additional stomata are found on the upper epidermis. This response can be observed within three days of a change in light conditions (Mohr and Schopfer 1995).

Table 4.4 A comparison of photosynthetic structures in shade and sun plants

Trait	Shade plants	Sun plants
Morphological	Mosaic leaf pattern; phototropic leaf petioles	
Anatomical	Large, thin leaves with thin cuticle; low root:shoot ratio; few palisade layers, short palisade cells, more inter-cellular spaces; horizontal leaf orientation	Small, thick leaves with thick cuticle; high root:shoot ratio; more palisade layers, long cells, tighter packing of cells; vertical leaf orientation
Ultrastructural	Large chloroplasts, several/ thick granal stacks	Small chloroplasts, fewer and smaller grana/thin granal stacks
Biochemical/ Physiological	High chl:RubisCO ratio; low Chl *a:b* ratio; low stomatal conductance; low photo-synthetic capacity; low light compensation/high light saturation point; low respiration rate	Low chl:RubisCO ratio; high Chl *a:b* ratio; high stomatal conductance; high photo-synthetic capacity; high light compensation/high light saturation point

In addition to their anatomical adaptations, leaves maximize light absorption by changing the spatial relationship between the leaf and the incident light. Many plants can extend the time during which the leaf blades are aligned perpendicular to sunlight by using a system of 'solar tracking' whereby the laminae follow the path of the sun during daylight hours. In some species, especially those growing in deserts, leaves turn away from the sun in order to minimize heat absorption and excess water loss. Light reception is also dependent on the orientation or angle of the leaf. Vertical orientation of leaves enhances light reception at low sun angles during early morning or late afternoon and thus reduces light reception at solar noon when radiation levels are highest. Leaves that are displayed horizontally receive light all day long, and especially at midday. Thus, leaves in a rainforest tend to be vertical at the tops of trees and horizontal in the understorey. Many plants change their leaf angles in response to a change in light; some do this in order to increase reception while others do it to avoid high light. An understorey herb, *Oxalis oregana*, found in redwood forests in the western USA provides a good example of optimizing light reception through leaf movement (Fig. 4.16). This plant can track sunlight on dull days but can change its leaf angle from horizontal to vertical in only 6 min when exposed to full sunlight (Björkman and Powles 1981). Such rapid responses can regulate light reception on a diurnal basis and help to protect the plant from excess light absorption. Slower-acting mechanisms such as changes in leaf-surface texture also regulate light reception, especially when there is a sustained change in the light environment. Many plants in high-light environments increase the reflectance of their leaves by acquiring a coat of leaf hairs or wax, as a means of external photoprotection.

Plants exhibit plasticity in their response to changes in light availability within a particular habitat. In low light plants need to absorb sufficient light for photosynthesis in order to survive. To do this plants need to maximize light absorption. In a high-light environment, however, the problem is reversed, with plants needing to utilize sufficient light energy for photsynthesis, while at the same time dealing with excess sunlight when photosynthetic capacity is exceeded. To deal with these variable light environments plants have evolved a variety of features that optimize light reception, absorption and processing (Table 4.4). Adaptation requires a genetically determined capacity to adapt to either sun or shade. Shade plants increase the light reception by producing larger leaves. Leaf size can vary within an individual plant, smaller leaves being produced near the top where irradiance is highest and larger leaves towards the interior and base where light levels are lower.

Light reception can also be regulated at the tissue and organelle level. Photosynthetic cells can be concentrated equally on both sides of a leaf (isobilateral) to maximize light absorbed by either side or concentrated on one side (dorsiventral) only, as is found in plants where leaves are predominantly horizontal. The density and location of chloroplasts within leaves can determine the capture of light energy: change of alignment from horizontal to vertical can greatly reduce the overall absorption of incident light.

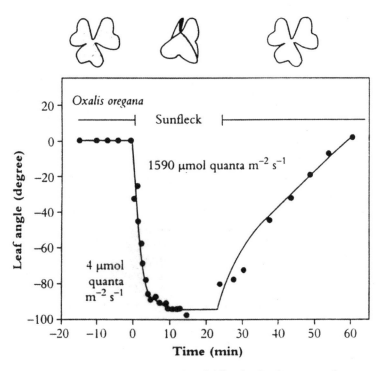

Figure 4.16 Time-course of leaf folding and unfolding in *Oxalis oregana* in response to arrival and departure of an intense sunfleck. (Björkman and Powles 1981)

Shade-grown plants and sun-grown plants process absorbed light energy in different ways. In high light there is a requirement for greater capacity in both light and dark reactions of photosynthesis; such differences are shown in light-response curves for shade and sun plants (Fig. 4.17). The initial slope of each light-response curve represents the quantum efficiency of photosynthesis. This is the same in shade and sun plants. The reason it does not change is that the efficiency of the light reactions is the same no matter how much light has been absorbed (i.e. eight photons are required for the evolution of one molecule of O_2 and the fixation of one molecule of CO_2). However, sun plants have a greater capacity for photosynthetic electron transport, since they possess a greater amount of carriers such as the cytochromes and plastoquinone. They also have a greater capacity for ATP synthesis per unit of chlorophyll compared with shade plants. Shade plants may contain four to five times more chlorophyll *a* and *b* per unit volume of chloroplast and have a higher chlorophyll *b/a* ratio than sun plants because their light-harvesting complexes are more elaborate. Shade enhances the capacity for light capture and energy transfer in the reaction centres. However, the capacity of the electron transport chain in shade plants is not increased, as there is relatively less (about one-fifth the amount) light-harvesting complexes (cytochrome *f*, plastoquinone, ferredoxin and carotenoid) per unit of chlorophyll than in sun plants. Shade plants have, therefore, more light-collecting apparatus but a smaller complement of electron carriers than sun plants. Since in dim light the rate of electron transport is limited by the number of photons falling on the leaf, there is no advantage for shade plants to have a large-capacity electron chain. Sun plants have less developed thylakoid systems, fewer granal stacks and fewer light-harvesting complexes so that they are less efficient at absorbing light energy at low photon flux than shade plants and so have lower quantum yields. The main differences between the two groups of plants is the capacity of the light-harvesting system and of electron transport. The light-absorbing system in shade plants makes them very effective at gathering the light energy available and passing it to the reaction centres, especially in dim light but they are limited in bright light by the rate of electron transfer; sun plants, in contrast, are very efficient at transporting electrons but not at gathering weak light. Synthesis of NADPH and ATP, rather than electron transport *per se*, determines CO_2 assimilation. Photosynthesis in shade plants may be limited by NADPH synthesis. Extensive granal stacking may be essential to obtain sufficient rates of electron flow to reduce $NADP^+$; cyclic electron flow could drive ATP synthesis to match NADPH production. Sun plants, on the other hand, probably have adequate $NADP^+$ reduction but may be limited by ATP supply; cyclic electron flow may then produce the required ATP without $NADP^+$ reduction.

Overall, sun plants can and do process more sunlight than shade plants. This higher capacity is matched by a greater investment in the biochemical machinery for CO_2 fixation (e.g. enzymes of the Calvin cycle), resulting in a higher light-saturation point and a higher maximum rate of photosynthesis. However, the higher photosynthetic capacity in sun plants is accompanied by higher respiration rates which increase the light-compensation point relative to shade plants. Higher respiration rates probably result from increased carbohydrate processing in high

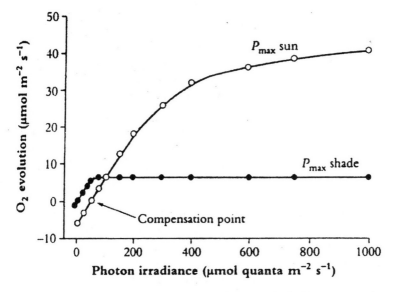

Figure 4.17 Photosynthetic-light response curve for typical shade and sun plants showing relationships between photosynthetic rate (measured as O_2 evolution) and light absorbed (expressed as photon irradiance). (Original data from Robinson *et al.* 1993)

light intensity, increased cost of constructing sun leaves and a higher cost of maintaining such leaves (Atwell *et al.* 1999). An equilibrium between photosynthesis and respiratory gas exchange can be achieved by varying the light. The light flux at which CO_2 uptake equals CO_2 release (or O_2 uptake equals O_2 release) for a leaf kept under constant CO_2 concentration (air), is called the *light compensation point* of photosynthesis. If a leaf, or plant, possesses a high light-compensation point it requires a relatively large light flux for photosynthesis to balance respiration. Conversely, a leaf or plant with a low light-compensation point can maintain a photosynthetically compensated carbon balance, even at relatively low light flux. Thus, this value characterizes the efficiency of a leaf with respect to the use of light and determines the minimum light flux required for the long-term survival of the photoautotrophic plant. The light compensation point varies within limits. Usually low values (around 100 lx) are measured for plants in light-depleted locations (shade plants), and high values (500–800 lx) for plants exposed to light (sun plants). Another cost of higher photosynthetic capacity is a higher rate of transpiration due to greater stomatal conductance. Sun plants often respond to this by increasing their root:shoot ratio. When water is limiting, however, stomatal conductance may be reduced, sacrificing photosynthesis in favour of slower transpiration (Atwell *et al.* 1999).

Absorption of excess light often leads to photoinhibition, a regulatory, photoprotective mechanism in photosynthesis whereby the amount of light reaching the reaction centres of PSII is reduced. The effect of photoinhibition is to reduce the quantum yield of photosynthesis. Photoprotection is normally sufficient to cope with light absorbed by leaves; photodamage only occurs when the capacity for photoprotection is exhausted. If excess light energy is absorbed, the plant employs photoprotective mechanisms to dissipate this excess energy as heat. The non-photochemical conversion of light energy is thought to occur in the PSII antennae and be mediated by xanthophyll pigments, including violaxanthin, antheraxanthin and zeaxanthin. These carotenoid pigments undergo interconversion in a process described as the 'xanthophyll cycle', in response to excess light; energy is dissipated in the process (Fig. 1.20). The potential value of the xanthophyll cycle is evident in the way plants alter their carotenoid content in response to changing environmental conditions. Total pool sizes of xanthophyll pigments increase with increasing exposure to excess light. Sun plants can have three- or four-fold larger pools of violaxanthin, antheraxanthin and zeaxanthin than shade plants. In addition, the carotenoid/chlorophyll ratio is typically higher in leaves exposed to full sunlight compared with shade-grown leaves (Demmig-Adams and Adams 1996). Photodamage is associated with damage to the D1 protein of PSII (Fig. 2.7), causing PSII reaction centres to become photochemically inactive. Non-functional PSII centres may then accumulate in the chloroplast grana and somehow increase thermal dissipation of excess light energy (Anderson and Aro 1994). Shade plants with their larger granal stacks may have greater capacity to accumulate non-functional PSII centres, whereas sun plants may rely on a higher rate of D1 turnover and larger xanthophyll pool sizes for their internal photoprotection.

Canopy structure is an important environmental factor affecting canopy light climate and energy conversion. Large leaf angles, with leaves close to the vertical, ensure good photosynthetically active radiation (PAR) penetration when the solar angle is high and a high proportion of leaves receive similar irradiation. An even distribution of PAR at leaf surfaces enhances canopy photosynthesis and improves light-use efficiency over canopies where upper, horizontal leaves intercept total sunlight, and light reaching lower leaves is greatly attenuated (down to 10%). Since leaves absorb little of the far-red light, understorey habitats are enriched with far-red light. As a result, plants growing in such environments are tall and spindly. While leaves of shade plants may be photodamaged on exposure to high photon fluxes, they may also benefit from the phenomenon of *sunflecks*. Sun patches and the more transient sunflecks occur when small and variable openings (possibly due to disturbances caused by a sudden breeze) in the canopy permit direct sunlight to penetrate to the forest floor, resulting in the familiar patchwork pattern of sunlight and shade. In a dense forest, sunflecks can, within seconds, increase ten-fold the photon flux density impinging on a shaded leaf. Sunflecks can contain up to 50% of the total light energy available during the day but it is available for only a few minutes' duration. Two components in the plant's photosynthetic physiology determine how it uses the light in a sunfleck and account for the variation in sunfleck-utilization efficiency. First, photosynthetic capacity will determine how much light a plant can use. Second, a few minutes' illumination are needed for the photosynthetic carbon reduction cycle intermediates to reach critical levels, and this 'induction requirement' of photosynthesis determines how quickly a leaf responds to an increase in irradiance. When a leaf that has been in low light for some time is exposed to an increase in irradiance, the rate of photosynthesis does not increase instantaneously to the new level: there is a gradual (10–60 min) increase in assimilation. This induction period varies according to the species of plant as well as the induction state of the leaf concerned. One consequence of generally low photosynthetic capacity in understorey plants is a limited ability to process the light energy absorbed during strong sunflecks. This limited ability is further impaired by a low induction state. Under these conditions, understorey plants need to dissipate excess energy in order to avoid photodamage.

Primary Production of the Biosphere

Biological processes on land and in the oceans contribute to the global carbon cycle. In both components of the biosphere, photosynthesis is responsible for virtually all of the biochemical production of organic matter. Early estimates of net primary production (NPP) and plant biomass (Whittaker 1975) are outdated; past syntheses of primary production from photosynthesis have focused on the terrestrial or ocean components separately, with the result that models of the global carbon cycle were compartmentalized, with limited opportunity for comprehensive analysis. Current integrated estimates of primary production are based on satellite measurements for both oceanic and terrestrial ecosystems (Field *et al.* 1998). This integrated approach builds from parallel data sets and gives a more accurate picture of the biosphere production.

The carbon cycle in terrestrial and ocean biomass involves both the production and turnover of organic matter. Photosynthesis and the biosynthesis of organic compounds, the processes that result in net primary production, are very similar. NPP is a major determinant of carbon sinks both on land and in the ocean. The major components of terrestrial plant biomass are roots and stems which respire but generally do not photosynthesize. The NPP of terrestrial ecosystems can be readily determined from incremental increases in biomass plus litter fall; below-ground measurements are not so easily measured. Since ocean NPP is dominated by phytoplankton, nearly all of the biomass is photosynthetic. NPP on land and in the oceans has been modelled with a variety of approaches. A common contemporary approach, developed independently for land and ocean models, calculates NPP as a function of the energy of photosynthesis, the absorbed photosynthetically active (400 to 700 nm) solar radiation (APAR), and an average light utilization efficiency (ε) (Morel 1991):

$$NPP = APAR \times \varepsilon$$

Models based on this approach are all strongly connected to global-scale observations. For the oceans, APAR can be related to satellite-derived measurements of surface chlorophyll (Morel 1991), and for terrestrial systems it can be determined from satellite-based estimates of vegetative greenness, usually the normalized difference of vegetation index (NDVI) (Sellers 1987). APAR depends on the amount and distribution of photosynthetic biomass and on the amount of available solar radiation (including the photosynthetically active wavelengths). ε is an effective photon yield for growth that converts the biomass-dependent variable (APAR) into a flux of organic compounds (NPP). For both terrestrial and oceanic models, ε cannot be measured directly and must be parameterized with field measurements. For marine systems, ε can be parameterized from thousands of ^{14}C-based field measurements of NPP (Behrenfeld and Falkowski 1997).

Terrestrial values are less readily obtained, mainly because ε depends on time-consuming determinations of NPP and APAR (Ruimy et al. 1994); uncertainty in ε is a major source of error in land and ocean NPP estimates. In general, ocean NPP models estimate ε solely as a function of sea-surface temperature (Antoine et al. 1996; Behrenfeld and Falkowski 1997); in terrestrial ecosystems, ε varies with ecosystem type.

Field et al. (1998) determined primary production of the biosphere by combining results from conceptually similar land and ocean NPP models – the Carnegie–Ames–Stanford approach (CASA) (Potter et al. 1993) for land and the Vertically Generalized Production Model (VGPM) ((Behrenfeld and Falkowski 1997) for the oceans. Both of these models are simple formulations using versions of the equation for NPP, above. Using the integrated CASA–VGPM biosphere model, Field et al.

Table 4.5 Annual and seasonal NPP of the major units of the biosphere, from CASA–VGPM. (From Field et al. 1998)

	Ocean NPP (Pg of C)		Land NPP (Pg of C)
Seasonal			
April–June	10.9		15.7
July–September	13.0		18.0
October–December	12.3		11.5
January–March	11.3		11.2
Biogeographic			
Oligotrophic	11.0	Tropical rainforests	17.8
Mesotrophic	27.4	Broadleaf deciduous forests	1.5
Eutrophic	9.1	Broad- and needleleaf forests	3.1
Macrophytes	1.0	Needleleaf evergreen forests	3.1
		Needleleaf deciduous forests	1.4
		Savannas	16.8
		Perennial grasslands	2.4
		Broadleaf shrubs with bare soil	1.0
		Tundra	0.8
		Desert	0.5
		Cultivation	8.0
Total	48.5		56.4

(1998) obtained an annual global NPP of 104.9 Pg of C (Table 4.5), with similar contributions from the terrestrial (56.4 Pg of C (53.8%)) and oceanic (48.5 Pg of C (46.2%)).

Spatial heterogeneity in NPP is comparable on land and in the oceans, with both regions exhibiting large regions of low productivity and smaller areas of high productivity. Maximal NPP is similar in the two systems (1000–$1500\,\mathrm{g\,C\,m^{-2}\,year^{-1}}$) but regions of high NPP are spatially more restricted in the oceans (essentially limited to estuarine and upwelling regions) than in terrestrial systems, for example, humid tropics (Field *et al.* 1998).

Seasonal fluctuations in ocean NPP are modest globally, even though regional seasonality can be very important. Ocean NPP ranges from 10.9 Pg of C in the Northern Hemisphere spring (April to June) to 13.0 Pg of C in the Northern Hemisphere summer (July to September) (Table 4.5). Spatial variation in NPP in both the terrestrial and ocean components of the biosphere model are driven mostly through variation in light capture by photosynthetic biomass or APAR and secondary variation in ε (Field *et al.* 1998). Spatial and seasonal variation in photosynthetic biomass is, in turn, largely controlled by the availability of other resources. Nitrogen, iron and light are critical in the oceans; on land water stress, temperature and nutrients such as phosphorus are important (Schimel *et al.* 1996). Consequently, regional and seasonal distributions of NPP reflect the interface between physical processes such as precipitation, PAR and ocean circulation and biological activities such as microbial activity and interaction between organisms.

Summary

The balance of CO_2 removal and release into the atmosphere is kept constant by the carbon cycle. Photosynthesis in the Carboniferous Age produced millions of tonnes of carbon that became trapped in plant tissues which are now vast reservoirs of oil and coal. Oceans provide another large reservoir of CO_2 and thus help to stabilize the atmospheric CO_2 balance; the CO_2 stored in sedimentary rocks is not immediately usable. In the short term, the largest contribution of CO_2 to the atmosphere is the respiration of plants, animals and bacteria and the burning of fossil fuels; in the long term, volcanic eruptions and emissions from soda springs would be the chief sources.

Photosynthesis is limited by a wide range of variables (light, CO_2, O_2, temperature, water, morphology of the leaf), all of which are involved in complex, mutual interaction. Only 5% of the solar energy reaching earth is used in photosynthesis and about 15% of the photosynthetically active radiation (400–700 nm) is lost through reflection and transmittance. Light-compensation point for normal C3 plants occurs at about 2% of normal sunlight. The rate of photosynthesis increases with increase in light up to light saturation at between a quarter and a half of full sunlight. Leaves of C4 species never reach saturation even in full sunlight and have maximum photosynthetic rates more than twice that of C3 plants.

Modification of photosynthetic capacity due to the environment is a complex process and requires the adaptation of morphological and functional properties of the leaf. Most plants are facultative shade or sun plants, but while sun plants can develop the anatomical and physiological characteristics of shade plants the latter do not adapt well to sunny habitats. CO_2 regulation is the more important of the two gas exchanges.

The absorption of radiant energy causes heating; conduction and convection currents help to lower temperature at the leaf surfaces. The temperature effects on biochemical reactions and responses are complex; effects on photosynthesis, in particular, depend on species and on environmental conditions. Optimal temperatures are correlated with genetical and physiological factors. Plant species can adapt to changes in temperature and adjust to seasonal change and different environments.

Increases in O_2 levels inhibit photosynthesis by causing oxidative photodestruction of proteins, nucleic acids and chlorophyll molecules. Biological electron transport systems are often linked to the formation of free radicals, especially oxygen radicals, which can be toxic.

Photosynthetic efficiencies differ immensely for plant species growing under a wide range of conditions. Differences in leaf photosynthetic capacity result mainly from variations in light, temperature, availability of water and nutrients. Highest capacities occur among desert annuals and grasses (C4 species) while slow-growing succulents (CAM species) have the lowest. C4 photosynthesis is successful due to improved water-use and nutrient-use efficiencies and to their relatively longer growth cycles. Water-use efficiency in CAM plants is much higher than that in C3 or C4 plants, which thus allows for diversification and speciation of CAM plants in water-limiting environments.

Shade plants and sun plants have evolved to optimize the use of sunlight over a wide diversity of habitats. Both types of plant have developed distinct morphological and photosynthetic characteristics. Adaptation requires a genetically determined capacity for either shade or sun. Shade-grown and sun-grown plants process absorbed light energy in different ways; sun plants process relatively more light energy. Canopy structure is an important factor affecting canopy light and energy conversion. Understorey habitats are enriched with far-red light and plants in these habitats are tall and spindly. Transient sun patches and sunflecks can rapidly increase the photon flux density impinging on shaded leaves.

Estimation of net primary productivity is difficult and evaluations by different researchers vary greatly. Approximately two-thirds of the total carbon fixed globally is accomplished by terrestrial ecosystems and the remaining one-third by the oceans. The greatest productivity occurs in ecosystems that provide an optimal combination of water, temperature and nutrients, such as is found in tropical rainforests. Water is a limiting factor in terrestrial ecosystems which thus have only moderate productivity. Poor photosynthetic efficiency of vegetation is due to an accumulation of losses arising from each energy transfer in the ecosystem.

Photosynthesis and organic compound biosynthesis make up net primary production (NPP). Present-day approaches calculate NPP as a function of the energy of photosynthesis, the absorbed photosynthetically active solar radiation (APAR) and average light utilization efficiency (ε). APAR is determined from satellite-based measurements.

References

Anderson, J.M. and Aro, E.-M. (1994). Grana stacking and protection of photosytem II in thylakoid membranes of higher plant leaves under sustained high irradiance: an hypothesis. *Photosynthesis Research* **41**: 315–326.

Antoine, D., André, J.M. and Morel, A. (1996). Oceanic primary production. 2. Estimation of global scale from satellite (coastal zone color scanner) chlorophyll. *Global Biogeochem. Cycles* **10**: 57–70.

Atwell, B.J., James, B., Kriedemann, P.E. and Turnbill, C.G. (1999). Sunlight: an all-pervasive source of energy, *Plant Physiology*, pp. 381–415. Macmillan Education Australia, South Yarra, Vic.

Bauer, A. and Martha, P. (1981). The CO_2 compensation point of C3 plants – a re-examination. I. Interspecific variability. *Zeitschrift für Pflanzenphysiologie* **103**: 445–450.

Behrenfeld, M.J. and Falkowski, A. (1997). A consumer's guide to phytoplankton primary production models. *Limnol. Oceanogr.* **42**: 1479–1491.

Berner, R.A., and Lasaga, A.C. (1989). Modelling the geochemical carbon cycle. *Scientific American* **260**: 74–81.

Björkman, O. (1991). Responses to different quantum flux densities. In: *Encyclopedia of Plant Physiology*, O.L. Lange, P.S. Nobel, C.B. Osmond and H. Zeigler (eds), New Series, **12A**: 57–107. Springer-Verlag, Berlin.

Björkman, O. and Powles, S.B. (1981). Leaf movement in the shade species *Oxalis oregano*. I. Response to light level and light quality. *Carnegie Year Book* **80**: 59–62.

Bowes, G. and Salvucci, M.E. (1989). Plasticity in the photosynthetic carbon metabolism of submerged aquatic macrophytes. *Aquat. Bot.* **34**: 233–266.

Calvin, M. (1982). Bioconversion of solar energy. In: *Trends in Photobiology*, C. Helene, M. Chalier, T.-M. Garestier and G. Laustriat (eds), pp. 645–659. Plenum, New York and London.

Cockburn, W., Whitelam, G.C., Broad, A. and Smith, J. (1996). The participation of phytochrome in the signal transduction pathway of salt stress responses in *Mesembryanthemum crystallinum* L. *J. Exp. Bot.* **47**: 647–653.

Corré, N.C. (1983). Growth and morphogenesis of sun and shade plants. I. The influence of light intensity. *Acta Botanica Neerlandica* **32**: 49–62.

Cushman, J.C. (2001). Crassulacean acid metabolism. A plastic photosynthetic adaptation to arid environments. *Plant Physiol.* **127**: 1439–1448.

Cushman, J.C. and Bohnert, H.J. (1999). Crassulacean acid metabolism: molecular genetics. *Ann. Rev. Plant Physiol. Plant Mol. Biol.* **50**: 305–332.

Cushman, J.C., Meyer, G., Michalowski, C.B., Schmitt, J.M. and Bohnert, H.J. (1999). Salt stress leads to the differential expression of two isogenes of phosphoenolpuruvate carboxylase during crassulacean acid metabolism in the common ice plant. *Plant Cell* **1**: 715–725.

Demmig-Adams, B. and Adams, W.W. (1996). The role of xanthophyll cycle carotenoids in the protection of photosynthesis. *Trends in Plant Science* **1**: 21–26.

Drennan, P.M. and Nobel, P.S. (2000). Responses of CAM species to increasing atmospheric CO_2 concentrations. *Plant Cell Environ.* **23**: 767–781.

Edwards, G.E., Furbank, R.T., Hatch, M.D. and Osmond C.B. (2001). What does it take to be C4? Lessons from the evolution of C4 photosynthesis. *Plant Physiol.* **125**: 46–49.

Ehleringer, J. and Pearcy, R.W. (1983). Variation in quantum yield for CO_2 uptake among C_3 and C_4 plants. *Plant Physiol.* **73**: 555–559.

Field, B.C., Behrenfeld, M.J., Randerson, J.T. and Falkowski, P. (1998). Primary production of the biosphere: integrating terrestrial and oceanic components. *Science* **281**: 237–240.

Gaastra, P. (1959). Photosynthesis of crop plants as influenced by light, carbon dioxide, temperature and stomatal diffusion resistance. *Mededelinger van de Landbouwhhogeschool Te Wageningen* **59**: 1–68.

Halliwell, B. and Gutteridge, J.M.C. (1985). The superoxide theory of oxygen toxicity, *Free Radicals in Biology and Medicine*, pp. 86–187. Clarendon Press, Oxford.

Haupt, W. (1986). Photomovement. In: *Photomorphogenesis in Plants*, R.E. Kendrick and G.H.M. Kronenberg (eds), pp. 415–441. Martinus Nijhoff, Dordrecht.

Hesketh, J.D. (1963). Limitations to photosynthesis responsible for differences among species. *Crop Science* **3**: 493–496.

McElwain, E.F., Bohnert, H.J. and Thomas, J.C. (1992). Light moderates the induction of phosphoenolpyruvate carboxylase by NaCl and abscisic acid in *Mesembryanthemum crystallinum*. *Plant Physiol.* **99**: 1261–1264.

Mohr, H. and Schopfer, P. (1995). The leaf as a photosynthetic system, *Plant Physiology*, pp. 225–243. Springer-Verlag, Berlin, Heidelberg and New York.

Morel, A. (1991). Light and marine photosynthesis: a spectral model with geochemical and climatological implications. *Prog. Oceanogr.* **26**: 263–306.

Orsenigo, M., Patrignoni, G. and Rascio, N. (1995). Ecophysiology of C3, C4 and CAM plants. In: *Handbook of Photosynthesis*, M. Pessarakli (ed.), pp. 1–25. Marcel Dekker, New York.

Osmond, C.B., Winter, K. and Ziegler, H. (1980). Functional significance of different pathways of CO_2 fixation of photosynthesis. *Encyclopedia of Plant Physiology* **12A**: 480–547.

Potter, C.S., Randerson, J.T., Field, C.B., Matson, P.A., Vitousek, P.M., Mooney, H.A. and Klooster, S.A. (1993). Terrestrial ecosystem production: a process-oriented model based on global satellite and surface data. *Global Biogeochem. Cycles* **7**: 811–842.

Reiskind, J.B., Madsen, T.V., van Ginkel, L.C. and Bowes, G. (1997). Evidence that inducible C4-type photosynthesis is a chloroplastic CO_2^- concentrating mechanism in *Hyrilla*, a submersed monocot. *Plant Cell Environ.* **20**: 211–220.

Robinson, S., Lovelock, C.E. and Osmond, C.B. (1993). Wax as a mechanism for protection against photoinhibition – a study of *Cotyledon orbiculata*. *Botanica Acta* **106**: 307–312.

Ruimy, A., Saugier, B. and Dedieu, G. (1994). Methodology for the estimation of net primary production from remotely sensed data. *J. Geophys. Res.* **99**: 5263–5283.

Sage, R.F. (2002). C4 photosynthesis in terrestrial plants does not require Kranz anatomy. *Trends Plant Sci.* **7**: 283–285.

Salisbury, F.B. and Ross, C.W. (1992). Photosynthesis: environmental and agricultural aspects, *Plant Physiology*, pp. 249–265. Wadsworth, Belmont, CA.

Schimel, D.S., Braswell, B.H., McKeown, R., Ojima, D.S., Parton, W.J. and Pulliam, W. (1996). Climate and nitrogen controls on the geography and timescales of terrestrial biogeochemical cycling. *Global Biogeochem. Cycles* **10**: 677–692.

Sellers, P.J. (1987). Canopy reflectance, photosynthesis, and transpiration II: The role of biophysics in the linearity of their interdependence. *Remote Sens. Environ.* **21**: 143–183.

Taiz, L. and Zeiger, E. (1991). Photosynthesis: carbon metabolism, *Plant Physiology*, pp. 219–248. Benjamin/Cummings, Redwood City, CA.

Ting, I.P. (1985). Crassulacean acid metabolism. *Ann. Rev. Plant Physiol.* **36**: 595–622.

Voznesenskaya, E.V., Franceschi, V.R., Kiirats, O., Freitag, H. and Edwards, G.E. (2001). Kranz anatomy is not essential for terrestrial C4 plant photosynthesis. *Nature* **414**: 543–546.

Whittaker, R.H. (1975). *Communities and Ecosystems*, 2nd edn. Macmillan, New York.

Winter, K. and Smith, J.A.C. (1996). An introduction to crassulacean acid metabolism. Biochemical and ecological diversity. In: *Crassulacean Acid Metabolism. Biochemistry, Ecophysiology and Evolution*, K. Winter and J.A.C. Smith (eds), pp. 1–13. Springer, Berlin and Heidelberg.

Zabka, G.G. and Chaturvedi, S.N. (1975). Water conservation in *Kalanchoe blossfeldiana* in relation to carbon dioxide dark fixation. *Plant Physiol.* **55**: 532–535.

5
Molecular Biology of Chloroplasts

Plastids

The main structural difference between prokaryotes and eukaryotes is the presence of subcellular organelles in the latter, particularly mitochondria (present in almost all eukaryotes) and plastids (present only in plants). Cytoplasmic organelles can be passed from one generation to the next by way of the egg (maternal inheritance), the sperm (paternal inheritance) or both (biparental inheritance). However, genetic evidence indicates that in angiosperms, plastids and mitochondria are transmitted maternally. Both of these organelles are most likely excluded from the sperm cells during male gametophyte development. Biparental inheritance of plastids and mitochondria has been reported in a few flowering plant genera, including *Pelargonium* (geranium), *Plumbago* (leadwort) and *Oenothera* (evening primrose). In *Pelargonium*, both male plastids and mitochondria differ ultrastructurally from those of the female, making it possible to establish that organelles from both are present in the embryo. *Plumbago* has dimorphic sperm – one of the male gametes is rich in mitochondria, the other in plastids and exhibits preferential fertilization, the plastids-rich sperm fusing most commonly with the egg.

Gymnosperms differ from angiosperms in that their plastids are usually transmitted through the sperm. Plastids multiply by binary fission of existing plastids, independently of cell division. In meristems, proplastid division keeps pace with cell division, so that daughter cells possess approximately the same number of plastids as the parent cells. Plastids have the capacity to differentiate and to dedifferentiate, resulting in interconversions that occur under certain environmental and developmental conditions (Fig. 5.1).

Photobiology of Higher Plants. By M. S. Mc Donald.
© 2003 John Wiley & Sons, Ltd: ISBN 0 470 85522 3; ISBN 0 470 85523 1 (PB)

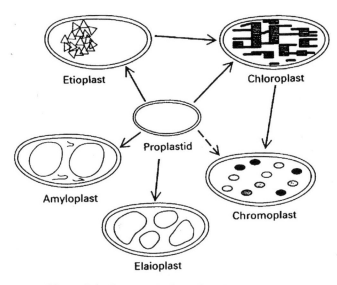

Figure 5.1   Interconversion of various plastid types

Proplastids

Proplastids are the precursors of other plastids and occur as discrete organelles in the cells of young developing leaves, shoots and roots, in meristems and in the embryo. Plastids are usually spherical or ovoid in shape and range in diameter from 0.2 to 1.0 μm. The matrix of the proplastid, called the stroma, is uniformly dense and granular. Embedded in the stroma are a small number of ribosomes, an electron-lucent nucleoid containing DNA fibrils and poorly developed lamellar membranes called thylakoids which appear to be continuous with the inner membrane of the surrounding envelope.

Amyloplasts

Amyloplasts are unpigmented plastids that contain starch grains. Their name is derived from amylose, a straight-chain polysaccharide composed of $\alpha(1-4)$ glucopyranosyl units, which, with amylopectin, constitutes the starch molecule. Amyloplasts occur mainly in storage tissues such as potato tubers. In the columella cells of root caps, amyloplasts serve as statoliths – they sediment in response to gravity and thereby initiate a gravitropic response.

Chloroplasts

Chloroplasts are green photosynthetic plastids containing chlorophyll and are responsible for the capture of light energy. The photosynthetic apparatus of the chloroplast is located in its extensive thylakoid system (Fig. 2.1); the dark reactions of photosynthesis take place in the stroma. In addition to their all-important role in providing a site for photosynthesis, chloroplasts carry out several other metabolic processes, including the biosynthesis of chlorophylls, carotenoids, purines, pyrimidines and fatty acids. They also play a role in the reduction of sulphate (SO_4^-) and of the inorganic ion nitrite (NO_2^-), the product of cytosolic nitrate reduction. When proplastids have been exposed to light for 3–4 minutes, enzymes are formed (or imported from the cytosol), light-absorbing pigments are synthesized and a well-developed thylakoid system is established. Ribosomes (70S), composed of 50S and 30S subunits, are found either free or bound to thylakoid membranes.

Chromoplasts

Chromoplasts are yellow, orange or red, depending on the particular combination of carotenoids present in the plastid. They are considered a degenerative form of plastid and are responsible for the colours of many fruits and flowers and of many roots such as carrots and sweet potatoes. Chromoplasts develop directly from proplastids or by redifferentiation of chloroplasts, as happens in the ripening of fruits. During the transition from chloroplast into chromoplast, the chlorophyll is broken down and the thylakoid lamellae and stroma become reorganized. Chromoplasts may ocasionally revert into chloroplasts as they do in the surface of carrot roots exposed to light or in some yellow or orange citrus fruits that regreen under appropriate light conditions. Chromoplast development is accompanied by the induction of enzymes that catalyze the biosynthesis of carotenoids in the thylakoid membranes which may carry the lipophilic precursors of carotenes and carotenes themselves.

Etioplasts

Etioplasts occur in plants that are grown in darkness. Etiolated tissues may appear yellow due to the presence of large amounts of the chlorophyll precursor protochlorophyllide in the etioplasts. Etioplasts are structurally simple, the most distinctive feature being a para-crystalline arrangement of lipid-storing membranes, called the *promellar body*. On exposure to light, etioplasts differentiate into chloroplasts; in the process protochlorophyllide is converted to chlorophyll and the promellar body reorganizes into grana and stromal lamellae.

Leucoplasts

Leucoplasts are colourless plastids found in epidermal and storage tissue; they are involved in the synthesis of monoterpenes, the volatile compounds contained in essential oils. 'Leucoplast' is merely a descriptive term and and is often used as a synonym for any colourless plastids such as proplastids, amyloplasts (which store starch) and elaioplasts (which store lipids and droplets of oil). Leucoplasts typically contain an extensive network of membranes which are involved in the synthesis of lipid molecules.

Organellar Genomes

Plant cells are unique in that they have three, distinct genomes – nuclear, plastidic and mitochondrial, each having its own genetic system and protein-synthesis machinery. It is now well established that the genomes of plastids and mitochondria contain a small number of genes that code for proteins that both function in the organelles and maintain their own genetic system. However, expression of organellar genomes is under the tight control of the nuclear genome. Regulatory mechanisms have also evolved that coordinate the expression of nuclear and organellar genes for proteins that function in the plant organelles.

It is generally believed that chloroplasts and mitochondria originated from endosymbiotic bacteria (Gray and Doolittle 1982). The first photosynthetic cells were most likely free-living prokaryotes, similar to present-day photosynthetic bacteria. In the absence of uncombined oxygen these organisms probably used H_2S as a substrate for the supply of electrons. At some point during the evolution of eukaryotes, these primitive cells were engulfed by nucleated prokaryotic cells and a symbiotic relationship was established (Reumann and Keegstra 1999). In the process, mitochondria and plastids retained their functional genomes but came under the biochemical and genetic control of the host cell. Plastids and mitochondria have a number of characteristics in common with free-living prokaryotes. For example, the organization of organellar and bacterial genomes is similar. In addition, chloroplasts carry out photosynthesis in much the same way as cyanobacteria (which contain chlorophyll and have photosystem I and II activity). The endosymbiotic hypothesis is also supported by the evidence that ribosomal RNA sequences established from organelles and from free-living prokaryotes suggest that plastids share a common ancestor with modern cyanobacteria and mitochondria with present-day proteobacteria that resemble the purple, sulphur types of eubacteria. Symbiosis between present-day photosynthetic organisms and non-photosynthetic hosts also support the endosymbiotic hypothesis. For example, *Cyanophora paradoxa*, a biflagellate protist, engulfs an endosymbiotic cyanobacterium, called a cyanelle, which functions as a chloroplast and supplies the host cell with photosynthetically reduced carbon.

There are many similarities but also many substantial differences between chloroplasts and cyanobacteria in their organization; differences appear in photosynthetic structures and metabolism and in the way in which their genetic material is organized. In prokaryotes, neither the genome nor the photosynthetic apparatus is confined within a membrane; in eukaryotes, the nucleus, chloroplast and mitochondrion, each with its own complement of DNA, are membrane-bound. Nuclear, chloroplastic and mitochondrial DNAs are referred to as nDNA, cpDNA and mtDNA, respectively.

The plastid genome

Plastids and mitochondria are semi-autonomous and cannot exist outside the eukaryotic cell without the supply of new proteins encoded in the nucleus. Despite the endosymbiotic relationship, very little DNA remains in either the chloroplast or the mitochondrion: during evolution, most of the DNA once present in these organelles was transferred to the nuclear genome. The DNA that remains within the organelles is mainly circular as in the genomes of bacteria and, unlike that of nuclear genomes of eukaryotes, does not form extensive supra-molecular complexes with proteins. Some organellar DNA sequences resemble the operons of prokaryotes in which genes encoding proteins in a common metabolic pathway are clustered together. Many genes in the plastid genomes are organized into polycistronic transcription units, that is, clusters of two or more genes that are transcribed by RNA-polymerase from a single promoter. For example, the genes of plastid rRNAs of most plants are organized into a cluster consisting of 16S, 23S, 4.5S and 5S rRNA genes. However, unlike the operons of prokaryotes, plastid genomes also contain polycistronic transcription units composed of distinct genes encoding proteins with diverse functions. Often, mRNAs from genes encoding proteins that are required during photosynthesis can be detected in non-photosynthetic plastids, such as amyloplasts in roots or chromoplasts in coloured fruits. Apparently these mRNAs are not translated into functional proteins, suggesting that plastid-gene expression may be under post-transcriptional control. The chloroplast genome of higher plants apparently does not contain any genetic sequences that are common to the mitochondrial or nuclear genome, apart from those sequences which have functional sequence homology, such as rRNAs and tRNAs.

The presence of DNA in chloroplasts was first demonstrated by Ris and Plaut (1962). They observed DNA-like filaments in a low electron-dense area within chloroplasts of *Chlamydomonas*. However, the first convincing evidence of the presence of chloroplast DNA in higher plants was provided by Tewari and Wildman in 1968; they isolated chloroplasts from leaves of tobacco (*Nicotiana tabacum*) and characterized some of its properties (Tewari 1987). The genome of the chloroplast is composed of a single, circular chromosome of double-stranded DNA and ranges in size from 120 to 160 kilobases (kb); it codes for about 140 products of which about 30 are involved in the process of photosynthesis (Tyagi 1998). Chloroplasts are highly polyploid organelles, containing many copies of the chloroplast DNA (plastome) per chloroplast. Molecules of the DNA are often aggregated and are attached to the thylakoid membranes forming 'nucleoids', resembling the organization of bacterial genomes. Most mature chloroplasts have 10–20 nucleoids with 2–24 DNA molecules in each. A cell with, say, 20 chloroplasts, each with 20 nucleoids containing 20 plastomes would have a total of 8000 copies of each single gene. The number of copies varies between species, the copy number varying with both the developmental stage of the chloroplast and the tissue in which it occurs. The high number of DNA molecules present in chloroplasts probably reflects the high demand for the expression of proteins active in photosynthesis; substantially fewer DNA molecules are found in amyloplasts or other non-photosynthetic plastids. The pea chloroplast from mature leaves normally contains about 14 copies per genome; there can be more than 200 copies of the genome per chloroplast in very young leaves (Dey *et al.* 1997). There are no histones associated with chloroplast DNA nor is there any 5-methyl cytosine, which is characteristic of nuclear DNA. Our understanding of the regulation of plastid-DNA replication is still very limited but it is now established that the replication of DNA in organelles is mostly independent of DNA replication in the nucleus. The process of plastid-DNA replication is probably similar to that of the small plasmid of budding yeast which contains specific replication origins (Sugiura and Takeda 2000). DNA sequences available from several different plants indicate that all the enzymes and proteins essential for plastid-DNA replication are encoded in the nuclear genome. The complete sequence of several chloroplast DNAs have been determined including *Marchantia polymorpha* (liverwort, 121 kb), *Oryza sativa* (rice, 135 kb) and *Nicotiana tabacum* (tobacco, 156 kb). The genome typically consists of four segments (Fig. 5.2), a large region of single-copy genes (LSC), a small region of single-copy genes (SSC) and two copies of an inverted repeat (IR_A and IR_B) that separate the single-copy regions. The large and small single-copy sequences contain the bulk of the rRNA, tRNA and the genes for the 'housekeeping' proteins of the chloroplast listed in Table 5.1.

Plastid DNA is generally homogeneous within a given plant species. However, just as all the somatic cells of an organism possess the same set of nuclear genes but may express those genes in different combinations, all the plastids in a single plant contain the same, identical DNA but may differ in gene expression. The functions of most plastid-encoded proteins have been identified but a few open-reading frames (ORFs) have, as yet, unknown functions. Variation of the length of the IR sequence, which ranges from 0.5 to 76 kb is responsible for the observed variation in size of plastid genomes. The presence or absence of IR sequences in plastid DNA has been used to characterize plastid genomes. While most plastid genomes contain IR regions, those of certain legumes and conifers do not, probably having lost the IR sequence during evolution.

Eukaryotes have complex interactions between the organellar genome products that they encode. Nuclear genes that encode proteins found exclusively in chloroplasts, for example, are transcribed in the nucleus and the mRNA species produced are transported to the cytosol and translated there; the proteins produced are transported into the chloroplast. Transport of chloroplast-coded proteins from the plastid into the cytosol or nucleus has not yet been demonstrated. In addition, the plastid genomes of non-photosynthetic plants (e.g. *Epifagus*) are small (50–73 kb) and have not many of the genes required for photosynthesis. On the other hand, algal plastid genomes are generally larger than the plastid DNAs of higher plants and contain many additional genes not found in plant plastids. For example, the red alga *Porphyra purpurea* contains 70 novel plastid genes that encode proteins; in higher plants these genes are found in the nuclear genome (Sugiura and Takeda 2000).

Major proteins synthesized in the chloroplast include the soluble large subunit of ribulose bisphosphate carboxylase (RubisCO) and a 32 kDa membrane protein associated with photosystem II. Most of the chloroplast genome encodes proteins involved in plastid transcription and translation or in photosynthesis. Thus the bulk of the chloroplast tends to be expressed in photosynthetically active cells.

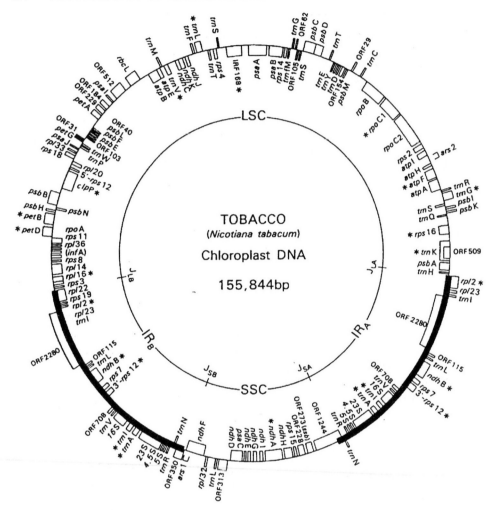

Figure 5.2   Gene map of chloroplast DNA of tobacco (*Nicotiana tabacum*). (From Sugiura 1992)

Interestingly, all plastid genomes contain protein-coding sequences homologous to the mitochondrial-encoded subunits of respiratory-chain NADH dehydrogenase. This gene is expressed highly in plastids and may point to the existence of a respiratory chain in chloroplasts (Dey *et al.* 1997).

Regulation of Gene Expression

Cell differentiation is a function of regulated gene expression. Some genes are strictly regulated by a developmental programme that directs the transition from juvenile seedlings to reproductive maturity. Such genes are active only in certain

Table 5.1   Plastid genes and their gene products

Ribosomal RNAs

rrn	16S, 23S and 5S rRNA operons

Ribosomal proteins

rps2	small subunit protein 2
rps12	small subunit protein 12

Transfer RNAs

trnK	tRNA^{lys}

RNA polymerase

rpoA	α subunit
rpoB	121 kd β subunit
rpoC1	78 kd β' subunit
rpoC2	154 kd β'' subunit

PSI

psaA	$\sim$65–70 kd chl a apoprotein (CP1)
psaB	$\sim$65–70 kd chl a apoprotein (CP1)

PSII

psbA	D1 reaction centre protein
psbB	50 kd chl a-binding protein
psbD	D2 reaction centre protein
psbH	10 kd phosphoprotein

Cytochrome b_6/f complex

petB	cytochrome b_6
petD	subunit IV

ATP synthase

atpA	CF$_1$ α subunit
atpBE	CF$_1$ β and ε subunits
atpF	CF$_0$ subunit I
atpH	CF$_0$ subunit III
atpI	27 kd hydrophobic protein (F$_0$a)

Ribulose-1,5-bisphosphate carboxylase

rbcL	large subunit

---

tissues or organs. Other genes are environmentally regulated, becoming active in response to stimuli such as light, wounding, stress and pathogens.

All genes contain a coding region composed of nucleotides that contain the codons for specific amino acids that make up the gene product and two non-coding regions at which transcription is initiated and terminated (Fig. 5.3).

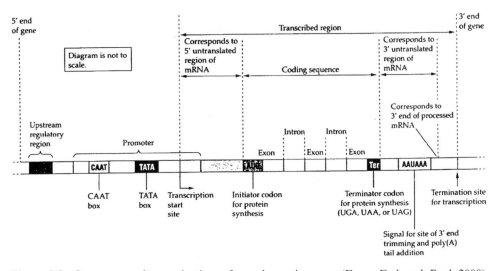

Figure 5.3   Structure and organization of a eukaryotic gene. (From Ferl and Paul 2000). Reproduced with permission

The sequence to be transcribed is 'read' from the template strand by the enzyme RNA polymerase, which polymerizes mRNA in a $5' \rightarrow 3'$ direction (downstream direction) (Fig. 5.4). Note that transcription proceeds in a $5' \rightarrow 3'$ direction but that the template strand is copied from the $3'$ end to the $5'$ end. Nuclear genes are transcribed by any one of three RNA polymerases. RNA polymerase I and III are used for the synthesis of rRNA and tRNA, respectively; RNA polymerase II is responsible for the transcription of structural genes that encode proteins. The

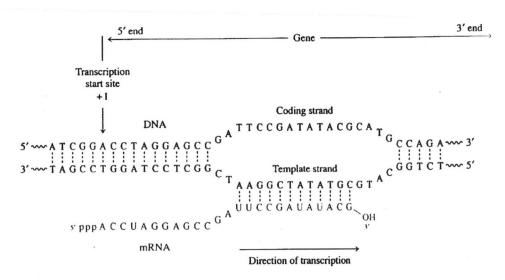

Figure 5.4   The sequence of a hypothetical gene and the RNA read from it

mitochondrial RNA polymerase is a monomeric enzyme encoded by the nuclear genome; its amino acid sequence is similar to that of RNA polymerase of certain bacteriophages, suggesting that these enzymes share a common ancestor. Chloroplasts, on the other hand, encode their own RNA polymerase. The genes encoding chloroplast RNA polymerase are similar in sequence to those for the RNA polymerase of cyanobacteria, which is consistent with the endosymbiotic hypothesis of chloroplast origin.

Three different kinds of RNA are transcribed from DNA: messenger RNA (mRNA), ribosomal RNA (rRNA) and transfer RNA (tRNA). Messenger RNA carries the protein code, rRNA is an integral part of ribosome structure and tRNA is an adapter molecule which aligns a specific amino acid to a particular codon on the mRNA. A gene coding for mRNA typically consists of three regions – promoter, protein coding and terminator (Fig. 5.3). Transcriptional control of gene expression is dependent on the role of initiation of mRNA synthesis, which, in turn, depends on access to the 5′ end of the gene. Much of the regulatory portion of plant genes is located upstream (5′ end) from the transcription start site and is referred to as the *promoter*. Within the promoter, DNA-binding proteins (so-called *trans*-acting, transcription factors) interact with DNA sequences (so-called *cis*-acting, regulatory elements) on the same strand as the coding region, to allow the RNA polymerase to operate. The most basic *cis*-element is the TATA box, which is found in most eukaryote genes and is located around position −30 (that is, 30 nucleotides upstream of the start codon). The TATA sequence is often juxtaposed to another *cis*-element, the CAAT box, which is located further upstream. TATA is responsible for positioning RNA polymerase II correctly, in order to initiate transcription. Regulatory elements can also be found far upstream from the TATA box. These elements typically act as *enhancers* and contribute to the efficiency of RNA polymerase in initiating transcription. Transcriptional control is based on the ability to regulate the rate of synthesis of mRNA for a specific gene product. This is achieved by altering the affinity of specific areas in the promoter region of the gene to enable binding the RNA polymerase. In prokaryotic cells, RNA polymerase is designated $\alpha_2 \beta\beta'\sigma$, that is, the holoenzyme is made up of two α-polypeptides plus one each of β, the related β' and σ; the core enzyme lacks σ and is designated $\alpha_2 \beta\beta'$. The key point about initiation of transcription is that RNA polymerase must transcribe whole genes rather than random pieces of DNA. This means that the RNA polymerase must bind to a DNA molecule at a specific position – within the promoter. The σ subunit of the RNA polymerase recognizes the promoter and so determines the binding of the enzyme to the DNA and ensures the correct initiation of transcription. When the promoter–RNA-polymerase complex has formed, the σ subunit dissociates and the holoenzyme is converted to the core enzyme. In chloroplasts, similar elements have been identified; however, since they are not identical to the prokaryote σ factor, they are called 'sigma-like factors'.

Another level of control is post-transcriptional processing of mRNA. In prokaryotes, the mRNA molecules that are translated are direct copies of the genes. In eukaryotes, most mRNAs undergo a series of modification and processing

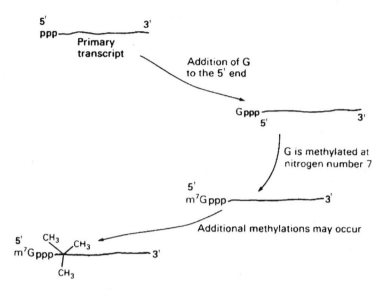

Figure 5.5   Capping of mRNA. (From Brown 1992)

steps before translation occurs. These include chemical modifications of the two ends of the mRNA molecule and removal of introns (splicing) and, in special cases, alteration of the nucleotide sequence of the mRNA (editing). These processes, which take place in the nucleus, are completed before the mRNA is transported into the cytoplasm. A primary transcript will have at the 5′ end the chemical structure pppNpN..., where N is the sugar-base component of the nucleotide and p represents the phosphate group (Fig. 5.5).

Mature eukaryotic mRNA has, at the 5′ end, the chemical structure m^7GpppNpN, where m^7G is the nucleotide carrying the modified base 7-methylguanine. This m^7G nucleotide is added to the mRNA molecule after transcription, by a two-step process, with methylation occurring only after a standard G has already been added; the process is called *capping*.

A second modification of most eukaryotic mRNAs is the addition to the 3′ end of the molecule, of a long stretch of up to 250 A-nucleotides, to produce a poly (A) tail (Fig. 5.6). This poly (A) tail is not encoded in the DNA, but is added post-transcriptionally before the mRNA is exported from the nucleus to the cytoplasm. Polyadenylation does not occur precisely at the 3′ end of the primary transcript. Instead, the final few nucleotides are removed by cleavage somewhere between 10 and 30 nucleotides downstream of the specific polyadenylation signal to produce a 3′ end to which the poly[A] tail is subsequently added by the enzyme poly(A) polymerase The poly (A) tail facilitates export of the mRNA from the nucleus, stabilizes the mRNA against exonuclease degradation and may have a role in the initiation of translation.

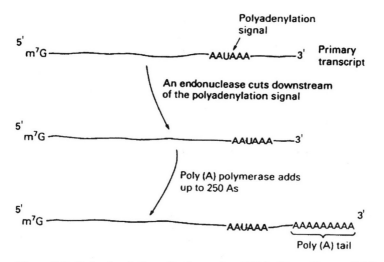

Figure 5.6   Polyadenylation of eukaryotic mRNA. (From Brown 1992)

Split Genes and RNA Splicing

The primary transcript of a eukaryotic mRNA typically contains coding regions, called *exons*, that are ultimately expressed, interspersed with non-coding regions called *introns*. Since the introns in DNA are transcribed, they are considered part of the gene. In a process called RNA splicing, the introns are removed from the primary transcript and the exons are joined to form a contiguous sequence specifying a functional polypeptide. Because of the loss of introns, mature mRNA is often only a fraction of the size of the original, primary transcript. For example, the gene for triose phosphate isomerase from maize contains nine exons and eight introns and spans over 3400 base pairs of DNA. The mature mRNA is only 1050 nucleotides long.

Plant nuclear pre-mRNA introns tend to be rich in adenosine and uridine ribonucleotides (Fig. 5.7). The nucleotide sequences at each exon–intron junction are highly conserved; the rest of the intron sequence is not conserved.

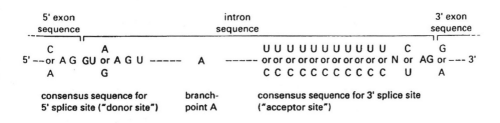
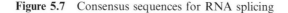

Figure 5.7   Consensus sequences for RNA splicing

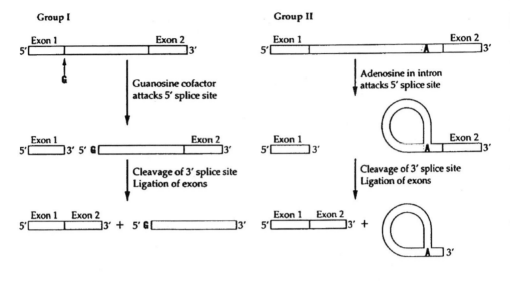

Figure 5.8   RNA splicing of group I and group II introns. (Sugiura and Takeda 2000)

In almost all of the nuclear introns in plants the boundary sequence at the 5′ splice site (donor site) and the 3′ splice site (acceptor site) consists of 5′ GU and AG 3′ (Fig. 5.7). This characteristic is known as the *GU–AG rule*. Another short consensus sequence, the branch site, is located 20 to 40 nucleotides upstream from the 3′ splice site. There are four distinct groups of introns. Group I and group II introns share the same key characteristics but differ in the details of the splicing mechanisms (Fig. 5.8).

In general, both splicing mechanisms involve two transesterification reactions. The group I intron sequences bind a free guanine nucleotide to a specific site. In the first esterification step the 3′-OH of the guanine attacks the phosphodiester bond at the 5′ splice site. In the second transesterification reaction, the 3′-OH of the 5′ exon attacks the phosphodiester bond at the 3′ splice site and displaces the 3′-OH group of the intron. The reaction entails cleavage at the 3′ splice site to generate a free intron-RNA and ligation of the two exons via a 3′-5′ phosphodiester bond. The group II intron sequences use an especially reactive adenosine nucleotide within the intron sequence itself. The 2′-OH of the specific adenosine attacks a phosphate that is 5′-linked to a guanine residue to form a lariat configuration. The lariat is formed via a 2′-5′ phosphodiester bond. Both group I and group II splicing mechanisms are normally aided by proteins, but the splicing is catalysed by the RNA itself, that is, the base-paired intron is a *ribozyme*, an RNA molecule that catalyses its own splicing reaction. In other words, group I and group II introns are self-splicing or autocatalytic. *In vivo*, however, at least in the case of group I introns, proteins probably facilitate splicing by promoting the correct folding of these large catalytic RNAs (Shaw and Lewin 1995; Weeks and Cech 1995). This phenomenon was first found in the genes for the rRNAs of certain protozoa, notably *Tetrahymena*. The

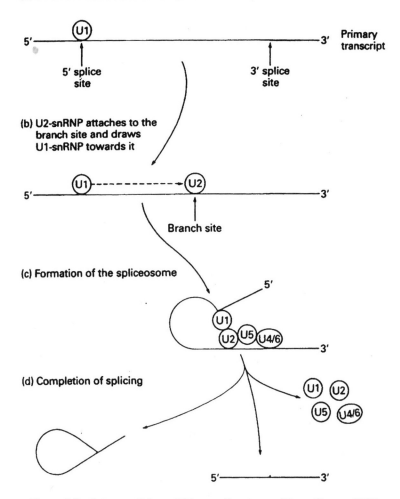

(a) U1-snRNP attaches to the 5' splice site

(b) U2-snRNP attaches to the branch site and draws U1-snRNP towards it

(c) Formation of the spliceosome

(d) Completion of splicing

Figure 5.9   Intron splicing within a spliceosome. (From Brown 1992)

introns lacked the GU–AG property; the introns fold up by intramolecular base-pairing into a complex tertiary structure in which the two splice sites are brought together and the intron itself catalyses its own splicing reaction.

A third group of introns, found in nuclear mRNA primary transcripts, are spliced by the same lariat-forming mechanism as the group II introns. However, they are not self-splicing. Splicing of this group is catalysed by a large, multi-unit, RNA–protein complex called a *spliceosome* (Fig. 5.9). It is composed of 45 proteins and 5 molecules of eukaryotic RNA; it contains about 5000 nucleotides, and is almost as large as a ribosome. The RNAs are U-RNAs (*uracil-rich*), called small-nuclear RNAs (snRNAs); five of them, designated U_1, U_2, U_4, U_5 and U_6 are involved in splicing reactions. These snRNAs occur in the nucleus where they are

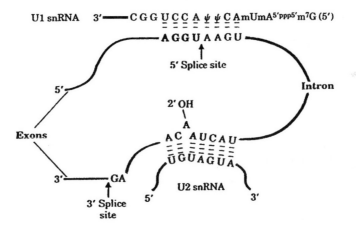

Figure 5.10 U_1-snRNA sequence complementary to 5′ splice site sequence. (From Lehninger *et al.* 1993)

complexed with proteins to form small nuclear ribonucleoproteins (snRNPs, referred to as 'snurps'). Each snRNP contains one or two snRNAs plus a number of proteins. Some of these proteins are common to all snRNAs; others are found in only one class of snRNP. The U_1-snRNA has a nucleotide sequence complementary to a sequence near the 5′ splice site of nuclear mRNA introns (Fig. 5.10). The U_1-snRNP binds to this region in the primary transcript and the U_2-snRNP attaches to the branch site. It is thought that the U_1- and U_2-snRNPs have an affinity for each other and that this draws the 5′ splice site towards the branch point. The other snRNPs (U_4-, U_5- and U_6-snRNP) then attach to the intron to form the spliceosome, within which the splicing occurs (Fig. 5.9).

A fourth group of introns, which are found in certain tRNAs, is distinguished from group I and group II introns in that its splicing mechanism requires ATP. In this reaction, a splicing endonuclease cleaves the phosphodiester bond at both ends of the intron and the two exons are joined together. A mechanism of post-transcriptional control occurs via mRNA degradation. Mature mRNA molecules are enzymatically degraded. The rate of degradation is specific for the transcript and can vary, in algal cells, from a few minutes to several hours (Mullet 1993; Sakamoto *et al.* 1993).

So-called RNA editing (Benne 1990) also provides post-transcriptional control. In this process the nucleotide sequene of an mRNA is altered by inserting new nucleotides or by deleting or changing existing ones. RNA editing was discovered in 1986 during studies of mitochondrial genes in the parasitic protozoan *Trypanosoma brucei* in which the primary transcripts are edited by extensive insertion of uracil nucleotides (Fig. 5.11). RNA editing results in the insertion of a different amino acid from that predicted from the DNA sequence. Post-transcriptional control of gene expression is a common feature of many chloroplast-encoded genes.

Control of gene expression also occurs at the level of translation. Translation is dependent on many factors – the availability of mRNA and ribosomes, the level of

(a) RNA editing in trypanosomes

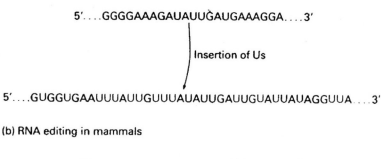

5′....GGGGAAAGAUAUUGAUGAAAGGA....3′

Insertion of Us

5′....GUGGUGAAUUUAUUGUUUAUAUUGAUUGUAUUAUAGGUUA....3′

(b) RNA editing in mammals

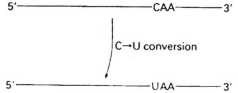

5′————————————CAA————3′

C→U conversion

5′————————————UAA————3′

Figure 5.11 RNA editing in (a) mitochondrial mRNA from *Trypanosoma brucei* and (b) in the mRNA of apolipoprotein-B. (From Brown 1992)

amino acid concentration, the availability of primed tRNA – all of which can limit the rate of translation of mRNA. Depression of cell growth due to low levels of nutrients, especially carbon and nitrogen (Herzig and Falkowski 1989) has been shown to be responsible for a reduction in translation rates.

Synthesized proteins are often modified before taking part in cell functions. Important post-translational modifications are achieved by protein phosphorylation and dephosphorylation – processes that are mediated by protein kinases and phosphatases, respectively.

Chloroplast Gene Expression

Plastids of higher plants contain their own DNA as well as the transcriptional and translational machinery necessary for its expression. Most of the genes located in the plastid genome encode products that are related either directly or indirectly to the chloroplast photosynthetic function. These include genes coding either for some of the proteins involved in photosynthesis (photosystems I and II, cytochromes, RubisCO and ATP synthase) or for products involved in their expression (ribosomal proteins, rRNAs, tRNAs, translation-initiation and elongation factors).

Chloroplasts are known to contain all of the chloroplast rRNA genes (3 to 5 genes), about 30 tRNA genes and about 100 genes encoding proteins; this fits the number of genes estimated for the size of the chloroplast genome. However, the plastid genes code for only a fraction of the proteins required for photosynthesis and protein synthesis; the remainder are encoded in the nucleus. The nuclear-encoded chloroplast proteins must be transported post-transcriptionally into the chloroplast where they can be assembled with the plastid-encoded proteins to form functional complexes. This process requires the coordinate expression of nuclear and plastid genes and provides a means by which the nucleus can regulate plastid function. Some of the major chloroplast genes that have been sequenced and their products are presented in Table 5.1; the gene nomenclature follows that proposed by Hallick (1989).

The progress of chloroplast DNA analysis has been rapid and exceeds that of chloroplast protein analysis. At present there are still a number of open-reading frames (ORFs) in the chloroplast genome which are potential polypeptide genes; homology searches of protein databases have predicted that some of the ORFs are protein genes. However, the final identification of chloroplast genes encoding polypeptides requires the identification of the translation products.

Genetic System Genes

Ribosomal RNA genes

Chloroplasts contain a 70S class of ribosome which is distinct from the 80S ribosome found in the cytoplasm. A 23S, a 5S and a 4.5S rRNA are all associated with the 50S subunit. The 4.5S rRNA has been found in higher-plant chloroplasts (Bowman and Dyer 1979) and it is homologous to the 3' end of the 23S rRNA of prokaryotes. Maize chloroplast rRNA genes were the first chloroplast genes cloned (Bedbrook *et al.* 1977). Sequencing of maize and tobacco chloroplast genes revealed a gene order of 16S–23S–4.5S–5S (which is similar to the order in *E. coli*) and an interspersion of tRNA genes within this cluster (Schwarz and Kössel 1980; Takaiwa and Sugiura 1980).

Transfer RNA genes

Hybridization of a total chloroplast tRNA fraction to chloroplast DNA has revealed the presence of 20–40 tRNA genes on the chloroplast genome (Haff and Bogorad 1976). Chloroplast genomes are thus believed to encode all the tRNA species used in chloroplast protein synthesis, although plant mitochondria take up some of their tRNAs from the cytoplasm. All the tobacco chloroplast DNA fragments that hybridized to total chloroplast tRNAs have been sequenced, revealing 30 tRNA genes (Wakasugi *et al.* 1986). The presence of introns in chloroplast tRNA genes was first demonstrated in maize *trn*I and *trn*A located in the long spacer DNA separating the 16S and 23S rDNAs (Koch *et al.* 1981). At least six chloroplast tRNA genes from land plants are now known to contain long, single introns (0.5–2.5 kb). In land plants, the tRNA genes are scattered over the chloroplast genome, while in *Euglena* most of the tRNA genes are clustered (Hallick *et al.* 1984). At least five *pseudo* tRNA genes have been found in rice chloroplasts; these have been located near the inversion endpoints and the involvement of tRNA genes in genome inversions during evolution has been proposed (Shimada and Sugiura 1989). All 61 possible codons are used in chloroplast genes which encode polypeptides. The minimum number of tRNA species required for translation of all 61 codons is 32, if normal wobble base-pairing occurs.

Ribosomal protein genes

Chloroplast ribosomes contain about 60 different protein components, one-third of which are thought to be encoded by chloroplast DNA (Eneas-Filho *et al.* 1981). Genes encoding chloroplast ribosomal proteins have been deduced through their homology with *E. coli* ribosomal protein genes (Sugita and Sugiura 1983). Twenty-one different ORFs potentially coding for polypeptides homologous to *E. coli* ribosomal proteins have been found in tobacco, liverwort and rice chloroplast genomes (Sugiura *et al.* 1991). Several nuclear-encoded chloroplast ribosomal proteins and their genes have been analysed; among them two have no similarity to any bacterial ribosomal proteins, indicating the uniqueness of chloroplast ribosomes (Gantt 1988; Johnson *et al.* 1990).

Photosynthetic System Genes

RubisCO subunit genes

Ribulose-1,5-bisphosphate carboxylase/oxygenase (RubisCO) is the major stromal protein in chloroplasts; it is composed of eight identical, large subunits (LS) of 55 kDa and eight identical, small subunits (SS) of 12 kDa. The LS is encoded by the chloroplast DNA (*rbc*L) and the SS is encoded by the nuclear DNA (*rbc*S) in higher plants and in algae. The maize chloroplast gene for the LS (*rbc*L) was the first chloroplast gene cloned and sequenced (McIntosh *et al.* 1980). The *rbc*L gene has become the most widely sequenced gene, enabling comparison for the determination of phylogenetic relationships among plant species (Zurawski and Clegg, 1987). The *rbc*L genes of higher plants and *Chlamydomonas* contain no introns while nine introns have been found in the *Euglena rbc*L gene (Koller *et al.* 1984). In the chloroplast genomes that contain it, *rbc*S is located downstream from and constitutes an operon with *rbc*L. No intron has been reported in chloroplast *rbc*S genes while the nuclear *rbc*S genes have one to three introns.

Photosystem II genes

Thylakoid membranes have four distinct complexes: photosystems I and II, cytochrome b_6/f and ATP synthase (Chapter 2). The genes encoding thylakoid proteins have usually been isolated and identified through protein analysis, in which a protein component is purified and its antibody synthesized. At least twelve components of PSII are encoded in the chloroplast genome. The gene for the 32 kDa protein Q_B or D1 protein (*psb*A) was the first photosystem gene sequenced in spinach and tobacco (Zurawski *et al.* 1982a). In land plants all of the PSII genes are continuous while some of the algal *psb* genes are split by one to six introns (Karabin *et al.* 1984). In higher plants the *psb*D gene overlaps *psb*C by about 50 base pairs (bp) (Alt *et al.* 1984), suggesting that chloroplasts must have a specific mechanism for producing the proper amount of each component of a given complex.

Photosystem I genes

Five components of PSI are encoded in the chloroplast genome. The genes for A1 and A2 of the P700 chlorophyll *a* apoprotein (*psa*A and *psa*B, respectively) were first sequenced from maize (Fish *et al.* 1985). The *psa*A and *psa*B genes in higher plants contain no introns, are situated tandemly and are about 45% homologous at the amino acid level. In *Chlamydomonas* the *psa*A gene is divided into three exons scattered around the chloroplast genome, while *psa*B is uninterrupted (Kück *et al.* 1987); the three distantly separated exons of *psa*A produce a functional mRNA by trans-splicing.

Cytochrome b_6/f-complex Genes

The cytochrome b_6/f complex consists of six components, four of which are encoded by the chloroplast genome (Heinemeyer *et al.* 1984; Wiley *et al.* 1984). The *pet*B and *pet*D genes are clustered with *psb*B and *psb*H in higher plants and they constitute a transcription unit (Herrmann *et al.* 1985). A transcription unit may be defined as a primarily transcribed region that can be determined by a combination of RNA hybridization (northern blotting), primer extension, ribonuclease protection and *in vitro* capping analysis. In higher plants both *pet*B and *pet*D contain single introns with short (6–8 bp) exons.

ATP synthase genes

ATP synthase consists of two parts, CF_1 and CF_0 (Chapter 2). CF_1 is composed of five different subunits and CF_0 is composed of four different subunits. The genes for six subunits are present in the chloroplast genome. Genes for the β and ε subunits (*atp*B and *atp*E) were first sequenced from maize and spinach (Krebbers *et al.* 1982; Zurawski *et al.* 1982b). The genes for the three CF_0 subunits (*atp*I, *atp*H, *atp*F) are clustered just before *atp*A (Bird *et al.* 1985). The deduced amino acid sequence of the six subunits show homology with their counterparts in *E. coli*.

ndh *genes*

Eleven chloroplast DNA sequences (*ndh*) whose predicted amino acid sequences resemble those of the genes of human mitochondrial respiratory-chain NADH dehydrogenase have been found in the chloroplasts of a variety of plants (Ohyama *et al.* 1986; Matsubayashi *et al.* 1987). Since most of these sequences are actively transcribed and the *ndh*A and *ndh*B are spliced rapidly, it is likely that these genes encode the components of a chloroplast NADH dehydrogenase (Matsubayashi *et al.* 1987). These observations suggest the existence of a respiratory chain in chloroplasts.

Transcription

The chloroplast genome contains over 120 genes and about 50 transcription units, suggesting that chloroplast genes are cotranscribed. The detection of primary transcripts has actually shown that most of the chloroplast genes are transcribed polycistronically. Polycistronic transcripts obtained from cotranscription consist of sequences for proteins and RNAs with related functions, such as photosynthesis, transcription or translation. Chloroplast polycistronic primary transcripts are generally processed into many overlapping, shorter RNA species, each of which accumulates at steady-state levels. Some of the shorter RNAs are monocistronic but others are not. For example, the *psb*B operon (PSII 47 kDa protein) is transcribed from a bacterial-type promoter as a tetracistronic precursor consisting of *psb*B, *psb*H, *pet*B and *pet*D. In the processing of the primary transcript to produce the functional, shorter RNA, many intermediate transcripts are produced: 17 different RNA species in spinach (Westhoff and Herrmann 1988) and at least 24 RNA species in maize (Barkan 1988) have been observed. Processing of the spinach primary transcript results ultimately in the formation of monocistronic mRNAs for *psb*B and *psb*H and a dicistronic mRNA for *pet*B and *pet*D (Westhoff and Herrmann 1988).

Most chloroplast transcription units contain short, inverted repeats at their 3' ends, which were originally thought to function as transcription terminators. It has been found, however, that these inverted repeats are ineffective as transcription terminators *in vitro* but serve as accurate and efficient RNA-processing signals (Stern and Gruissem 1987). The stability of RNAs containing inverted repeats at their 3' ends is greatly enhanced. Investigations in spinach (Stern *et al.* 1989), barley (Gamble and Mullet 1989; Sexton *et al.* 1990) and mustard (Nickelsen and Link 1989) of the stability of chloroplast mRNAs and protein interaction with their 3'-end inverted repeats suggest that nuclear-encoded proteins function in chloroplast mRNA maturation and differential mRNA stability.

Transcription rates and steady-state RNA levels are generally not coincident; many chloroplast genes are known to be constitutively transcribed, suggesting that post-transcriptional RNA processing of primary transcripts represents an important step in the control of chloroplast gene expression (Deng and Gruissem 1987; Mullet and Klein 1987). Chloroplast RNA processing includes endonucleolytic cleavage, 3'-end trimming, *cis*/*trans*-splicing and RNA editing (Gruissem and Tonkyn 1993; Sugiura 1991; Rochaix 1996). All of these steps are believed to be accomplished by specific proteins and RNA factors in higher-plant chloroplasts.

Some regions of chloroplast DNA are strikingly similar to those of the DNA of present-day bacteria, a fact that reflects the endosymbiotic origin of chloroplasts.

One segment of chloroplast DNA encodes eight proteins that are homologous to eight *E. coli* ribosomal proteins; the order of these is the same in the two DNAs. Liverwort chloroplast DNA has some genes that are not detected in tobacco chloroplast DNA and *vice versa*. Since the chloroplasts from these two different organisms contain virtually the same set of proteins, it appears that some genes are present in chloroplast DNA in one species and in the nuclear DNA of the other. This indicates that some exchange of genes between chloroplast and nucleus has occurred during evolution. The topology of the plastome has also been shown to influence transcription: the rate of *in vitro* transcription was shown to be higher with the super-coiled form than with the relaxed form (Stirdivant *et al.* 1985). In addition, existence of gyrase and topoisomerases in plastids has been demonstrated (Lam and Chua 1987), supporting the concept that plastome topology and accessibility to transcription machinery are important in the regulation of plastid gene expression. However, the chloroplast genome is not large enough to contain all the regulatory genes. Chloroplast gene expression is coordinated with the expression of nuclear genes, which provides most of the proteins required for chloroplast function. The structures of plastid and mitochondrial mRNAs differ from those of cytoplasmic mRNA. Organellar mRNAs generally lack the 5′ cap and the 3′ poly (A) tail. As a result of differences in post-transcriptional processing, mitochondrial mRNAs commonly retain the 5′ triphosphate group, while the 5′ ends of most plastid mRNAs carry a monophosphate group. Similar to prokaryotic mRNAs, the untranslated sequences that occur in both the 5′ and the 3′ ends of most organellar mRNAs can form stem loops that have regulatory and stabilizing functions (Fig. 5.12).

These mRNA structures often interact with proteins and serve as signals that affect the processing, translation and nucleolytic degradation of the mRNA (Gruissem and Tonkyn 1993). A small fraction of plastid mRNAs contains a short polyA sequence at the 3′ end. These poly[A] sequences are added on to the plastid mRNAs after endonuclease-catalysed cleavage at the 3′ end, at internal sites within the transcript. Unlike the poly[A] tail of cytoplasmic mRNAs, the poly[A] sequences of plastid mRNAs appear to promote efficient degradation of the modified mRNA. While an RNA-binding protein responsible for processing the 3′

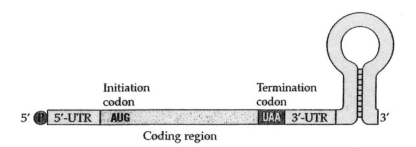

Figure 5.12   Structure of a typical, mature chloroplast mRNA. (Sugiura and Takeda 2000)

end of the chloroplast mRNA has been isolated (Schuster and Gruissem 1991) and related proteins containing RNA-binding domains have also been reported (Li and Sugiura 1990; Ye *et al.* 1991), all the molecular components involved in these processes have not been discretely identified. Some direct correlation has been found between the 5′ UTRs (untranslated regions) and RNA stability; the 3′ IR may only be a signal for the formation of the 3′ end rather than being involved in the control of RNA stability.

Chloroplast gene expression is highly regulated and regulation can occur at the transcriptional, post-transcriptional, translational and post-translational levels. Expression of plastid genes must be differentially regulated under different conditions. During the development of chloroplasts from protoplastids the large subunits of RubisCO must be synthesized in the chloroplast and assembled with the nuclear-encoded small subunits and the photosynthetic complexes must be assembled and inserted into the thylakoid membranes (Fig. 5.15). During the differentiation of chloroplasts to chromoplasts, on the other hand, the photosynthetic complexes must be disassembled as carotenoid pigments are synthesized. Chloroplasts themselves can be altered in response to varying environmental stimuli, notably light. Some photosynthetic proteins do not accumulate until plants are illuminated and expression of others can be altered by different light intensities in order to optimize the rate of photosynthesis. In addition, the activity of the photosynthetic apparatus can exert feedback regulation on the expression of photosynthetic genes. In particular, redox signals have been shown to be important in the expression of both plastid and nuclear genes; light-regulated translation of chloroplast mRNAs has been shown to be dependent on redox potential (Danon and Mayfield 1994b). Light-intensity regulation of *cab* gene expression is signalled by the redox state of the plastoquinone pool (Christopher *et al.* 1997; Grover *et al.* 1999).

The transcription–translation apparatus of chloroplasts has a number of prokaryote-like features (Gruissem and Tonkyn 1993; Igloi and Kössel 1992; Sugiura 1992). For example, chloroplast ribosomes are 70S in size, chloroplast rRNAs and tRNAs are very similar in sequence to their *Escherichia coli* counterparts and chloroplast mRNAs have triphosphates at their 5′ ends and are believed to lack poly (A) tails. Chloroplasts have a complete system for the transcription and translation of chloroplast genes, yet, most of the components of the chloroplast genetic system are not coded for by chloroplast DNA but instead are nuclear-encoded. While half of the total leaf complement of ribosomes is in the chloroplast and half of the total leaf protein is synthesized there, formation of the chloroplast genetic system requires the coordinated expression of nuclear and chloroplast genes.

RNA Polymerase

The structure of chloroplast genes and their modes of expression as currently understood exhibit both prokaryotic and eukaryotic features. This implies the presence of multiple RNA polymerase species; multiple RNA polymerase activities have been found in chloroplasts (Gruissem and Tonkyn 1993). For example, two distinct RNA polymerase activities (peak A and peak B), both of which correspond to large multi-subunit complexes, have been found in plastids of mustard seedlings (Pfannschmidt and Link 1994). This plastid-encoded RNA polymerase (PEP) encodes the eubacteria-like core α,β,β' subunits. The peak A enzyme corresponds to a soluble complex composed of about 14 protein subunits (Gruissem and Tonkyn 1993; Mullet 1993) and is thought to transcribe primarily the mRNA and tRNA genes (Greenberg et al. 1984). The peak B enzyme, referred to as 'transcriptionally active chromosome' (TAC) is isolated as a complex that is tightly bound to and actively transcribing its DNA template; this enzyme preferentially transcribes plastid rRNA. PEP may exist as a eubacteria-like RNA polymerase (enzyme B) in etioplasts or as part of a larger complex (enzyme A) in chloroplasts (Pfannschmidt and Link 1997). In other words, the A and B forms of plastid RNA polymerase may transcribe the same genes at different developmental stages of the chloroplast.

A clue to the existence of a second plastid RNA polymerase (NEP – nuclear-encoded plastid RNA polymerase) was found through sequencing the plastid genome of *Epifagus virginiana*, a non-photosynthetic parasitic plant. The *E. virginiana* plant lacks all but one of the plastid-encoded PEP genes yet it still maintains transcription of at least a subset of plastid genes (Morden et al. 1991). Therefore, transcription of *E. virginiana* plastids may be via the NEP enzyme or via the eubacterial PEP encoded in the RNA polymerase genes transferred to the nucleus as part of its adaptation to parasitism. This novel NEP resembles the single-peptide RNA polymerase of bacteriophages T3 and T7.

Transcript accumulation in ribosome-deficient plastids of a barley mutant *albostrians*, provides additional evidence for one or more functional non-chloroplast-encoded RNA polymerases. Transcription of a subset of plastid genes can be detected in this mutant, although translation of plastid-encoded RNA polymerase subunits is absent (Hess et al. 1993). Similar results were obtained by the deletion of the *rpo*B (RNA polymerase subunit β') gene of plastids from the tobacco mutant ΔrpoB (Allison et al. 1996). Transcript mapping of the actively transcribing genes of ΔrpoB revealed the existence of a class of genes that have novel promoter elements besides the canonical -10 and -35 sequences of eubacteria-like plastid promoters (Hajdukiewicz et al. 1997). In both of these plants the residual polymerase activity must be caused exclusively by a nuclear-encoded protein.

It is generally believed that NEP and PEP act sequentially, forming a cascade during chloroplast development. The nuclear-encoded RNA polymerase (NEP) is imported into the proplastids of meristematic cells where it transcribes the plastid *rpo* genes (*rpo*A–C, RNA polymerase subunits genes). The newly transcribed *rpo* RNAs are translated and the resultant RNA polymerase (PEP) takes over the transcription of the housekeeping genes and initiates transcription of ribosomal proteins, tRNAs and photosynthetic proteins in developing chloroplasts. (Hajdukiewicz et al. 1997; Hess et al. 1993; Kapoor et al. 1997). Meanwhile the imported RNA polymerase (NEP) forms the TACs which transcribe rRNAs. The nuclear-encoded components of the ribosomes are imported and assembled with the plastid-encoded components to form more ribosomes in order to cope with the high levels of protein synthesis required to translate the mRNAs encoding photosynthetic proteins. An alternative to the cascade model would be a parallel one in which there is no hierarchical relationship between the NEP and PEP (Maliga 1998). Accordingly, both RNA polymerases would be available in sufficient quantities in all plastid types at all times. The activity of individual promoters would be regulated by gene-specific transcription factors and general regulatory mechanisms such as phosphorylation of the components of the plastid transcription machinery (Baginsky et al. 1997).

Chloroplast genomes from several plant species have been sequenced (Igloi and Kössel 1992), revealing rpoA, rpoB and rpoC genes which encode proteins homologous to the α, β and β' subunits of bacterial RNA polymerase (rpoC is usually split into rpoC1 and rpoC2, which encode β' and β'' subunits respectively). In other words, the principal RNA polymerase of plastids resembles that of E. coli. The E. coli RNA polymerase consists of four core subunits α_2, β and β', plus one of several regulatory σ (sigma) subunits. However, chloroplast genomes lack a gene encoding the expected σ subunit although it has been detected immunologically in chloroplasts. Presumably the σ subunit is encoded by a nuclear gene in plants as it is in the unicellular red alga Cyanidium caldarium and in Galdieria sulphuraria (Tanaka et al. 1996; Liu and Troxler 1996). Bacterial sigma-like polypeptides are not encoded in the chloroplast genome although they have been found to occur during plastid differentiation in several higher plant species (Lerbs et al. 1988; Tiller and Link 1993b; Troxler et al. 1994). Genes for these sigma-like factors of chloroplast-RNA polymerase have been isolated and have been shown to be regulated by light and chloroplast development (Maliga 1998), supporting the concept that light is instrumental in initiating additional transcriptional activity in plastids (Tozawa et al. 1998). There is evidence that such an enzyme is involved in the regulation of differential activity of photosynthesis-related genes (Link 1996). Sensitivity to light is known to vary depending on the age of the plant and both the developmental-state-dependent and the light-dependent cues contribute to the establishment of steady-state levels of plastid gene expression (Kapoor et al. 1994). The observed light-dependent changes in mRNA levels may be due either to enhanced rate of transcription or to altered stability of the transcripts. In vitro transcription studies have shown that the rates of transcription change during the development of chloroplasts from etioplasts as well as from proplastids. Also, genes of different organs, as well as of different plastids, show variable rates of transcription (Rapp et al. 1992). Light influences steady-state levels of transcripts by both its intensity and its quality; red/far-red wavelengths mediated by phytochrome and blue-light acting via cryptochrome appear to be the most important photoreceptors. Phosphorylation/dephosphorylation of sigma-like polypeptides has been reported to affect transcription initiation (Tiller and Link 1993a). In mustard chloroplasts, a σ-like factor is phosphorylated in the dark and dephosphorylated in light. This reversible phosphorylation could provide an efficient mechanism for differential transcription of chloroplast genes in response to changing light conditions (Tiller and Link 1993a). In addition to the sigma-like factors, other DNA-binding proteins have been reported. A sequence-specific DNA-binding factor required for the transcription of the barley chloroplast blue-light-responsive *psb*D-*psb*C (PSII D1 protein) promoter has been identified (Kim and Mullet 1995). In short, multiple promoters and RNA-polymerase activities appear to be involved in transcription in plastids.

The Ca^{2+}/calmodulin pathway and G-proteins have been shown to be involved in light-regulated plastid-gene (*psa*A, PSI, *psb*A, PSII and *rbc*L) expression in rice (Grover *et al.* 1999). Phosphorylation and redox status influence chloroplast transcription and these in turn may be controlled by a light-signal transduction pathway using kinases and phosphatases or by alterations in redox status brought about by light-dependent electron transfer (Link 1996). The processing of plastid gene transcripts by endonucleases (Sugita and Sugiura 1996), has been reported to be influenced by light. mRNA editing of *atp*A (subunit CF_1 of H^+-ATPase) has been shown to be light-dependent (Hirose *et al.* 1996), thereby emphasizing light-regulated control of plastid gene expression.

Promoters

Unlike nuclear genes, which typically have only one promoter that is recognized by RNA polymerase, transcription of many plastid genes and operons is initiated either by multiple promoters or at multiple sites. The upstream regions of many transcription initiation sites of the genes of the chloroplast genome contain consensus sequences similar to the -10 and -35 sequences (TATAAT and TTGCACA, respectively) of prokaryotes (Sugiura 1992; Gruissem and Tonkyn 1993). Sequences similar to the canonical prokaryotic -10 and -35 elements were recognized when chloroplast genes were first isolated from plants. They were later verified, using *in vitro* transcription systems, to be *bona fide* promoter elements (Gruissem and Zurawski 1985a,b). In addition, it has been possible to analyse plastid gene promoters in *E. coli*. For example, mutations in the maize *atp*B (ATPase subunit β) promoter have the same effects either in *E. coli* or in a chloroplast *in vitro* transcription system (Bradley and Gatenby 1985). The *psb*A (PSI, D1 protein) and several other photosynthesis-related genes in higher-plant chloroplasts contain both the prokaryote-type -10 and -35 sequences and between them a sequence motif similar to the TATA box of nuclear genes (Sugita and Sugiura 1984). Prokaryote-like sequences are necessary for proper transcription in chloroplasts and the TATA box-like region is also needed for correct *psb*A transcription *in vitro* (Eisermann *et al.* 1990). On the other hand, several genes, including some tRNA genes, lack the prokaryotic sequences (Gruissem and Tonkyn 1993) while the *rps*16 gene (30S ribosomal protein) from mustard possesses the -10 motif but lacks the -35 sequence (Neuhaus *et al.* 1989). Several *Chlamydomonas* chloroplast promoters have been identified by their ability to direct transcription of reporter genes in transformed chloroplasts and alignment of these and additional sequences have revealed *E. coli*-like promoters upstream of chloroplast genes *rrn*16 (ribosomal RNA 16S) and *rbc*L (RubisCO large subunit) from *Chlamydomonas* (Klein *et al.* 1992). The $-10/-35$ promoter type is widespread in chloroplast genomes and in most cases has sequence requirements indistinguishable from the *E. coli* promoters. Multiple transcription initiation sites are known especially for chloroplast genes *psb*D (PSII, D2 protein) (Berends *et al.* 1990) and 16S rDNA (Vera and Sugiura 1995). One of the several transcription initiation sites of *psb*D is directed by a light-responsive promoter whose activity is regulated differentially by chloroplast-specific protein factors (Wada *et al.* 1994). Also, one of the multiple promoters of 16S rDNA lacks any conserved sequence motifs and is active, irrespective of the developmental stage of the chloroplast, suggesting that it is responsible for the constitutive transcription of these genes (Vera and Sugiura 1995).

The promoter can affect transcription in two ways. The first is by its strength, which is a function of its DNA sequence and can be considered to be the efficiency with which the promoter engages the RNA polymerase to initiate transcription. Stronger promoters lead to more frequent initiation. For example, the transcription rates of ten plastid genes of spinach vary up to 25-fold (Deng and Gruissem 1987), while those of fifteen barley plastid genes vary up to 300-fold (Baumgartner *et al.* 1993). Secondly, regardless of strength, different promoters can be selectively utilized under varying conditions to differentially modulate the transcription of individual transcripts. For example, the transcription rates of some plastid genes can be differentially altered during chloroplast development, as in the case of *psb*A (PSI P700) and *rbc*L (RubisCO large subunit) in barley and maize (Klein and Mullet 1990) *trn*K (lys tRNA) and *rpl*2 (50S r-protein) in spinach (Deng and Gruissem 1987) and *rpo* (RNA polymerase subunit) and *rps* (30S r-protein) genes in barley (Baumgartner *et al.* 1993). Differential modulation of transcription rates is also observed in spinach in response to different light qualities (Deng *et al.* 1989) or photo-oxidative stress (Tonkyn *et al.* 1992).

The organization of genes into operons can play a role in the regulation of gene expression. Most plastid genes are arranged in operons of two types – those which contain genes for proteins with related function and those for proteins with different functions. For example, the *rpo*B-*rpo*C1-*rpo*C2 operon encodes subunits of RNA polymerase, and *psb*I-*psb*K-*psb*D-*psb*C operon encodes components of the PSII complex. On the other hand, some operons encode proteins with different functions. The *psa*A-*psa*B-*rps*14 operon, for example, encodes subunits of PSI and a ribosomal protein while the *psb*B-*psb*H-*pet*B-*pet*D operon encodes subunits of both PSII and the cytochrome b_6/f complex. The grouping of genes into an operon does not necessarily lead to fixed ratios of transcription rates of the individual genes. This concept is borne out by the results of analysis of the ORF31–*pet*E–ORF42 gene cluster in maize which showed that a monocistronic *pet*E transcript is produced from an alternative promoter during light-induced plastid maturation in addition to the bi- and tri-cistronic transcripts that are produced constitutively from the proximal promoter (Haley and Bogorad 1990).

Chloroplast RNA Splicing

Several chloroplast genes are interrupted by introns. In the tobacco chloroplast genome, 18 genes contain introns – 6 tRNA genes and 12 protein-encoding genes. Introns found in chloroplast genes can be classified into four groups on the basis of their intron boundary sequences and on possible secondary structures (Christopher and Hallick 1989; Shinozaki *et al.* 1986). Characterization of chloroplast introns (Rochaix 1992) revealed that most belong to the group I and group II intron classes, originally identified as self-splicing introns in the protozoan *Tetrahymena* and fungal mitichondrial genes. These introns contain sets of conserved sequences characteristic of each intron type. Chloroplast group I introns can be folded with a secondary structure typical of that of group I introns of fungal mitochondrial genes (Michel and Dujon 1983). The introns of *trn*L (Leu-tRNA), and of the 23S rDNA belong to this group. Chloroplast group II introns, which include those of *trn*I (Ile-tRNA) and *trn*A (Ala-tRNA) in the 16S–23S spacer can form a secondary structure with six helical domains radiating from a central core, which is the structure of group II introns of fungal mitochondrial genes. The majority of introns from land-plant chloroplasts fall into this category. A third class of intron, designated group III, was discovered in *Euglena* (Christopher and Hallick 1989). Chloroplast introns in this class have degenerate versions of the group II intron consensus boundary sequences but lack the highly conservative, secondary structural features characteristic of group II introns. Chloroplast group III introns include those of protein-encoding genes *trn*V-UAC (Val-tRNA), *trn*G-UCC (Gly-tRNA) and *trn*K-UUU (Lys-tRNA) from higher plants. The similarity of the postulated secondary structures of group II and group III introns has led to the two groups being sometimes combined. A fourth intron group, found in genes encoding many 30S r-proteins, has been described (Christopher and Hallick 1989).

In some genes, exons encoding different parts of a single protein may be located distant from each other on the plastid genome. In such cases, *trans*-splicing of split genes plays an important role in the maturation process of the functional mRNA encoding the entire gene product. Investigations of the mechanism of *trans*-splicing suggest the involvement of group II introns (Jarrell *et al.* 1988; Kohchi *et al.* 1988) and a small plastid RNA encoded by the *tsc*A (*trans*-splicing in the chloroplast) locus (Goldschmidt-Clermont *et al.* 1990, 1991). The tobacco gene for a ribosomal protein CS12 is divided into one copy of 5'-*rps*12 (30S r-protein) which contains exon 1 and two copies of 3'-*rps*12 which contain exons 2 and 3. The 5'- and 3'-*rps*12 segments are separated by 28 kb and are transcribed independently. The two transcripts are spliced in *trans* to produce a mature mRNA for CS12 (Zaita *et al.* 1987). Liverwort *rps*12 is divided into two parts; the 5'- and 3'-*rps*12 segments are present in single copies and are located on opposing DNA strands (Ohyama *et al.* 1986). The mature mRNA is also produced by *trans*-splicing (Kohchi *et al.* 1988).

The *psa*A gene of *Chlamydomonas reinhardtii*, which encodes one of the major reaction-centre proteins of photosystem I, consists of three exons that are widely separated on the chloroplast genome (Kück *et al.* 1987; Turmel *et al.* 1995). All three exons are flanked by consensus intron boundary sequences typical of group II introns. The fact that exons 1 and 2 are directed in opposite orientations on the chloroplast genome implies that transcription of these exons is discontinuous. The three exons are transcribed independently as precursors and the synthesis of mature *psa*A-mRNA involves *trans*-assembly of these three separate transcripts. In addition, an internal segment of intron I is encoded at another site in the chloroplast genome, the *tsc*A locus, which is transcribed to produce a small RNA (Goldschmidt-Clermont *et al.* 1991). Thus, at least three RNAs must be brought together for the splicing of *psa*A intron I. This mode of splicing shows a strong dependence on protein factors, since self-splicing has not been shown to occur for any *trans*-spliced introns. Apart from the *tsc*A of *Chlamydomonas*, the *spr*A gene of tobacco chloroplast genome has been found to encode a small-plastid RNA (Vera and Sugiura 1994). Part of this small-plastid RNA sequence exhibits the potential for base-pairing with a segment of the leader sequence of pre-16S-rRNA, suggesting a role for small-plastid RNA in chloroplast ribosome biogenesis, that is, 16S-rRNA biogenesis (Vera and Sugiura 1994). As the tobacco chloroplast genome still includes about ten spacer regions of about 1 kb which do not harbour any significant ORFs, additional small-plastid RNA species may be encoded in these spacers.

Chloroplast RNA Editing

RNA editing occurs in a number of chloroplast genes (review: Kössel *et al.* 1993). Multiple site-specific conversions from C to U (or, rarely, U to C) occur in primary transcripts from both plastid and plant mitochondrial genomes. The frequency of editing is organelle- or gene-specific; the total number of editing sites has been estimated at 1200 in wheat mitochondria but only about 30 in the chloroplasts of flowering plants (Sugiura and Takeda 2000). Most nucleotide conversions change the first or second base of the codon, thereby altering the amino acid sequence of the protein; amino acid changes often result in a more highly conserved protein. Editing of C to U in *rpl*2 mRNA (50S r-protein) of maize plastid (Hoch *et al.* 1991) and *psb*L mRNA (PSII L-protein) of tobacco (Kudla *et al.* 1992) converts the sequence of the first codon from ACG to AUG, thus restoring the conventional initiating codon. Editing of chloroplast RNA transcripts also occurs at locations internal to the protein coding sequence. For example, four C-to-U conversions in maize *ndh*A mRNA (NADH dehydrogenase) restores amino acids that are conserved in the *ndh*A-encoded peptides of chloroplasts of other species (Maier *et al.* 1992).

In addition to amino acid changes, editing can shorten ORFs predicted from DNA sequences or produce new ORFs. The editing of glutamine codons (CAA and CAG) and the arginine codon (CGA) yields termination codons (UAA, UAG and UGA, respectively), thereby introducing a premature termination of the ORF in both mitochondria and plastids. These stop codons that result from editing generally convert the predicted C-terminal extension of a protein to the conserved protein size. Likewise, a threonine codon (ACG) can be converted to the initiating codon AUG. Thus, C-to-U editing can create both an initiation and a termination codon within the same transcript, producing a new ORF. It is clear from these examples that DNA sequencing alone is not always sufficient to predict the products of organellar genes.

Editing processes in plant mitochondria and chloroplasts share certain features such as C-to-U conversions. In addition, both organelles carry common sequence motifs around the editing sites. Thus it is probable that the editing systems of mitochondria and plastids share components or even mechanisms. RNA editing in chloroplasts does not require translation, indicating that all protein components necessary for editing are encoded by the nuclear genome and imported into the chloroplast (Sugiura and Takeda 2000).

Chloroplast RNA Stability

It is well established that the levels of chloroplast transcripts vary considerably during plastid development and differentiation in higher plants. The amount of transcript available for translation depends on both the rate of transcription and the rate of degradation, which together determine the amount of a particular mRNA. The level of mRNA does not always mirror the transcription rate; control of the degradation of transcripts has an important role to play in the regulation of mRNA levels. For example, the steady-state levels of *psb*A and *atp*B transcripts increase significantly during greening of etiolated spinach cotyledons, while the transcription rates remain relatively constant (Deng and Gruissem 1987). A study of the half-lives of several chloroplast genes during plastid development in barley provides direct evidence that the stability of some transcripts is selectively increased and modulated during chloroplast development (Kim *et al.* 1993). In mature barley chloroplasts, as well, enhanced levels of *psb*A mRNA are due primarily to selective stabilization (Baumgartner *et al.* 1993). In contrast, differential accumulation of *psb*D mRNA relative to *rbc*L mRNA occurs as a result of light-stimulated transcription of *psb*D (Baumgartner *et al.* 1993). These two regulatory mechanisms probably act together to optimize the expression of plastid genes. Expression of a single gene can be controlled by a combination of transcription initiation rate and transcript stability. A specific example is the complex interplay between transcription and mRNA stability that controls the levels of transcripts from the large ATP synthase gene cluster in spinach plastids during light-induced development. During the initial 24 h of light-induced development of etioplast to chloroplast, transcription decreases concomitantly with increased stability; transcriptional activity of this cluster then increases with a concomitant decrease in the stability of the transcripts. As the chloroplast matures, the transcripts from this cluster again become markedly more stable and the rate of transcription declines (Green and Hollingsworth 1992).

Most chloroplast RNA transcripts contain inverted repeat (IR) sequences at their 3' untranslated regions that can potentially form stem-loop structures (Fig. 5.12). Similar structures are found in prokaryotes, where they play an important role in transcript termination. However, it has been demonstrated that the IR sequences at untranslated regions of chloroplast RNA transcripts act as processing signals rather than termination signals (Stern and Gruissem 1987). The 3′ IR sequences appear to be *cis*-acting elements required for processing and stabilizing the mRNA 3′ ends. When the IR sequences are removed, by either deletion mutagenesis (Stern *et al.* 1989) or by endonucleolytic cleavage (Hsu-Ching and Stern 1991), the resulting RNAs are rapidly degraded *in vitro*. Transformation of *Chlamydomonas* chloroplasts causing partial or total deletion of the *atp*B 3′ IR and resulting in a 60–80% loss of mRNA, provides more direct evidence for the stabilizing function of the 3′ IR sequence (Stern *et al.* 1991).

Translation

Protein synthesis in plastids is similar to bacterial translation, occurring on 70S ribosomes and mediated by similar initiation and elongation factors. The rRNAs and some ribosomal proteins are encoded in the plastid genome and the remainder of the ribosomal proteins are encoded in the nucleus. Sequences resembling the prokaryotic ribosome binding sites precede most plastid open-reading frames and translation begins with the incorporation of a formyl methionine residue. Chloroplast-encoded protein genes show a general sequence requirement for translation consistent with the prokaryotic nature of the plastid translation apparatus.

Control of gene expression at the translational level allows an organism to adapt rapidly to changes in environmental conditions by altering the preferential translation of mRNAs without concomitant adjustment in transcript level. This can be especially important for regulating the synthesis of highly expressed genes when continued translation of existing mRNA under unfavourable conditions could unnecessarily consume limited metabolites. Translation of chloroplast proteins can be affected by ATP levels (Michaels and Herrin 1990), by light conditions and by the metabolic state of the organelles (Scheibe 1990). Translational control of gene expression in plastids is well documented. In *Euglena* cells transferred to darkness, synthesis of RubisCO large subunits stopped immediately even as levels of the *rbc*L transcripts increased transiently (Reinbothe *et al.* 1991). Control of RubisCO synthesis was determined to be translationally mediated during the cell-cycle in *Euglena* (Brandt *et al.* 1989) and during light induction of chloroplast development in amaranth (Berry *et al.* 1990) and barley (Klein 1991). Expression of the plastid *psb*A gene can also be modulated at the translational level (Kuchka *et al.* 1988). Translation of the *psb*A transcript remains high in leaves of senescing bean despite a sharp decrease in the level of transcript. This is in contrast to other photosynthetic proteins whose translation decreases with decreasing transcript levels (Bate *et al.* 1990). In tobacco, *psb*A mRNA is present in all tissues, while its product accumulates tissue-specifically and in response to light. Investigations with a chimeric *uid*A reporter gene, encoding β-glucuronidase (GUS) under the control of the *psb*A 5′- and 3′-regulatory regions integrated into the tobacco genome, showed that this pattern of expression is mediated at the translational level by the *psb*A untranslated region (Staub and Maliga 1993).

Several proteins binding to the 5'-end untranslated regions have been shown to be important in translation in *Chlamydomonas* (Mayfield *et al.* 1995). In higher plants also, the leader region of *psb*A has been shown to control light-dependent accumulation of D1 polypeptide (Staub and Maliga 1993, 1994). It seems that the 5'-end untranslated region interacts with proteins variably depending on phosphorylation status (Danon and Mayfield 1994a) or redox state (Danon and Mayfield 1994b), which can be controlled by light. Thus translation of mRNAs related to photosynthesis can be linked to the photosynthetic process and chlorophyll biosynthesis *per se*. In addition, light-dependent activity of chloroplast elongation factor *Tu*, or differential organ-specific expression of its genes, as in the case of tobacco (Sugita *et al.* 1994), may play a significant role in translational control of plastid gene expression.

Another mechanism of translational control is mediated by ribosome stalling at hairpin structures in a leader peptide, which leads to premature termination of translation. This mechanism would be controlled by the presence or absence of ribosomes translating a leader peptide. The presence of hairpin structures in the leader region of spinach chloroplast operon supports the concept that this operon could be regulated by such a mechanism (Laboure *et al.* 1988).

Chloroplast Proteins

Polysomes fall into two classes – free and membrane-bound; each class translates a different set of proteins. Proteins synthesized on free polysomes diffuse into the cytosol and plastid stroma and are directed to their final destinations by in-built protein sequences that act as targeting signals. On the other hand, proteins synthesized on membrane-bound polysomes are inserted directly into the membrane during translation; this process is termed *cotranslational translocation*. Although chloroplasts contain their own DNA and ribosomes, most of their proteins are encoded in nuclear DNA and imported from the cytoplasm. Chloroplast proteins encoded by nuclear DNA are translated on ribosomes in the cytoplasm (80S ribosomes) and by chloroplast DNA (70S ribosomes) in the plastid stroma (Fig. 5.13).

Transport of the nuclear-encoded proteins to the correct plastid compartment is critical for the biogenesis and function of chloroplasts (Keegstra and Froehlich 1999; Bauer *et al.* 2000); these proteins can be divided into several groups based on their final destinations. Chloroplasts are separated from the cytoplasm by two distinct membranes – the outer and inner membranes of the chloroplast envelope. In addition, the chloroplast contains an internal system of thylakoid membranes. Thus the chloroplast contains five functionally distinct locations – the outer and inner envelope membranes, the intermembrane space, the thylakoid membrane and

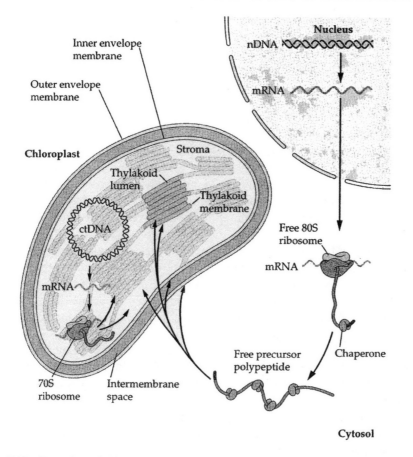

Figure 5.13   Targeting of chloroplast proteins to five different locations in the organelle. (From Raikhel and Chrispeels 2000)

the thylakoid lumen. Each location has a unique set of proteins which must be directed not only into the plastid but also to the correct location.

Major complexes such as photosystems I and II and ATP synthase must be assembled on the thylakoid membranes during chloroplast development. Synthesis of many chloroplast protein complexes is greatly enhanced by light during the light-induced greening of plastids. Genetic and biochemical evidence suggests that some nuclear gene products act as regulators of chloroplast protein synthesis, often affecting the translation of a single species of chloroplast mRNA by interaction with its 5′ UTR (Spremulli 2000). These protein–mRNA complexes increase in light and are thought to play a role in light-dependent activation of translation. The binding of these *trans*-acting factors to the 5′ UTRs of the chloroplast mRNAs must be coordinated with the photosynthetic activity of the chloroplast. Photosynthetic electron transfer generates the reducing power ($NADPH^+$) and the energy-rich ATP. The redox environment of the chloroplast and the availability of ATP are thought to influence the formation of the multiprotein complex that binds the 5' UTR of the *psb*A–mRNA (PSII) in *Chlamydomonas* (Spremulli 2000). Formation of this mRNA–protein complex is promoted by light and correlates with a 50- to 100-fold enhancement of *psb*A translation.

Targeting Proteins to the Plastids

Chloroplast proteins encoded in the nucleus are synthesized in the cytosol by free ribosomes. These proteins are synthesized as precursor proteins, with a cleavable, N-terminal extension of 40–50 amino acids, termed a *presequence* or *transit peptide*. Transit peptides are both necessary and sufficient to mediate chloroplast recognition, translocation across the chloroplast envelope, productive import of passenger proteins and targeting of these proteins to specific locations in the organelle (Cline and Henry 1996). Proteins that lack a transit peptide cannot be imported. If a transit peptide is added to the N-terminus of a protein that is foreign to the chloroplast, for example a reporter protein, then the chimeric precursor is imported. Chloroplast precursor proteins are imported at defined contact sites – proteinaceous channels where the inner and outer membranes meet. After their translocation through the chloroplast envelope, a peptidase removes the transit peptides of stromal precursor proteins. Stroma-targeting domains of transit peptides have little overall amino acid homology although the types and distribution of amino acids show a distinctive pattern: transit peptides are rich in basic, hydroxylated and hydrophobic amino acids but generally lack acidic residues. A cytosolic protein kinase which recognizes transit sequences of chloroplast precursor proteins, catalyses the phosphorylation of one specific serine or threonine residue within the stroma-targeting, envelope-transfer domain of the transit sequence (Waegemann and Soll 1996). Phosphorylated and non-phosphory-lated precursor proteins bind with similar efficiency to the organelle surface but translocation of the former into the chloroplast is inhibited. Dephosphorylation, catalysed by a protein phosphatase, is required to allow complete import of the precursor into the stroma. This phosphorylation–dephosphorylation cycle could be a communicative circuit between the chloroplast and the rest of the cell.

Transport across the two envelope membranes is a common step for all proteins that enter plastids, regardless of their ultimate destination in the organelles. Most of the proteins studied to date are imported via a single transport system, which is sometimes referred to as the *general transport system* (Cline and Henry 1996). Studies using cell-free systems suggest a two-step model for chloroplast protein import. The process employs chaperones on both sides of the chloroplast envelope and a group of proteins collectively termed the *protein apparatus* that spans both the outer and inner membranes.

The outer-membrane proteins form the Toc complex (Toc is the acronym for translocon at the outer membrane of the chloroplast). Three outer-membrane proteins have been identified – Toc86, Toc75 and Toc34; the numbers designate the molecular mass of the given proteins in kilodaltons (Schnell *et al.* 1997). Toc86 is a

GTP-binding protein that is postulated to function as an import-receptor (Kessler *et al.* 1994). Toc86 is readily degraded when intact chloroplasts are subjected to mild protease activity, indicating that it is exposed on the exterior chloroplastic surface (Fig. 5.14). Toc34, also a GTP-binding protein, has most of its hydrophilic domain exposed at the cytoplasmic surface of the outer membrane (Chen and Schnell 1997). Toc75, the most abundant protein in the outer membrane, is postulated to form the channel through which precursor proteins are transported across the outer membrane (Schnell *et al.* 1997). Toc70 is deeply embedded in the membrane and is highly resistant to proteolysis from the outer surface (Tranel *et al.* 1995). Evidence supporting its proposed function as a channel comes from experiments demonstrating that highly purified Toc75, produced in *E. coli*, can form a voltage-gated channel when reconstituted in a phospholipid membrane (Hinnah *et al.* 1997).

Inner-membrane proteins form the Tic complex (translocon on the inner membrane of the chloroplast). Several proteins have been identified as putative components of the Tic complex; the best studied is Tic110 which has been shown to have a hydrophilic domain in the stroma (Jackson *et al.* 1998). At least three other proteins have been identified as inner-membrane translocation components. Tic55 has been found in association with other translocon proteins (Caliebe *et al.* 1997). This protein is unusual among translocation proteins in that it contains an iron–sulphur centre; no role for a putative redox centre in protein translocation has been found. Tic20 and Tic22 have been shown to be associated with precursors undergoing translocation. Tic20 is believed to be an integral membrane protein, possibly part of the inner-membrane translocation channel, whereas Tic22 is a peripheral-membrane protein and possibly serves to connect the inner and outer membrane complexes (Kouranov *et al.* 1998). In addition, three different chaperones appear to associate peripherally with the stromal side of the Tic complex – Hsp100, Hsp70 and chaperonin 60 (Nielsen *et al.* 1997). These stromal chaperones are believed to drive protein import into the stroma by pulling precursors into chloroplasts through repeated cycles of binding and release. The general import pathway of chloroplast precursor proteins can be divided into three stages (Fig. 5.14).

Molecular chaperones are employed to unfold the precursor proteins in order to cross the membranes and to refold them in the stroma. Chaperones are proteins that interact specifically and reversibly with newly synthesized peptides, preventing misfolding of the molecule and ensuring the correct structure of the mature protein. Heat-shock proteins (Hsp) may be classed as chaperones which have the function of recognizing the increased exposure of interacting protein surfaces to high temperatures and protecting them. The translocon at the outer-envelope membrane contains two associated chaperones – Hsp70 and Com70 (chloroplast outer-envelope membrane protein). Com70 is found on the cytoplasmic surface of the outer membrane and has been postulated to play a role in the early stages of protein import (Kourtz and Ko 1997), possibly by holding the precursor in an unfolded or partially folded state. A different Hsp70, located on the inner side of the outer membrane may be responsible for the unfolding activity associated with the outer

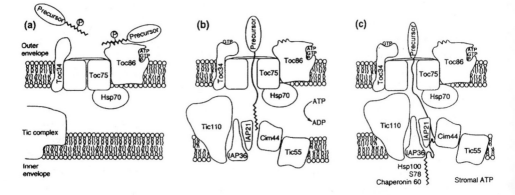

Figure 5.14 A model of protein import into chloroplasts across the outer- and inner-envelope membranes. (a) The phosphorylated precursor protein binds to the receptor Toc86 at the cytosol surface of the outer membrane. (b) Translocation of the precursor protein to the translocation pore in Toc75 is accompanied by dephosphorylation of the precursor and ATP hydrolysis in the intermembrane space. (c) Translocation of the precursor occurs simultaneously across the outer and inner membranes. (From Heins *et al.* 1998)

membrane (Guéra *et al.* 1993). This intermembrane-space Hsp70 may provide the driving force for precursor transport across the outer membrane and may account for the observed distinction between the translocation machineries of the two membranes (Scott and Theg 1996).

In the first stage the precursors associate with the chloroplast surface by binding to the receptor Toc80 in a reversible, energy-independent step. The next two stages can be distinguished by their different energy requirements. In the presence of guanosine triphosphate (GTP), or the hydrolysis of low concentrations of ATP ($10\text{--}100\,\mu M$), precursor proteins bind (dock) to the import machinery (Waegemann and Soll 1996). Transfer of the precursor from the chloroplast surface to the translocation pore in Toc75 requires phospholipids on the cytosol side of the outer membrane. This step necessitates the dephosphorylation of the precursor and ATP hydrolysis in the intermembrane space. The conformation of Toc34 is probably altered by binding GTP and as a consequence Toc34 possibly affects the gating properties of Toc75. When the precursor inserts into the outer-membrane component of the import machinery, components of the inner membrane can be isolated. This supports the contention that the import apparatus of both the outer and the inner membranes are physically close together and that precursor translocation proceeds simultaneously across both membranes (Nielsen *et al.* 1997). Complete translocation into the organelle is achieved only in the presence of higher ATP concentrations ($0.1\text{--}3\,mM$) in the stroma (Theg *et al.* 1989). Ribulose-1,5-bisphosphate carboxylase (RubisCO) provides an excellent example of how proteins are imported into the chloroplast stroma (Fig. 5.15). The RubisCO enzyme is made up of eight, identical, large (L) subunits that are encoded on the chloroplast DNA and eight, identical, small (S) subunits that are encoded on the nuclear DNA.

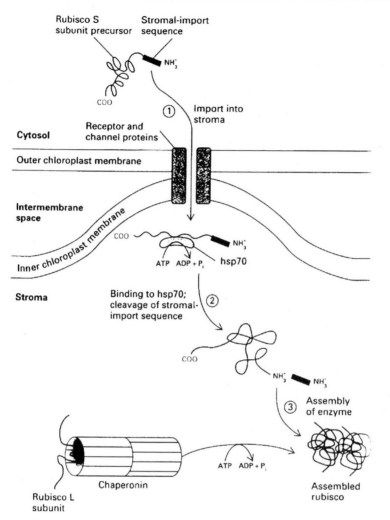

Figure 5.15   Import of RubisCO small (S) subunits into the chloroplast stroma and assembly of small and large (L) subunits into the mature RubisCO enzyme. (From Lodish *et al.* 1995)

The small subunit is synthesized in a precursor form that traverses both the outer and inner chloroplast membranes to reach the stroma where it combines with the large subunit to form the holoenzyme.

A receptor protein located at the point of contact between the outer- and inner-envelope membranes mediates translocation from the cytosol to the stroma. The receptor binds to the stromal-import sequence and the precursor then passes through the channel protein into the stroma. The precursor binds transiently to a 70 kDa heat-shock protein (Hsp70), and the stromal-import sequence is cleaved. The large subunits are synthesized in the stroma by chloroplast ribosomes and are stored complexed to a 60 kDA (Hsp60) protein called a *chaperonin*. The large and small subunits combine to form the mature RubisCO enzyme.

Targeting Proteins to the Envelope Membranes

Outer-membrane proteins

The proteins transported to the outer-envelope membrane can be divided into two groups: those with (Tranel *et al.* 1995; Muckel and Soll 1996) and those without (Wu and Ko 1993; Li and Chen 1996) a cleavable N-terminal transit peptide. Thus, at least two pathways exist for targeting proteins into the outer-envelope membrane. The first is used for the transport of various small proteins such as Toc34 that have no cleavable targeting N-terminal peptide (Chen and Schnell 1997). Targeting and efficient insertion of these proteins into the outer membrane does not require the hydrolysis of ATP nor does it depend on proteins of the membrane. Insertion occurs directly into the membrane and is dependent on a C-terminal hydrophilic domain within these proteins (Li and Chen 1996).

Most of the outer-envelope membrane proteins lack the N-terminal transit peptide (Wu and Ko 1993; Muckel and Soll 1996) and only a few of them, such as Toc75 and Toc86, have been shown to possess it (Tranell *et al.* 1995; Muckel and Soll 1996). A second pathway is used by the precursors of Toc75 and Toc86, which are part of the protein-import machinery of the chloroplast. In contrast to Toc34, insertion of these proteins into the outer membrane requires the hydrolysis of ATP. The precursor of Toc75 has a bipartite transit peptide (Tranel and Keegstra 1996). The N-terminal stroma-targeting domain of the transit peptide engages the components of the general import pathway and is removed by stromal proteases. The resulting intermediate, still containing the second domain of the original bipartite transit peptide, is hydrolysed to serve as a 'stop-transfer' signal to halt translocation within the membrane while the Toc75 assumes its mature conformation (Tranel and Keegstra 1996). The second domain of the bipartite transit peptide is then removed by a stromal protease. The presequence of Toc86, on the other hand, contains no chloroplast-targeting information; insertion into the outer membrane is achieved by a C-terminal fragment of Toc86 (Muckell and Soll 1996).

Inner-membrane proteins

Proteins destined for the inner-envelope membrane carry cleavable transit peptides containing envelope transfer, stroma-targeting information, which, in most cases, is indistinguishable from presequences of stromal or thylakoid proteins. Current evidence suggests that inner-membrane proteins use the Toc and Tic complex and targeting to the inner membrane is probably mediated by hydrophobic, transmembrane domains of the mature protein. There are two possible pathways. In general, proteins of the inner membrane first enter the stroma because of their stroma-targeting sequence and the processed, mature protein is subsequently re-exported from the stroma into the inner membrane. Alternatively, at the same time as translocation, a 'stop-transfer' membrane-insertion signal is recognized by the Tic complex. Regardless of which pathway is used, an open question remains as to how chloroplasts differentiate between the membrane-anchoring or signalling

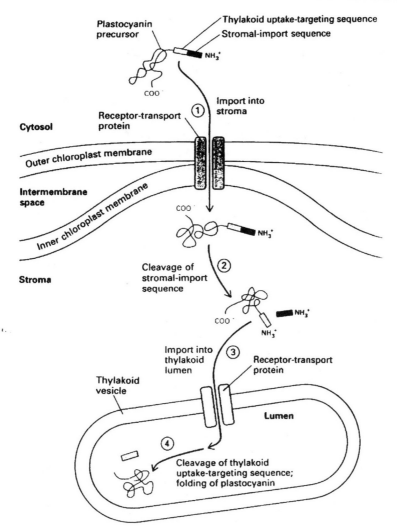

Figure 5.16   Transport of plastocyanin from cytosol to thylakoids. (Lodish *et al.* 1995)

function of the inner membrane and hydrophobic, membrane-anchor domains in the thylakoid.

Thylakoid proteins

Proteins destined for the thylakoid membrane or the thylakoid lumen must travel through the stroma and either be inserted into the thylakoid membrane or cross the membrane and enter the thylakoid lumen. Proteins targeted to the thylakoid lumen, such as plastocyanin, require and contain the successive action of two targeting sequences – they contain a bipartite transit pepide (Fig. 5.16). The first, like that of the RubisCO S-subunit, targets the protein to the stroma where a stromal protease removes it. This exposes a second transit peptide. The lumen-targeting domain of the transit peptide subsequently directs the protein across the thylakoid membrane into the lumen where a second protease cleaves it from the peptide. The role of these sequences has been demonstrated by cell-free experiments using mutant proteins generated by recombinant DNA techniques. For instance, when the thylakoid-uptake targeting sequence is deleted, plastocyanin accumulates in the stroma and is not transported into the lumen. It is believed that hydrophobic regions of the mature protein itself, rather than a cleavable transit peptide, are responsible for targeting proteins into the thylakoid lumen.

Import of proteins into chloroplasts requires energy. Experimental evidence suggests that import into the stroma depends on ATP hydrolysis catalysed by chaperones in the stroma. While import of proteins into chloroplasts does not require an electrochemical potential across the inner membrane, transport of some proteins into the thylakoid does require a pH difference across the thylakoid membrane. There are at least two pathways for transporting proteins across the thylakoid membrane into the lumen. The translocation of luminal proteins such as plastocyanin and OE33 (a component of the oxygen-evolving complex) requires ATP and a soluble protein and is stimulated by a pH difference between the chloroplast stroma and the thylakoid lumen. The soluble protein factor is a carrier mechanism, called SecA, which is a stromal prokaryotic component of bacterial translocation systems. This thylakoid pathway is termed the *SEC pathway* because of its similarity to the bacterial secretory process.

A second pathway is ATP-independent. Thylakoid luminal proteins are transported by an ATP-independent pathway that requires only a pH gradient across the membrane. This pathway is termed the *ΔpH pathway*. The structure of the transit sequences provides the targeting signals for the two different mechanisms, despite the similarity of the two groups of proteins. The transit sequence is prokaryotic, with three distinct domains – an amino-terminal (N-) domain enriched in basic and hydroxylated amino acids and a few acidic amino acids for entry into the stroma, a hydrophobic core (H-) domain enriched in serine, threonine, lysine and arginine, and a hydrophilic carboxy-terminal (C-) domain for cleavage by the thylakoidal processing protease (Chaal *et al.* 1998). The only difference appears to be the twin-arginine motif located between the N- and H-domains of the transit peptide which is essential for targeting proteins via the ΔpH pathway. In many respects the dual pathway of moving proteins from the stroma into the thylakoid resembles the process for the secretion of bacterial proteins across the bacterial plasma membrane, a similarity consistent with the evolution of chloroplasts from ancestral photosynthetic prokaryotes. The thylakoid membrane corresponds to the plasma membrane of the ancestral bacterium; thus the chloroplast stroma corresponds to the bacterial cytoplasm. The secretion of bacterial proteins occurs post-translationally and requires binding of precursor proteins to cytoplasmic chaperones, just as in the post-translational uptake of proteins from the stroma to the thylakoid. In addition, the thylakoid targeting sequences of plastocyanin and other thylakoid luminal proteins resemble the sequences that target bacterial proteins to cross the bacterial plasma membrane. In both, a sequence of 20 amino acids contains a continuous sequence of about 5 hydrophobic amino acids near the N-terminus, usually preceded by a lysine or arginine residue.

A third pathway is used to transport integral-thylakoid-membrane proteins, such as the light-harvesting chlorophyll *a/b* protein (LHCP). The LHCP, with three-membrane spanning sequences, is synthesized with only the envelope-transit sequence so that the signals required for correct insertion into the membrane are derived from the mature protein. This pathway requires GTP rather than ATP and is stimulated by ΔpH (Keegstra and Cline 1999). This pathway also requires a stromal protein factor – a chloroplast homologue of the signal recognition particle (SRP), a ribonucleoprotein involved in targeting proteins to the endoplasmic reticulum (ER). It is therefore called the *SRP pathway*. The stromal protein required for the pre-LHCP import is believed to be CP54, a homologue of the SRP54-kD protein, as its depletion from the stroma prevents insertion (Li *et al.* 1995). The complex recognizes and binds to a sequence of hydrophobic amino acids specific to the protein (and so not recognized by the AT- and Sec system-proteins) and allows the protein to enter across the water-membrane interface and insert into the membrane lipid. Thus, similar to the ER and bacterial SRPs, chloroplast SRP54 appears to function as an essential chaperone or pilot for targeting preproteins to the thylakoid membrane (Keegstra and Cline 1999).

A fourth system is used by proteins with a single-membrane span, such as certain subunits of PSII and CF_0 subunit II, which has an N-terminus in the lumen. While such proteins have similarities in the transit peptide to SEC-transported proteins, these proteins integrate with SEC, SRP and either GTP or ATP. Probably the structure of the protein allows insertion into the hydrophobic membrane, followed by reorganization of the secondary/tertiary form and removal of the signal domain by specific peptidases (Lawlor 2001).

Several luminal proteins are encoded on the chloroplast genome. Targeting of these proteins to the thylakoid membrane meets the same problem as nuclear-encoded proteins that have reached the stroma. One such chloroplast-encoded protein is cytochrome *f*. This protein is anchored in the thylakoid membrane with a C-terminal membrane-spanning domain. The bulk of the protein is in the lumen. There is an N-terminal cleavable signal sequence which acts as the lumen-targeting domain; as for plastocyanin, the targeting to the thylakoid follows the SEC-mediated pathway.

It is not clear why plants need multiple pathways for the translocation and integration of thylakoid proteins. Although it is possible that such redundancy provides a back-up system, it is more likely that the different pathways exist to accommodate the specific assembly problems of the different groups of proteins that follow each pathway (Keegstra and Cline 1999). For example, the chloroplast SRP system provides a means of keeping membrane proteins soluble during trans-stromal transport. In prokaryotes, the SRP system appears to be primarily dedicated to the assembly of cytoplasmic membrane proteins (Ulbrandt *et al.* 1997). Similarly, for the lumenal proteins, the SEC pathway appears incapable of transporting protein substrates of the ΔpH pathway. It has been suggested that ΔpH passenger proteins, such as OE23, are tightly folded and that the ΔpH pathway, but not the SEC pathway, can transport folded proteins (Creighton et al. 1995). The possibility that large, folded domains can indeed be transported via the ΔpH pathway has important implications for the mechanism of ΔpH transport because translocation of ΔpH substrates does not open ion-permeable channels across the thylakoid membrane (Teter and Theg 1998). It has been proposd that proteins traversing the envelope do not pass through stromal intermediates and perhaps those intermediates that are detected are rare escapes (Hoober et al. 1994). This alternative view is not widely accepted but is consistent with the large amounts of vesicular traffic and contacts between thylakoid and envelopes observed in chloroplasts active in thylakoid biogenesis (Hoober et al. 1994). Direct transfer of lipophilic proteins such as the 22-kD thylakoid heat-shock protein and the LHCP, which do not contain simple thylakoid signal sequences, from the envelope to the thylakoid membrane would obviate the need for stromal proteins such as CP54. Alternatively, CP54 may mediate envelope to thylakoid transfer. While most in vitro experiments are clear and concise and support the concept of multi-functional signal sequences that produce stromal intermediates, these alternative concepts remain to be tested.

Summary

Plastids have the capacity to differentiate and dedifferentiate, resulting in interconversions to form amyloplasts, chloroplasts, chromoplasts, etioplasts and leucoplasts. Chloroplasts and mitochondria are believed to have originated from endosymbiotic bacteria.

Plant cells have three different genomes – nuclear, plastidic and mitochondrial, each with its own genetic system and protein synthesis machinery. Chloroplasts are highly polyploid and contain many copies of chloroplast DNA (plastome) per chloroplast. Most of the chloroplast genome encodes proteins involved in plastid transcription and translation and in photosynthesis. In addition to their own DNA, plastids possess the transcriptional and translational machinery necessary for its expression. They also contain all of the chloroplast rRNA, approximately 30 tRNA genes and about 100 genes encoding proteins. Chloroplast genes are interrupted by introns. RNA-editing occurs in some chloroplast genes, the frequency being organelle- or gene-specific. Editing processes in plant chloroplasts and mitochondria share certain features with prokaryotes.

Chloroplast gene expression is highly regulated and regulation can occur at the transcriptional, post-transcriptional, translational and post-translational levels. Chloroplast genes and their mode of expression exhibit both prokaryotic and eukaryotic features. Chloroplast genomes for many plant species have been sequenced; the principal RNA polymerase of plastids has been shown to resemble that of E. coli. Transcription sites of chloroplast genes are similar to those of prokaryotes.

Protein synthesis in plastids is similar to bacterial translation, occurring on 70S ribosomes and mediated by similar initiation and elongation factors. Chloroplast proteins synthesized on free polysomes diffuse into the stroma and are targeted to their destinations by targeting signals; proteins synthesized on membrane-bound polysomes are inserted directly into the membrane during translation. Chloroplast proteins encoded in the nucleus occur as precursors with a cleavable N-terminal transit peptide. Transit peptides mediate chloroplast recognition, translocation across the chloroplast envelope, and import and targeting of proteins to specific locations in the organelle. The transit apparatus uses chaperones on both sides of the chloroplast envelope.

References

Allison, L.A., Simon, L.D. and Maliga, P. (1996). Deletion of *rpo*B reveals a second distinct transcription system in plastids of higher plants. *EMBO J.* **15**: 2802–2809.

Alt, J., Morris, J., Westhoff, P. and Herrman, R.G. (1984). Nucleotide sequence of the clustered genes for the 44 kd chlorophyll *a* apoprotein and the '32 kd'-like protein of the photosystem II reaction center in the spinach plastid chromosome. *Curr. Genet.* **8**: 597–606.

Baginsky, S., Tiller, K. and Link, G. (1997). Transcription factor phosphorylation by a protein kinase associated with chloroplast RNA polymerase from mustard (*Sinapsis alba*). *Plant Mol. Biol.* **34**: 181–189.

Barkan, A. (1988). Proteins encoded by a complex chloroplast transcription unit are each translated from both monocistronic and polycistronic mRNAs. *EMBO J.* **7**: 2637–2644.

Bate, N.J., Strauss, N.A. and Thompson, J.E. (1990). Expression of chloroplast photosynthesis genes during leaf senescence. *Physiol. Plant.* **80**: 217–225.

Bauer, J., Chen, K., Hiltbunner, A., Wehrli, E., Eugster, M., Schnell, D. and Kessler, F. (2000). The major protein import receptor of plastids is essential for chloroplast biogenesis. *Nature* **403**: 303–307.

Baumgartner, B.J., Rapp, J.C. and Mullet, J.E. (1993). Plastid genes encoding the transcription/translation apparatus are differentially transcribed early in barley (*Hordeum vulgare*) chloroplast development: evidence for selective stabilization of *psb*A mRNA. *Plant Physiol.* **101**: 781–791.

Bedbrook, J.R., Kolodner, R. and Bogorad, L. (1977). *Zea mays* chloroplast ribosomal RNA genes are part of a 22,000 base pair inverted repeat. *Cell* **11**: 739–749.

Benne, R. (1990). RNA editing in trypanosomes: is there a message? *Trends in Genetics* **6**: 1770–1781.

Berends, T., Jones, J.T. and Mullet, J.E. (1990). Sequence and transcription analysis of the barley ctDNA region upstream of *psb*D-*psb*C encoding *trn*K (UUU), *rps*16, *trn*Q (UUG), *psb*K, *psb*I and *trn*S (GCU). *Curr. Genet.* **17**: 445–454.

Berry, J.O., Breiding, D.E. and Klessig, D.F. (1990). Light-mediated control of translational initiation of ribulose-1,5-bisphosphate carboxylase in Amaranth cotyledons. *Plant Cell* **2**: 795–803.

Bird, C.R., Koller, B., Auffret, A.D., Hutty, A.K., Howe, C.J., Dyer, T.A. and Gray, J.C. (1985). The wheat chloroplast gene for CF_0 subunit I of ATP synthase contains a large intron. *EMBO J.* **4**: 1381–1388.

Bowman, C.M. and Dyer, T.A. (1979). 4.5S ribonucleic acid, a novel ribosome component in the chloroplasts of flowering plants. *Biochem. J.* **183**: 605–613.

Bradley, D. and Gatenby, A.A. (1985). Mutational analysis of the maize chloroplast ATPase-beta subunit gene promoter: the isolation of promoter mutants in E. coli and their characterization in a chloroplast *in vitro* transcription system. *EMBO J.* **4**: 3641–3648.

Brandt, P., Breidenbach, E., Prestin, B. and Boschetti, A. (1989). Transcriptional, translational and assembly control of the large subunit of ribulose-1,5-bisphosphate carboxylase/oxygenase during the cell cycle of *Euglena gracilis*. *J. Plant Physiol.* **134**: 420–426.

Brown, T.A. (1992). Types of RNA molecule: mRNA, *Genetics – a Molecular Approach*, pp. 95–112. Chapman and Hall, London.

Caliebe, A., Grimm, R., Kaiser, G., Lübeck, J., Soll, J. and Howe, C.J. (1997). The chloroplastic protein import machinery contains a Rieske-type iron-sulfur cluster and a mononuclear iron-binding protein. *EMBO J.* **16**: 7342–7350.

Chaal, B.K., Mould, R.M., Babrook, A.C., Gray, J.C. and Howe, C.J. (1998). Characterization of a cDNA encoding the thylakoid processing peptidase from *Arabidopsis thaliana* – implications for the origin and catalytic mechanism of the enzyme. *J. Biol. Chem.* **273**: 2715–2722.

Chen, D.D. and Schnell, D.J. (1997). Insertion of the 34-kDA chloroplast protein import component, IAP34, into the chloroplast outer membrane is dependent on its intrinsic GTP-binding capacity. *J. Biol. Chem.* **272**: 6614–6620.

Christopher, D.A. and Hallick, R.B. (1989). *Euglena gracilis* chloroplast ribosomal protein operon: a new chloroplast gene for ribosomal protein L5 and description of a novel organelle intron category designated group III. *Nucl. Acids Res.* **17**: 7591–7608.

Christopher, D.A., Xinli, L., Kim, M. and Mullet, J.E. (1997). Involvement of protein kinase and extraplastidic serine/threonine protein phosphatases in signalling pathways regulating plastid transcription and the *psb*D blue-light responsive promoter in barley. *Plant Physiol.* **113**: 1273–1282.

Cline, K. and Henry, R. (1996). Import and routing of nucleus-encoded chloroplast proteins. *Ann. Rev. Cell Dev. Biol.* **12**: 1–26.

Creighton, A.M., Hulford, A., Mant, A., Robinson, D. and Robinson, C. (1995). A monomeric, tightly folded stromal intermediate on the delta pH-dependent thylakoid protein transport pathway. *J. Biol. Chem.* **270**: 1663–1669.

Danon, A. and Mayfield, S.P. (1994a). ADP-dependent phosphorylation regulates RNA-binding *in vitro*: implications in light-modulated translation. *EMBO J.* **13**: 2227–2235.

Danon, A. and Mayfield, S.P. (1994b). Light-regulated translation of chloroplast messenger RNAs through redox potential. *Science* **266**: 1717–1719.

Deng, X.-W. and Gruissem, W. (1987). Control of plastid gene expression during development: the limited role of transcriptional regulation. *Cell* **49**: 379–387.

Deng, X.-W., Tonkyn, J.C., Peter, G.F., Thornber, J.P. and Gruissem, W. (1989). Post-transcriptional control of plastid mRNA accumulation during adaptation of chloroplasts to different light quality environments. *Plant Cell* **1**: 645–654.

Dey, P.M., Brownleader, M.D. and Harborne, J.B. (1997). The plant, the cell and its molecular components. In: *Plant Biochemistry*, P.M. Dey and J.B. Harborne (eds), pp. 1–47. Academic Press, London.

Eisermann, A., Tiller, K. and Link, G. (1990). *In vitro* transcription and DNA binding characteristics of chloroplast and etioplast extracts from mustard (*Sinapsis alba*) indicate differential usage of the *psb*A promoter. *EMBO J.* **9**: 3981–3987.

Eneas-Filho, J., Hartley, M.R. and Mache, R. (1981). Pea chloroplast ribosomal proteins; characterization and site of synthesis. *Mol. Gen. Genet.* **184**: 484–488.

Ferl, R. and Paul, A.L. (2000). Genome organization and expression. In: *Biochemistry and Molecular Biology of Plants*, B.R. Buchanan, W. Gruissen and R.L. Jones, (eds), pp. 312–356. American Society of Plant Physiologists, Rockville, MD.

Fish, L.E., Kück, U. and Bogorad, L. (1985). Two partially homologous adjacent light-inducible maize chloroplast genes encoding polypeptides of the P700 chlorophyll *a*-protein complex of photosystem I. *J. Biol. Chem.* **260**: 1413–1421.

Gamble, P.E. and Mullet, J.E. (1989). Blue light regulates the accumulation of two *psb*D-*psb*C transcripts in barley chloroplasts. *EMBO J.* **8**: 2785–2794.

Gantt, J.S. (1988). Nucleotide sequences of cDNAs encoding four complete nuclear-encoded plastid ribosomal proteins. *Curr. Genet.* **14**: 519–528.

Goldschmidt-Clermont, M., Girard-Bascou, J., Choquet, Y. and Rochaix, J.-D. (1990). *Trans*-splicing mutants of *Chlamydomonas reinhardtii. Mol. Gen. Genet.* **223**: 417–425.

Goldschmidt-Clermont, M., Choquet, Y., Girard-Bascou, J., Michel, F. and Rochaix, J.-D. (1991). A small chloroplast RNA may be required for *trans*-splicing in *Chlamydomonas reinhardtii. Cell* **65**: 135–143.

Gray, M.W. and Doolittle, W.F. (1982). Has the endosymbiont hypothesis been proven? *Microbiol. Rev.* **46**: 1–42.

Green, C.D. and Hollingsworth, M.J. (1992). Expression of the large ATP synthase gene cluster in spinach plastids during light-induced development. *Plant Physiol.* **100**: 1164–1170.

Greenberg, B.M., Narita, J.O., Deluca-Flaherty, C., Gruissem, W., Rushlow, K.A. and Hallick, R.B. (1984). Evidence for two RNA polymerase activities in *Euglena gracilis* chloroplasts. *J. Biol. Chem.* **259**: 14880–14887.

Grover, M., Dhingra, A., Sharma, A.K., Maheshwari, S.C. and Tyagi, A.K. (1999). Involvement of photoreceptor(s), Ca^{2+} and phosphorylation in light-dependent control of transcript levels for plastid genes (*psb*A, *psa*A and *rbc*L) in rice (*Oryza sativa* L.). *Physiologia Plantarum* **105**: 701–707.

Gruissem, W. and Tonkyn, J.C. (1993). Control mechanisms of plastid gene expression. *Crit. Res. Plant Sci.* **12**: 19–55.

Gruissem, W. and Zurawski, G. (1985a). Analysis of promoter regions for the spinach chloroplast *rbc*L, *atp*B and *psb*A genes. *EMBO J.* **4**: 3375–3383.

Gruissem, W. and Zurawski, G. (1985b). Identification and mutational analysis of the promoter for a spinach chloroplast transfer RNA gene. *EMBO J.* **4**: 1637–1644.

Guéra, A., America, T., Van Waas, M. and Weisbeck, P.J. (1993). A strong protein unfolding activity is associated with the binding of precursor chloroplast proteins to chloroplast envelopes. *Plant Mol. Biol.* **23**: 309–324.

Haff, L.A. and Bogorad, L. (1976). Hybridization of maize chloroplast DNA with transfer ribonucleic acids. *Biochem. J.* **15**: 4105–4109.

Hajdukiewicz, P.T.J., Allison, L.A. and Maliga, P. (1997). The two RNA polymerases encoded by the nuclear and the plastid compartments transcribe distinct groups of genes in tobacco plastids. *EMBO J.* **16**: 4041–4048.

Haley, J. and Bogorad, L. (1990). Alternative promoters are used for genes within maize chloroplast polycistronic transcription units. *Plant Cell* **2**: 323–333.

Hallick, R.B. (1989). Proposals for the naming of chloroplast genes. *Plant Mol. Bio. Rep.* **7**: 266–275.

Hallick, R.B., Hollingsworth, M.J. and Nickoloff, J.A. (1984). Transfer RNA genes of *Euglena gracilis* chloroplast DNA. *Plant Mol. Biol.* **3**: 169–175.

Heinemeyer, W., Alt, J. and Herrmann, R.G. (1984). Nucleotide sequence of the clustered genes for apocytochrome *b6* and subunit 4 of the cytochrome *b/f* complex in the spinach plastid chromosome. *Curr. Genet.* **8**: 543–549.

Heins, L., Collinson, I. and Soll, J. (1998). The protein translocation apparatus of chloroplast envelopes. *Trends Plant Sci.* **3**: 56–61.

Herrmann, R.G., Westhoff, P., Alt, J., Tittgen, J. and Nelson, N. (1985). Thylakoid membrane proteins and their genes. In: *Molecular Form and Function of the Plant Genome*, L. van Vloten-Doting, G.S.P. Groot and T.C. Hall (eds), pp. 233–256. Plenum Press, New York.

Herzig, R. and Falkowski, P.G. (1989). Nitrogen limitation of *Isochrysis galbana*. I. Photosynthetic energy conversion and growth efficiencies. *J. Phycol.* **25**: 462–471.

Hess, W.R., Prombona, A., Fieder, B., Subramanian, A.R. and Borner, T. (1993). Chloroplast *rps*15 and the *rpo*B/C1/C2 gene cluster are strongly transcribed in ribosome-deficient plastids: evidence for a functioning non-chloroplast-encoded RNA polymerase. *EMBO J.* **12**: 563–571.

Hinnah, S.C., Hill, K., Wagner, R., Schlicher, T. and Soll, J. (1997). Reconstitution of a chloroplast protein import channel. *EMBO J.* **16**: 7351–7360.

Hirose, T., Fan, H., Suzuki, J.Y., Wakasugi, T., Kossel, H. and Sugiura, M. (1996). Occurrence of silent RNA editing in chloroplasts: its species specificity and the influence of environmental and developmental conditions. *Plant Mol. Biol.* **30**: 667–672.

Hoch, B., Maier, R.M., Appel, K., Igloe, G.L. and Kössel, H. (1991). Editing of a chloroplast mRNA by creation of an initiation codon. *Nature* **353**: 178–180.

Hoober, J., White, R., Marks, D. and Gabriel, J. (1994). Biogenesis of thylakoid membranes with emphasis on the process in *C. reinhardtii. Photosyn. Res.* **39**: 15–31.

Hsu-Ching, C. and Stern, D.B. (1991). Specific ribonuclease activities in spinach chloroplasts promote mRNA maturation and degradation. *J. Biol. Chem.* **266**: 24205–24211.

Igloi, G.L. and Kössel, H. (1992). The transcriptional apparatus of chloroplasts. *Crit. Rev. Plant Sci.* **10**: 525–558.

Jackson, D.T., Froehlich, J.E. and Keegstra, K. (1998). The hydrophilic domain of Tic 110, an inner membrane component of the chloroplastic protein translocation apparatus faces the stromal compartment. *J. Biol. Chem.* **273**: 16583–16588.

Jarrell, K.A., Dietrich, R.C. and Perlman, P.S. (1988). Group II intron domain 5 facilitates a *trans*-splicing reaction. *Mol. Cell Biol.* **8**: 2361–2366.

Johnson C.H., Kruft, V. and Subramanian, A.R. (1990). Identification of a plastid-specific ribosomal protein in the 30S subunit of chloroplast ribosomes and isolation of the cDNA clone encoding its cytoplasmic precursor. *J. Biol. Chem.* **265**: 12790–12795.

Kapoor, S., Masheshwari, S.C. and Tyagi, A.K. (1994). Developmental and light-dependent cues interact to establish steady-state levels of transcripts for photosynthesis-related genes (*psb*A, *psb*D, *psa*A and *rbc*L) in rice (*Oryza sativa* L.). *Curr. Genet.* **25**: 362–366.

Kapoor, S., Suzuki, J.Y. and Sugiura, M. (1997). Identification and functional significance of a new class of non-consensus-type plastid promoters. *Plant J.* **11**: 327–337.

Karabin, G.D., Farley, M. and Hallick, R.B. (1984). Chloroplast gene for M_R 32,000 polypeptide of photosystem II in *Euglena gracilis* is interrupted by four introns with conserved boundary sequences. *Nucl. Acids Res.* **12**: 5801–5812.

Keegstra, K. and Cline, K. (1999). Protein import and routing systems of chloroplasts. *Plant Cell* **11**: 557–570.

Keegstra, K. and Froehlich, J.E. (1999). Protein import into chloroplasts. *Curr. Opin. Plant Biol.* **2**: 471–476.

Kessler, F., Blobel, G., Patel, H.A. and Schnell, D.J. (1994). Identification of two GTP-binding proteins in the chloroplast import machinery. *Science* **266**: 1035–1039.

Kim, J. and Mullet, J.E. (1995). Identification of a sequence-specific DNA binding factor required for the transcription of the barley chloroplast blue light-responsive *psb*D-*psb*C promoter. *Plant Cell* **7**: 1445–1457.

Kim, J., Christopher, D.A. and Mullet, J.E. (1993). Direct evidence for selective modulation of *psb*A, *rpo*A, *rbc*L and 16S RNA stability during barley chloroplast development. *Plant Mol. Biol.* **22**: 447–463.

Klein, R.R. (1991). Regulation of light-induced chloroplast transcription and translation in eight-day-old dark-grown barley seedlings. *Plant Physiol.* **97**: 335–342.

Klein, R.R. and Mullet, J.E. (1990). Light-induced transcription of chloroplast genes: psbA transcription is differentially enhanced in illuminated barley. *J. Biol. Chem.* **265**: 1895–1902.

Klein, R.R., De Camp, J.D. and Bogorad, L. (1992). Two types of chloroplast gene promoters in *Chlamydomonas reinhardtii*. *Proc. Natl. Acad. Sci. USA* **89**: 3453–3457.

Koch, W., Edwards, K. and Kössel, H. (1981). Sequencing of the 16S-23S spacer in a ribosomal RNA operon of *Zea mays* chloroplast DNA reveals two split tRNA genes. *Cell* **25**: 203–213.

Kohchi, T., Umesono, K., Ogura, Y., Komine, Y., Nakahigashi, K., Komano, T., Yamada, Y., Ozeki, H. and Ohyama, K. (1988). A nicked group II intron and *trans*-splicing in liverwort, *Marchantia polymorpha*, chloroplasts. *Nucl. Acids Res.* **16**: 10025–10036.

Koller, B., Gingrich, J.C., Stiegler, G.L., Farley, M.A., Delius, H. and Hallick, R.B. (1984). Nine introns with conserved boundary sequences in the *Euglena gracilis* chloroplast ribulose-1,5-bisphosphate carboxylase gene. *Cell* **36**: 545–553.

Kössel, H., Hoch, B., Maier, R.M., Igloi, G.L., Kudla, J., Zeltz, P., Freyer, R., Neckermann, K. and Ruf, S. (1993). RNA editing in chloroplasts of higher plants. In: *Plant Mitochondria: with Emphasis on RNA Editing and Cytoplasmic Male Sterility*, A. Brennicke and U. Küch (eds), pp. 93–102. VCH Publishers, New York.

Kouranov, A., Chen, X., Fuks, B. and Schnell, D.J. (1998). Tic20 and Tic22 are new components of the protein import apparatus at the chloroplast inner envelope membrane. *J. Cell Biol.* **143**: 991–1002.

Kourtz, L. and Ko, K. (1997). The early stage of chloroplast protein import involves Com 70. *J. Biol Chem.* **272**: 2808–2813.

Krebbers, E.T., Larrinua, I.M., McIntosh, L. and Bogorad, L. (1982). The maize chloroplast genes for the β and ε subunits of the photosynthetic coupling factor CF_1 are fused. *Nucl. Acids Res.* **10**: 4985–5002.

Kuchka, M.R., Mayfield, S.P. and Rochaix, J.-D. (1988). Nuclear mutations specifically affect the synthesis and/or degradation of the chloroplast-encoded D2 polypeptide of Photosystem II in *Chlamydomonas reinhardtii*. *EMBO J.* **7**: 319–324.

Kück, U., Choquet, Y., Schneider, M., Dron, M. and Bennoun, P. (1987). Structural and transcriptional analysis of two homologous genes for the P700 chlorophyll *a*-apoproteins in *Chlamydomonas reinhardtii*: evidence for *in vivo* trans-splicing. *EMBO J.* **6**: 2185–2195.

Kudla, J., Igloe, G.L., Metzlaff, M., Hagemann, R. and Kössel, H. (1992). RNA editing in tobacco chloroplasts leads to the formation of a translatable *psb*L mRNA by a C to U substitution within the initiation codon. *EMBO J.* **11**: 1099–1103.

Laboure, A.-M., Lescure, A.-M. and Briat, J.-F. (1988). Evidence for a translation-mediated attenuation of a spinach chloroplast rDNA operon. *Biochimie* **70**: 1343–1352.

Lam, E. and Chua, N.H. (1987). Chloroplast DNA gyrase and *in vitro* regulation of transcription by template topology and novobiocin. *Plant Mol. Biol.* **8**: 415–424.

Lawlor, D.W. (2001). Molecular biology of the photosynthetic system, *Photosynthesis*, pp. 247–280. BIOS Scientific Publishers, Oxford.

Lerbs, S., Brautigam, E. and Mache, R. (1988). DNA-dependent RNA polymerase of spinach chloroplasts: characterization of α-like and σ-like polypeptides. *Mol. Gen. Genet.* **211**: 459–464.

Li, H.M. and Chen, L.J. (1996). Protein targeting and integration signal for the chloroplastic outer envelope membrane. *Plant Cell* **8**: 2117–2126.

Li, X., Henry, R., Yuan, J., Cline, K. and Hoffman, N.E. (1995). A chloroplast homologue of the signal recognition particle subunit SRP54 is involved in the posttranslational integration of a protein into thylakoid membranes. *Proc. Natl. Acad. Sci. USA* **92**: 3789–3793.

Li, Y. and Sugiura, M. (1990). Three distinct ribonucleoproteins from tobacco chloroplasts: each contains a unique amino terminal acidic domain and two ribonucleoprotein consensus motifs. *EMBO J.* **9**: 3059–3066.

Link, G. (1996). Green life: control of chloroplast gene transcription. *Bio Essays* **18**: 465–471.

Liu, B. and Troxler, R.F. (1996). Molecular characterization of a positively photoregulated nuclear gene for a chloroplast RNA polymerase sigma factor in *Cyanidium caldarium*. *Proc. Natl. Acad. Sci. USA* **93**: 3313–3318.

Lodish, H., Baltimore, D., Berk, A., Zipursky, S.L., Matsudaira, P. and Darnell, J. (1995). Organelle biogenesis: the mitochondrion, chloroplast, peroxisome and nucleus, *Molecular Cell Biology*, pp. 809–849. Scientific American Books, New York.

Maier, R.M., Hoch, B., Zeltz, P. and Kössel, H. (1992). Internal editing of the maize chloroplast *ndh*A transcript restores codons for conserved amino acids. *Plant Cell* **4**: 609–616.

Maliga, P. (1998). Two plastid RNA polymerases of higher plants: an evolving story. *Trends Plant Sci.* **3**: 4–6.

Matsubayashi, T., Wakasugi, T., Shinozaki, K., Shinozaki, K.Y., Zaita, N., Hidaka, T., Meng, B.Y., Ohto, C., Tanaka, M., Kato, A., Murayama, T. and Sugiura, M. (1987). Six chloroplast genes (*ndh*A to F) homologous to human mitochondrial genes encoding components of the respiratory-chain NADH dehydrogenase are actively expressed: determination of the splice sites in *ndh*A and *ndh*B pre-mRNAs. *Mol. Gen. Genet.* **210**: 385–393.

Mayfield, S.P., Yohn, C.B., Cohen, A. and Danon, A. (1995). Regulation of chloroplast gene expression. *Ann. Rev. Plant Physiol. Plant Mol. Biol.* **46**: 147–166.

McIntosh, L., Poulsen, C. and Bogorad, L. (1980). Chloroplast gene sequence for the large subunit of ribulose bisphosphate carboxylase of maize. *Nature* **288**: 556–560.

Michaels, A. and Herrin, D.L. (1990). Translational regulation of chloroplast gene expression during the light-dark cell cycle of *Chlamydomonas*: evidence for control by ATP/energy supply. *Biochem. Biophys. Res. Comm.* **170**: 1082–1088.

Michel, F. and Dujon, B. (1983). Conservation of RNA secondary in two intron families including mitochondrial-, chloroplast- and nuclear-encoded members. *EMBO J.* **2**: 33–38.

Morden, C.W., Wolfe, K.H., dePamphilis, C.W. and Palmer, J.D. (1991). Plastid translation and transcription genes in a non-photosynthetic plant: intact, missing and pseudo genes. *EMBO J.* **10**: 3281–3288.

Muckel, E. and Soll, J. (1996). A protein import receptor of chloroplasts is inserted into the outer envelope membrane by a novel pathway. *J. Biol. Chem.* **271**: 23846–23852.

Mullet, J.E. (1993). Dynamic regulation of chloroplast transcription. *Plant Physiol.* **103**: 309–313.

Mullet, J.E. and Klein, R.R. (1987). Transcription and RNA stability are important determinants of higher plant chloroplast RNA levels. *EMBO J.* **6**: 1571–1579.

Neuhaus, H., Scholz, A. and Link, G. (1989). Structure and expression of a split chloroplast gene from mustard (*Sinapsis alba*): ribosomal protein gene *rps*16 reveals unusual transcription features and complex RNA maturation. *Curr. Genet.* **15**: 63–70.

Nickelsen, J. and Link, G. (1989). Interaction of a 3′ RNA region of the mustard *trn*K gene with chloroplast proteins. *Nucl. Acids Res.* **17**: 9637–9648.

Nielsen, E., Akita, M., Davila-Aponte, J. and Keegstra, K. (1997). Stable association of chloroplastic precursors with protein translocation complexes that contain proteins from both envelope membranes and a stromal hsp100 molecular chaperone. *EMBO J.* **16**: 935–946.

Ohyama, K., Fukuzawa, H., Kochi, T., Shirai, H., Sano, T., Sano, S., Umesono, K., Shiki, Y., Takeuchi, M., Chang, Z., Aota, S., Inokuchi, H. and Ozeki, H. (1986). Chloroplast gene organization deduced from complete sequence of liverwort *Marchantia polymorpha* chloroplast DNA. *Nature* **322**: 572–574.

Pfannschmidt, T. and Link, G. (1994). Separation of two classes of DNA-dependent RNA polymerases that are differentially expressed in mustard (*Sinapsis alba* L.) seedlings. *Plant Mol. Biol.* **25**: 69–81.

Pfannschmidt, T. and Link, G. (1997). The A and B forms of plastid DNA-dependent RNA polymerase from mustard (*Sinapis alba* L.) transcribe the same genes in a different developmental context. *Mol. Gen. Genet.* **257**: 35–44.

Raikhel, N. and Chrispeels, M.J. (2000). Protein and vesicle traffic. In: *Biochemistry and Molecular Biology of Plants*, B.B. Buchanan, W. Gruissem and R.L. Jones (eds), pp. 160–201. American Society of Plant Physiologists, Rockville, MD.

Rapp, J.C., Baumgartner, B.J. and Mullet, J.E. (1992). Quantitative analysis of transcription and RNA levels of 15 barley chloroplast genes: transcription rates and mRNA levels vary over 300-fold; predicted mRNA stabilities vary 30-fold. *J. Biol. Chem.* **267**: 21404–21411.

Reinbothe, S., Reinbothe, C., Krauspe, R. and Parthier, B (1991). Changing gene expression during dark-induced chloroplast dedifferentiation in *Euglena gracilis*. *Plant Physiol. Biochem.* **29**: 309–318.

Reumann, S. and Keegstra, K. (1999). The endosymbiotic origin of the protein import machinery of chloroplast envelope membranes. *Trends Plant Sci.* **4**: 302–307.

Ris, H. and Plaut, W. (1962). The ultrastructure of DNA-containing areas in the chloroplasts of *Chlamydomonas*. *J. Cell Biol.* **13**: 383–391.

Rochaix, J.-D. (1992). Post-transcriptional steps in the expression of chloroplast genes. In: *Annual Review of Cell Biology, Vol. 8*, G.E. Palade (ed.), pp. 1–28. Annual Reviews Inc., Palo Alto, CA.

Rochaix, J.-D. (1996). Post-transcriptional regulation of chloroplast gene expression in *Chlamydomonas reinhardtii*. *Plant Mol. Biol.* **32**: 327–341.

Sakamoto, W., Kindle, K.L. and Stern D.B. (1993). *In vivo* analysis of *Chlamydomonas* chloroplast *pet*D gene expression using stable transformation of β-glucuronidase translational fusions. *Proc. Natl. Acad. Sci. USA* **90**: 497–501.

Scheibe, R. (1990). Light/dark modulation: regulation of chloroplast metabolism in a new light. *Bot. Acta* **103**: 327–334.

Schnell, D.J., Blobel, G., Keegstra, K., Kessler, F., Ko, K. and Soll, J. (1997). A consensus nomenclature for the protein-import components of the chloroplast envelope. *Trends Cell Biol.* **7**: 303–304.

Schuster, G. and Gruissem, W. (1991). Chloroplast mRNA 3' end processing requires a nuclear-encoded RNA-binding protein. *EMBO J.* **10**: 1493–1502.

Schwarz, Z. and Kössel, H. (1980). The primary structure of 16S rRNa from *Zea mays* chloroplast is homologous to *E. coli* 10S rRNA. *Nature* **283**: 739–742.

Scott, S.V. and Theg, S.M. (1996). A new chloroplast protein import intermediate reveals distinct translocation machineries in the two envelope membranes: energetics and mechanistic implications. *J. Cell Biol.* **132**: 63–75.

Sexton, T.B., Christopher, D.A. and Mullet, J.E. (1990). Light-induced switch in barley *psb*D-*psb*C promoter utilization: a novel mechanism regulating chloroplast gene expression. *EMBO J.* **9**: 4485–4494.

Shaw, L.C. and Lewin, A.S. (1995). Protein induced folding of a Group I intron in Cytochrome *b* pre-mRNA. *J. Biol. Chem.* **270**: 21552–21562.

Shimada, H. and Sugiura, M. (1989). Pseudogenes and short repeated sequences in the rice chloroplast genome. *Curr. Genet.* **16**: 293–301.

Shinozaki, K., Deno, H., Sugita, M., Kuramitzu, S. and Sugiura, M. (1986). Intron in the gene for the ribosomal proteins S16 of tobacco chloroplast and its conserved boundary sequences. *Mol. Gen. Genet.* **202**: 1–5.

Spremulli, L. (2000). Protein synthesis, assembly, and degradation. In: *Biochemistry and Molecular Biology of Plants*, B. Buchanan, W. Gruissem and R. Jones (eds), pp. 412–454. American Society of Plant Physiologists, Rockville, MD.

Staub, J.M. and Maliga, P. (1993). Accumulation of D1 polypeptide in tobacco plastids is regulated *via* the untranslated region of the *psb*A mRNA. *EMBO J.* **12**: 601–606.

Staub, J.M. and Maliga, P. (1994). Translation of *psb*A mRNA is regulated by light *via* the 5' untranslated region in tobacco plants. *Plant J.* **6**: 547–553.

Stern, D.B. and Gruissem, W. (1987). Control of plastid gene expression: 3' inverted repeats act as mRNA processing and stabilizing elements, but do not terminate transcription. *Cell* **51**: 1145–1157.

Stern D.B., Jones, H. and Gruissem, W. (1989). Function of plastid mRNA 3' inverted repeats: RNA stabilization and gene-specific protein binding. *J. Biol. Chem.* **264**: 18742–18750.

Stern, D.B., Radwanski, E.R. and Kindle, K.L. (1991). A 3' stem/loop structure of the *Chlamydomonas* chloroplast *atp*B gene regulates mRNA acumulation *in vitro*. *Plant Cell* **3**: 285–297.

Stirdivant, S.M., Crossland, L.D. and Bogorad, L. (1985). DNA supercoiling affects *in vitro* transcription of two maize chloroplast genes differently. *Proc. Natl. Acad. Sci. USA* **82**: 4886–4890.

Sugita, M., Murayama, Y. and Sugiura, M. (1994). Structure and differential expression of two distinct genes encoding the chloroplast elongation factor Tu in tobacco. *Curr. Genet.* **25**: 164–168.

Sugita, M. and Sugiura, M. (1983). Putative gene of tobacco chloroplast coding for ribosomal protein similar to *E. coli* ribosomal protein S19. *Nucl. Acids Res.* **11**: 1913–1918.

Sugita, M. and Sugiura, M. (1984). Nucleotide sequence and transcription of the gene for the 32,000 dalton thylakoid membrane protein from *Nicotiana tabacum*. *Mol. Gen. Genet.* **195**: 308–313.

Sugita, M. and Sugiura, M. (1996). Regulation of gene expression in chloroplasts of higher plants. *Plant Mol. Biol.* **32**: 315–326.

Sugiura, M. (1991). Transcript processing in plastids: trimming, cutting, splicing. In: *The Molecular Biology of Plastids. Cell Culture and Somatic Cell Genetics of Plants.* **7A**: 125–137. L. Bogorad and I.K. Vasdil (eds). Academic Press, San Diego, CA.

Sugiura, M. (1992). The chloroplast genome. *Plant Mol. Biol.* **19**: 149–168.

Sugiura, M. and Takeda,Y. (2000). Nucleic acids. In: *Biochemistry and Molecular Biology of Plants*, B. Buchanan, W. Gruissem and R. Jones (eds), pp. 260–310. American Society of Plant Physiologists, Rockville, MD.

Sugiura, M., Torazawa, K. and Wakasugi, T. (1991). Chloroplast genes coding for ribosomal proteins in land plants. In: *The Translational Apparatus of Photosynthetic Organelles*, R. Mache, E. Stutz and A.R. Subramanian (eds), pp. 59–69. Springer-Verlag, Berlin.

Takaiwa, F. and Sugiura, M. (1980). The nucleotide sequence of 4.5S and 5S ribosomal RNA genes from tobacco chloroplasts. *Mol. Gen. Genet.* **180**: 1–4.

Tanaka, K., Oikawa, K., Ohta, N., Kuroiwa, T. and Takahashi, H. (1996). Nuclear encoding of a chloroplast RNA polymerase sigma subunit in a red alga. *Science* **272**: 1932–1935.

Teter, S.A. and Theg, S.M. (1998). Energy-transducing thylakoid membranes remain highly impermeable to ions during protein translocation. *Proc. Natl. Acad. Sci. USA* **95**: 1590–1594.

Tewari, K.K. (1987). Replication of chloroplast DNA. In: *DNA Replication in Plants*, J.A. Bryant and V.L. Dunham (eds), pp. 69–116. CRC Press, Boca Raton, FL.

Tewari, K.K. and Wildman, S.G. (1968). Function of chloroplast DNA. I. Hybridization studies involving nuclear and chloroplast DNA with RNA from cytoplasmic (80S). *Proc. Natl. Acad. Sci. USA* **59**: 569–576.

Theg, S., Baurle, C., Olsen, L., Selman, B. and Keegstra, K. (1989). Internal ATP is the only energy requirement for the translocation of precursor. *J. Biol. Chem.* **264**: 6730–6736.

Tiller, K. and Link, G. (1993a). Phosphorylation and dephosphorylation affect functional characteristics of chloroplast and etioplast transcription systems from mustard *Sinapsis alba* L. *EMBO J.* **12**: 1745–1753.

Tiller, K. and Link, G. (1993b). Sigma-like transcription factors from mustard (*Sinapsis alba* L.) etioplast are similar in size to, but functionally distinct from their chloroplast counterparts. *Plant Mol. Biol.* **21**: 503–513.

Tonkyn J.C., Deng, X.-W. and Gruissem, W. (1992). Regulation of plastid gene expression during photo-oxidative stress. *Plant Physiol.* **99**: 1406–1415.

Tozawa, Y., Tanaka, K., Takahashi, H. and Wakasa, K. (1998). Nuclear encoding of a plastid sigma factor in rice and its tissue- and light-dependent expression. *Nucl. Acids Res.* **26**: 415–419.

Tranel. P.J. and Keegstra, K. (1996). A novel, bipartite transit peptide targets OEP75 to the outer membrane of the chloroplastic envelope. *Plant Cell* **8**: 2093–2104.

Tranel, P.J., Froehlich, J., Goyal, A. and Keegstra, K. (1995). A component of the chloroplastic protein import apparatus is targeted to the outer envelope membrane *via* a novel pathway. *EMBO J.* **14**: 2436–2446.

Troxler, R.F., Zhang, F., Hu, J. and Bogorad, L. (1994). Evidence that s factors are components of chloroplast RNA polymerase. *Plant Physiol.* **104**: 753–759.

Turmel, M., Choquet, Y., Goldschmidt-Clermont, M., Rochaix, J.-D., Otis, C. and Lemieux, C. (1995). The *trans*-spliced intron I in the *psa*A gene of the *Chlamydomonas* chloroplast: a comparative analysis. *Curr. Genet.* **27**: 270–279.

Tyagi, A.K. (1998). Regulation of plastid gene expression. In: *Concepts in Photobiology: Photosynthesis and Photomorphogenesis*, G. Singhal, G. Renger, S. Sopory, K.-D. Irrgang and Govindjee (eds), pp. 717–729. Narosa, New Delhi.

Ulbrandt, N.D., Newitt, J.A. and Bernstein, H.D. (1997). The *E. coli* signal recognition particle is required for the insertion of a subset of inner membrane proteins. *Cell* **88**: 187–196.

Vera, A. and Sugiura, M. (1994). A novel RNA gene in the tobacco plastid genome: its possible role in the maturation of 16S rRNA. *EMBO J.* **13**: 2211–2217.

Vera, A. and Sugiura, M. (1995). Chloroplast rRNA transcription from structurally different tandem promoters: an additional novel-type promoter. *Curr. Genet.* **27**: 280–284.

Wada, T., Tunoyama, Y., Shiina, T. and Toyoshima, T. (1994). *In vitro* analysis of light-induced transcription in the wheat *psb*D/C gene cluster using plastid extracts from dark-grown and short-term illuminated seedlings. *Plant Physiol.* **104**: 1259–1267.

Waegemann, K. and Soll, J. (1996). Phosphorylation of the transit sequence of chloroplast precursor proteins. *J. Biol. Chem.* **271**: 6545–6554.

Wakasugi, T., Ohme, H., Shinozaki, K. and Sugiura, M. (1986). Structure of tobacco chloroplast genes for tRNA-Ile (CAU), tRNA-Leu (CAA), tRNA-Cys (GCA), tRNA-Ser (UGA) and tRNA-Thr (GGU): a compilation of tRNA genes from tobacco chloroplasts. *Plant Mol. Biol.* **7**: 385–392.

Weeks, K.M. and Cech, T.R. (1995). Protein facilitation of group I intron splicing by assembly of the catalytic core and the 5' splice site domain. *Cell* **82**: 221–230.

Westhoff, P. and Herrmann, R.G. (1988). Complex RNA maturation in chloroplasts: the psbB operon from spinach. *Eur. J. Biochem.* **171**: 551–564.

Wiley, D.L., Aufrett, A.D. and Gray, J.C. (1984). Structure and topology of cytochrome f in pea chloroplast membranes. *Cell* **36**: 555–562.

Wu, C. and Ko, K. (1993). Identification of an uncleavable targeting signal in the 70-kilodalton spinach chloroplast outer envelope membrane protein. *J. Biol. Chem.* **268**: 19384–19391.

Ye, L., Fukami-Kobayashi, K., Go, M., Konishi, T., Watanabe, A. and Sugiura, M. (1991). Diversity of a ribonucleoprotein family in tobacco chloroplasts: two new chloroplast ribonucleoproteins and a phylogenetic tree of ten chloroplast RNA-binding domains. *Nucl. Acids Res.* **19**: 6485–6490.

Zaita, N., Torazawa, K., Shinozaki, K. and Sugiura, M. (1987). *Trans*splicing *in vivo*: joining of transcripts from the 'divided' gene for ribosomal protein S12 in the chloroplasts of tobacco. *FEBS Lett.* **210**: 153–156.

Zurawski, G. and Clegg, M.T. (1987). Evolution of higher-plant chloroplast DNA-encoded genes: implications for structure-function and phylogenetic studies. *Ann. Rev. Plant Physiol.* **38**: 391–418.

Zurawski, G., Bohnert, H.J., Whitfeld, P.R. and Bottomly, W. (1982a). Nucleotide sequence of the gene for the M_R 32,000 thylakoid membrane protein from *Spinacia oleracea* and *Nicotiana debneyi* predicts a totally conserved primary translation product of M_R 38,950. *Proc. Natl. Acad. Sci. USA* **79**: 7699–7703.

Zurawski, G., Bottomly, W. and Whitfeld P.R. (1982b). Structure of the genes for the β and ε subunit of spinach ATPase chloroplast indicates a distronic mRNA and an overlapping translation stop/start signal. *Proc. Natl. Acad. Sci. USA* **79**: 6260–6264.

6
Photomorphogenesis

Introduction

Light is an important environmental factor that controls plant growth and development. Living organisms can utilize light in two ways: light can be used as a source of energy to drive a process such as photosynthesis; alternatively, light can be used as a source of information, as in *photomorphogenesis*. It has long been established that light can influence the structural development of plants under conditions that exclude significant levels of photosynthesis. Seedlings grown in darkness have a pale, spindly appearance (*skotomorphogenesis*). This etiolated condition is dramatically different from the compact, green appearance of light-grown seedlings (Fig. 6.1). Given the central role of photosynthesis in plant metabolism, one would be tempted to attribute this contrast to the difference in the availability of light energy. However, very little time or light is required to transform the etiolated seedling to the green form. Within about 10 min of applying a flash of relatively dim light to the etiolated seedling, one can observe the beginning of hook straightening, decrease in the rate of stem elongation and initiation of pigment synthesis. Photosynthesis cannot be responsible for this transformation because chlorophyll is not present. Light has acted as a signal to induce a change in the form of the seedling in a rapid light response which is called photomorphogenesis.

For light to serve as a source of information to a plant, it must first be absorbed by a pigment which thereby becomes photochemically active. Such a pigment is referred to as a *photoreceptor*. By absorbing at different wavelengths, the photoreceptor 'reads' the information contained in the light and interprets it in the form of a primary action for the cell. Primary action usually involves some form of chemical transduction, such as a conformational change in a protein, or a redox reaction. Absorption of light by the photoreceptor initiates a cascade of biochemical events, called a signal transduction chain (Chapter 9), which ultimately induces a response to the light stimulus. Most photomorphogenic responses in higher plants appear to be under the control of one of three kinds of signal-transducing photoreceptors:

Photobiology of Higher Plants. By M. S. Mc Donald.
© 2003 John Wiley & Sons, Ltd: ISBN 0 470 85522 3; ISBN 0 470 85523 1 (PB)

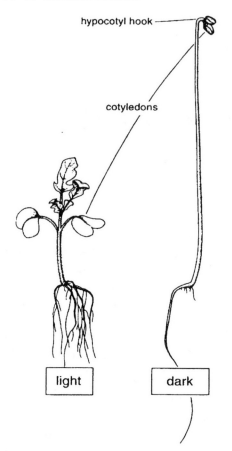

Figure 6.1   Photomorphogenesis and skotomorphogenesis in mustard seedlings (*Sinapsis alba*). (From Mohr and Schopfer 1995)

(i)   *phytochrome*, which absorbs mainly in the red (R) and far-red (FR) and also in the blue regions of the spectrum;

(ii) *cryptochrome*, a group of similar, as yet unidentified, pigments which absorb blue and long-wave ultraviolet light (the UV-A region, 320–400 nm);

(iii) *UV-B photoreceptors*, a group of yet unidentified pigments which absorb ultraviolet radiation between about 280 and 320 nm.

In addition, the photoreceptor protochlorophyllide *a* absorbs light energy in the red and blue parts of the spectrum and becomes reduced to chlorophyll *a*, the key pigment in photosynthesis. The main photoresponses in higher plants and their wavelengths of maximum sensitivity are shown in Figure 6.2. Photoreceptors control morphogenic processes in plants right through from seed germination to the formation of flowers and new seeds. While light itself, or indeed its photoreceptor, is unlikely to carry morphogenic information, the competence of cells to respond to

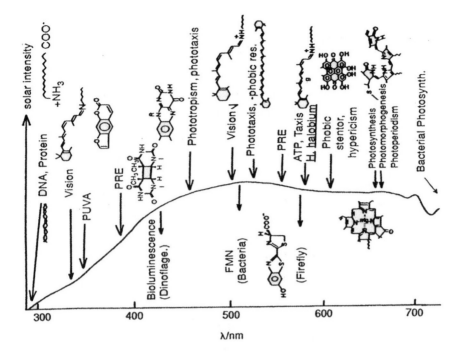

Figure 6.2   The photobiological spectrum, solar spectrum (solid line) and photoresponses in higher plants. (From Kohen *et al.* 1995). Reprinted with permission of Elsevier Science

light and to amplify this stimulus is the main factor that controls photomorphogenesis.

Discovery of Phytochrome

Most of the research leading to the discovery and isolation of phytochrome was carried out at the US Department of Agriculture, Beltsville, Maryland, between 1945 and 1960. Earlier work at Beltsville by H.A. Garner and W.W. Allard (1920) had already established the relative lengths of light and darkness that control flowering in certain plants. Then, in 1938, K.C. Hamner and J. Bonner, also at USDA, discovered that cocklebur (*Xanthium strumarium*), which requires long nights to flower (a short-day plant, SDP), is prevented from flowering by a brief interruption of the long dark period by a flash of light ('night break'). Red light was found to be the most effective wavelength for interrupting the long night. The action spectra for the prevention of flowering in the short-day plants *Xanthium strumarium* and *Glucine max* cv. Biloxi, soybean and for promotion of flowering in the long-day plants (LDP) *Hordeum vulgare*, barley (Borthwick *et al.* 1948) and *Hyoscyamus niger*, henbane (Parker *et al.* 1950) were remarkably similar, especially in the red wavelengths, with maximum effect between 600 and 660 nm (Fig. 6.3).

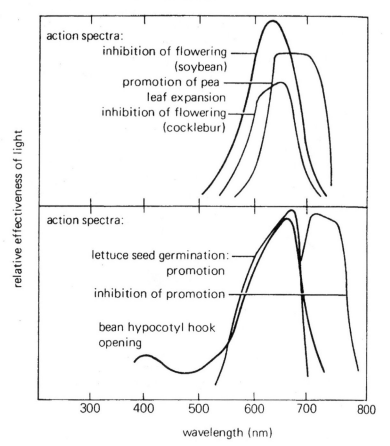

Figure 6.3 Action spectra for the inhibition of flowering, promotion of leaf expansion, promotion and inhibition of lettuce seed germination and bean hypocotyl hook opening. (From Vierstra and Quail 1983; Parker *et al.* 1949; Withrow *et al.* 1957; Borthwick *et al.* 1954)

Action spectra for other light-dependent growth responses were also determined: promotion of leaf growth in etiolated pea seedlings (Parker *et al.* 1949), the inhibition of stem growth in dark-grown seedlings of barley (Borthwick *et al.* 1951) and the promotion of germination in light-sensitive lettuce seeds cv. Grand Rapids (Borthwick and Parker 1952). The similarity of the action spectra indicated that the same photoreceptor was involved in all of these growth responses. During the 1950s, H.A. Borthwick, a botanist, and S.B. Hendricks, a physical chemist, began to study the action spectra of such diverse phenomena, at USDA. They turned their attention to earlier work by L.H. Flint and E.D. McAlister (1935) who had shown that red light promoted germination in the light-sensitive Grand Rapids lettuce seeds and that germination was suppressed by light of wavelengths longer than 700 nm (far-red). They demonstrated that the response to a few minutes of red light could be nullified if it was followed immediately by a similar duration of far-red light (FR). By repeatedly alternating the R and FR treatments they found that the

Table 6.1 Germination of lettuce seeds, cv. Grand Rapids, following consecutive exposures to red (R) and far-red (FR) light. (From Borthwick and Parker 1952)

Treatment	% Germination
Dark control	8.5
R	98.0
R,FR	54.0
R,FR,R	100.0
R,FR,R,FR	43.0
R,FR,R,FR,R	99.0
R,FR,R,FR,R,FR	54.0
R,FR,R,FR,R,FR,R	98.0

radiation last applied determined whether the seed germinated or not, red light promoting and far-red nullifying that promotion (Table 6.1).

They then tested this R/FR reversibility in flowering and found that the night-break reaction controlling flowering in short-day plants showed the same reversibility characteristics as did germination, that is, inhibition of flowering in short-day plants due to a night break of red light was nullified by immediately following the red light with far-red light (Borthwick and Parker 1952). In other words, the radiation used last in the sequence determines the response of the plants (Table 6.2).

These observations led the Beltsville group to propose the existence of a novel pigment system, later called *phytochrome*. They predicted that this hypothetical pigment would exist in two forms: a blue form that absorbed red light (P_r) and an olive-green form that absorbed far-red light (P_{fr}). They also deduced that the pigment must be photochromic, i.e. it exists in two interconvertible forms.

Table 6.2 Effect of daily interruptions of the dark period with consecutive interruptions of red (R) and far-red (FR) on flower initiation in cocklebur and soybean. (From Downs 1956)

Treatment	Mean stage of floral development in cocklebur	Mean number of flowering nodes in Bioloxi soybean
Dark control	6.0	4.0
R	0.0	0.0
R,FR	5.6	1.6
R,FR,R	0.0	0.0
R,FR,R,FR	4.2	1.0
R,FR,R,FR,R	0.0	0.0
R,FR,R,FR,R,FR	2.4	0.6
R,FR,R,FR,R,FR,R	0.0	0.0
R,FR,R,FR,R,FR,R,FR	0.6	0.0

Absorption of red light by P_r would convert the pigment to the far-red absorbing form (P_{fr}), while subsequent absorption of far-red light by P_{fr} would convert the pigment back to the red-absorbing form:

$$P_r \xrightleftharpoons[\text{FR}]{\text{R}} P_{fr}$$

Based on simple physiological experiments, the Beltsville investigators predicted several other characteristics of this hypothetical pigment. They deduced that P_{fr} formed by the absorption of red light, was the physiologically active form which mediated the observed growth responses and that phytochrome must be synthesized in the dark as P_r, which was the inactive form. On the other hand, P_{fr} was apparently unstable and was either destroyed or could revert back to P_r in darkness by means of a non-photochemical, temperature-dependent reaction. They also suggested that the pigment acted catalytically and was therefore most likely a protein.

Since the hypothetical pigment could not be seen in dark-grown, chlorophyll-free tissues, where it was probably present in very low concentrations, and also because of its unique photoreversible property, the existence of phytochrome was generally met with scepticism (Sage 1992). All the contemporary ideas and predictions were based on physiological and spectrophotometric studies using seeds and etiolated plants, and needed to be verified by physical evidence of the existence of phytochrome, that is, its isolation and characterization *in vitro*. Hendricks and his co-workers also predicted, from the action spectra for the responses and their reversal, that the absorbance properties of P_r and P_{fr} must be significantly different: P_r absorbing mainly in the red part of the spectrum and P_{fr} mainly in the far-red part. Photoconversion of P_r to P_{fr} and back again should therefore produce photoreversible changes in the relative absorbance at both of these wavelengths in tissues or in extracts that contained the pigment. A special kind of spectrophotometer, capable of measuring very small absorbance changes in dense, light-scattering tissue samples, was required. Using an instrument called a 'ratiospec', which measures absorbance at two different wavelengths, Butler and his co-workers (1959) were able to demonstrate the predicted photoreversible absorbance changes in etiolated maize seedling tissue following alternating treatments with R and FR light. Within hours of the first detection of phytochrome *in vivo* by this method, they were able to demonstrate similar absorbance changes in aqueous extracts of the same tissue. This spectral analysis was the first physical confirmation that phytochrome really existed. A few years later, Siegelman and Firer (1964) successfully isolated and purified phytochrome from dark-grown oat seedlings and demonstrated its photoreversible character *in vitro*. Mumford and Jenner (1966) also purified phytochrome from oat seedlings and showed it to be a chromoprotein. Phytochrome has since been shown to be ubiquitous and has been found in algae, bryophytes and higher plants. In addition to the clear-cut morphological effects in etiolated seedlings, phytochrome also

Table 6.3   Selected examples of phytochrome-mediated responses

Phototropic floral induction	Chloroplast development
Nyctinastic leaf movement	Chlorophyll synthesis
Phototropic sensitivity	Carotenoid synthesis
Seed germination	Anthocyanin synthesis
Stem elongation	Protein synthesis
Plumular hook opening	mRNA synthesis
Leaf and cotyledon expansion	Enzyme activation

mediates a wide range of other biochemical and morphological changes in higher plants (Table 6.3).

Photoconversions of Phytochrome

Phytochrome accumulates in the stable, red-absorbing, P_r form in seeds and dark-grown seedlings. Dark-grown (etiolated) seedlings have been used in nearly all phytochrome studies for two reasons. First, dark-grown seedlings accumulate large amounts of phytochrome compared with seedlings grown in light. Second, they contain no chlorophyll that absorbs blue and red light that interferes with spectrophotometric analysis of phytochrome. The main diagnostic property of phytochrome is that it exists in two, photo-interconvertible forms, P_r and P_{fr}, which have different absorbance properties. P_r has a major absorbance band at about 665 nm and a secondary maximum at about 380 nm; P_{fr} has a major peak at 730 nm and a secondary maximum at about 400 nm (Fig. 6.4). Both forms absorb weakly in the violet (390–420 nm) and blue (420–490 nm) parts of the spectrum, but these wavelengths are not in the range of physiological interest attributed to phytochrome. Neither form of phytochrome absorbs green (490–540 nm) light, which is therefore commonly used as a 'safe' light for phytochrome studies; in some very sensitive responses, mentioned later, not even green light is safe.

When radiation of any wavelength is absorbed by either P_r or P_{fr}, it drives the phototransformation of the molecule to the alternative form. A pulse of red light converts P_r to P_{fr}, while a pulse of far-red light immediately following the red-light treatment converts the pigment back to the P_r form. Far-red light thus negates the effect of the red light and cancels any physiological response. The absorption spectra for P_r and P_{fr} have a considerable overlap in the red part of the spectrum – the spectral range of physiological interest (Fig. 6.4).

Because red light is absorbed by both forms, when P_r molecules are irradiated with red light most of them will absorb it and be converted to P_{fr}, but some of the P_{fr} molecules will also absorb the red light and be converted back to P_r. In other words, due to the overlap, it is not possible to convert 100% of the pigment to either form. Instead, irradiation with visible light establishes a photoequilibrium: a photostationary state in which photoconversion takes place simultaneously in both directions and the rate of conversion of P_r to P_{fr} is balanced by the reverse reaction.

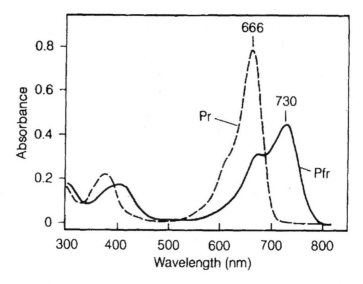

Figure 6.4   The absorption spectra of phytochrome. (From Vierstra and Quail 1983)

Table 6.4   Typical phytochrome photoequilibra (Φ) established by blue, red and far-red light

Light	λ (nm)	Φ
Blue	450	0.4
Red	660	0.8
Far-red	720	0.03
Far-red	756	0.01

Photoequilibrium (Φ) is a dynamic process involving a constant cycling between P_r and P_{fr}; it is dependent on the wavelengths of light applied and is given by

$$\Phi = \frac{[P_{fr}]}{[P_{tot}]}$$

where $P_{tot} = [P_r] + [P_{fr}]$. The proportion of phytochrome present in the P_{fr} form after saturation with red light (660 nm) is approximately 80% of the total phytochrome, i.e. $\Phi_r = 0.8$; the very small amount of far-red (720 nm) absorbed by P_r achieves a photoequilibrium of 97% P_r and 3% P_{fr} i.e. $\Phi_{fr} = 0.03$ (Table 6.4).

Since phytochrome responses are induced by red light, they could, in theory, be due to either the appearance of P_{fr} or the disappearance of P_r. A quantitative relationship has been shown to exist between the extent of the physiological response and the amount of P_{fr} generated by irradiation; no such relationship has been established for the amount of P_r generated on irradiation. Evidence such as this has led to the conclusion that P_{fr} is the physiological form. Where a quantitative relationship between the amount of P_{fr} generated and the magnitude of the response does not hold, it is suggested that the ratio of P_{fr} to P_r determines the response.

The phototransformations of $P_r \rightarrow P_{fr}$ and of $P_{fr} \rightarrow P_r$ are not one-step processes: a number of short-lived intermediates exist between the two (Kendrick and Spruit 1977). There is convincing evidence for the presence of multiple discrete species of phytochrome with different apoproteins harbouring the same chromophore, which may well interact to drive diverse plant responses (Kendrick and Kronenberg 1994). Photoconversion of the phytochrome molecule can be separated into two phases – an initial nanosecond photochemical reaction which leads to a rearrangement of the chromophore and a second millisecond reaction which results in a conformational change in the protein component. At low temperatures, the intermediates formed on irradiation can be stabilized and identified on the basis of their absorption spectra. A different set of intermediates occurs in the conversion of $P_r \rightarrow P_{fr}$ compared with those from the conversion of $P_{fr} \rightarrow P_r$ (Rüdiger 1986). When irradiated with the full range of visible wavelengths, both P_r and P_{fr} are excited and phytochrome cycles continuously between these two forms. Under such visible light conditions the intermediate forms of phytochrome accumulate and make up a considerable steady-state fraction of the total phytochrome and contribute to the photoequilibrium established under continuous illumination. The role of these intermediates is not known. It is possible that, if they accumulate to physiologically significant levels, they may play a role in initiating or amplifying phytochrome responses under conditions of natural sunlight.

Regulation of Phytochrome

Light is not the only factor regulating the interconversions of P_r and P_{fr}. Both *in vivo* and *in vitro*, the P_{fr} form can spontaneously revert to P_r in darkness, in a reaction called dark reversion (Fig. 6.5). The rate of the reaction is both pH- and temperature-dependent. Phytochrome extracted from cereals exhibits dark reversion only if it is already partly degraded by proteases; *in vitro* reactions are accelerated by the addition of reducing agents and can also be induced by interaction with certain monoclonal antibodies (Lumsden *et al.* 1985). It is possible that dark reversions *in vivo* are also dependent on partial proteolysis. Dark reversion is of particular interest in photoperiodism where the timed disappearance of P_{fr} may have a role in the regulation of flowering. However, the significance of reversion is still a puzzle as some dark-grown members of the Gramineae show photoperiodic responses even though reversion does not occur.

Both synthesis and destruction are major factors in determining the biological effects of phytochrome. In etiolated seedlings, phytochrome is synthesized in darkness as P_r until synthesis is balanced by destruction. Because of the relatively long half-life (approximately 100 h) P_r builds up to a high level and is practically stable in darkness, especially in meristematic cells. Both forms of phytochrome are degraded by proteolytic enzymes and are broken down in the normal turnover

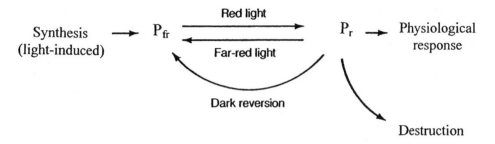

Synthesis   ⟶ P_{fr} ⇄ P_r ⟶ Physiological
(light-induced)

Red light

Far-red light

Dark reversion

Destruction

response

Figure 6.5 A summary of the process regulating the turnover of phytochrome in cells. (From Taiz and Zeiger 1991)

process in the cell. Exposure to light and consequent photoconversion of P_r to P_{fr} accelerates the breakdown of phytochrome, suggesting that P_{fr} is more prone to proteolysis than P_r. Thus, the same light stimulus that converts P_r to P_{fr} and hence accelerates its degradation also decreases its synthesis. While the mechanism responsible for the rapid destruction of phytochrome is not known, some evidence (Shanklin *et al*. 1987; Cherry and Vierstra 1994) suggests that it occurs through a process known as *ubiquitination*. Ubiquitin is a small polypeptide (76 amino acids) found in nearly all eukaryotes. It binds to proteins by means of peptide bonds, thus tagging them for destruction by proteases in the cell. Ubiquitin has been found attached to some of the P_{fr} forms of phytochrome after etiolated tissues have been exposed to red light. Attachment of ubiquitin to P_{fr} or to other proteins, targets them for degradation in a process that requires ATP and at least three enzymes. Proteases can then recognize and hydrolyse the targeted protein and release free ubiquitin. The proteolytic breakdown and disappearance of P_{fr} by ubiquitination is quite distinct from dark reversion. A summary of the regulation of phytochrome turnover is given in Fig. 6.5.

The Phytochrome Family

Most of the biochemical information on phytochrome has resulted from studies on etiolated cereal seedlings. Methods of purifying phytochrome from plant tissues were developed through research pioneered by H.W. Siegelman. Some initial success was reported by Siegelman and Firer (1964) who purified phytochrome from dark-grown oat seedlings. Although on a fresh-weight basis there is about 50 times more phytochrome in etiolated tissues than in the equivalent green tissue, the overall amount in etiolated tissues is still only about 0.2% of the total extractable protein. Thus, a minimum of a kilogram fresh weight of plant material is required to purify even milligram quantities of phytochrome. Etiolated tissues are also rich reservoirs of proteases and phytochrome is susceptible to proteolysis. To minimize enzymatic breakdown, extractions are best performed rapidly, in safe green light, at low temperatures and in buffer solutions containing protease inhibitors.

Assay methods for phytochrome were developed in parallel with purification techniques. The ratiospec method formed the basis for a dual-wavelength spectrophotometric assay of total phytochrome and of the proportion present in the physiologically active P_{fr} form. In this technique, the difference in absorbance at two, preselected, wavelengths (usually 660 and 720 nm) is measured before and after irradiation of the tissue sample with alternating R and FR light. The relative amounts of phytochrome can be calculated from the reversible change in absorbance which is a function of the concentration of the phytochrome present. This method is highly reliable for dark-grown tissues but has limited usefulness for green tissues where chlorophyll, which absorbs strongly in the red part of the spectrum, interferes greatly with the assay. Alternative methods were needed to study phytochrome in light-grown, green tissues, the normal condition of plants. Immunochemical methods were developed since phytochrome is a protein molecule capable of eliciting antibody production. However, when the antibodies raised against phytochrome extracted from dark-grown plants were tested against phytochrome extracted from light-grown plants, little or no reaction was obtained (Shimazaki et al. 1983; Shimazaki and Pratt 1985; Thomas et al. 1984; Tokuhisha et al. 1985). This led to the conclusion that the phytochrome protein in light-grown plants was not the same as that from dark-grown plants, even though they are spectrophotometrically (i.e. in their R/FR reversibility) similar. This finding prompted a wide range of investigations into the differences between phyto-chromes from light-grown and dark-grown plants (Jordan et al. 1986; Furuya 1989). These studies were greatly helped by the availability of *monoclonal antibodies* against phytochrome. Monoclonal antibodies are clones of single antibodies which recognize specific chemical groups (*epitopes*) within a macro-molecule (*antigen*) to which a given antibody binds. Specific monoclonal antibodies recognize different epitopes and thus provide molecular probes for comparing proteins from different sources. From such studies it became evident that there are antigenic differences between phytochromes extracted from light-grown and dark-grown tissues and that, in fact, a number of different phytochromes exist (Furuya 1989; Wang et al. 1991; Pratt 1995; Tokuhisha and Quail 1989; Cordonnier 1989). Confirmation of the existence of a range of phytochromes came from phytochrome gene studies. Southern blot analysis (a technique for detecting specific DNA sequences) of *Arabidopsis* genome cDNA has identified five genes encoding phytochrome. The cDNAs of three of them, designated PhyA, PhyB and PhyC have been isolated and sequenced (Sharrock and Quail 1989; Quail 1991; Vierstra 1993); later, two other genes, PhyD and PhyE, were isolated and sequenced (Quail 1991; Clack et al. 1994). Phytochrome genes have now been isolated and sequenced from a number of species, including monocotyledons, dicotyledons, ferns, mosses and algae. That five distinct genes for phytochrome have been found in *Arabidopsis*, a plant with one of the smallest and least complex genomes, suggests that, at least in angiosperms, all species will have a family containing at least five genes. If the expressed products of these genes are different, it is possible that each specific phytochrome may function differently in different physiological responses.

Phytochrome Structure

The chromophore of phytochrome is an open-chain tetrapyrrole similar to the photosynthetic phycobilin pigment found in red algae and cyanobacteria. The chromophore (*prosthetic group*) is attached via a sulphur bridge to a cysteine residue in the protein (*apoprotein*) part of the phytochrome molecule (Fig. 6.6). Phytochrome is believed to exist *in vivo* and *in vitro* as a homodimer complex with one chromophore per monomer. Published values for the molecular mass of phytochrome have varied considerably, due largely to the difficulty in isolating the protein in undegraded form. The molecular mass is in the range 120 kDa (courgette) to 127 kDa (maize). The molecular mass of oat phytochrome, which has been the most intensively studied, is 124 kDa (reviewed by Vierstra and Quail 1986). The amino acid sequences of phytochrome from several species have been deduced by cDNA sequencing techniques (Hershey *et al.* 1985; Sharrock and Quail 1989) and all have been found to include a unique 11-amino acid sequence at the NH_2-terminal of the protein (Fig. 6.6).

The protein folds loosely into three domains, with the chromophore attached to a 10 kDa hydrophobic region in the amino-terminal domain. In solution, P_r aggregates to form a dimer and this interaction occurs through the 55 kDa carboxy-terminal domain of the protein. The third domain (chromophore domain) consists of a residual 60 kDa core which is protease-resistant and contains the single chromophore (Vierstra *et al.* 1984; Jones *et al.* 1985). The absorption of light by the chromophore (not the protein) results in phytochrome-mediated responses. It is generally agreed that P_r is biologically inactive and that photoconversion to P_{fr} brings about the physiological responses. Photoconversion of P_r to P_{fr} by red light causes a *cis–trans* isomerization of the chromophore (Rüdiger 1986). The absorption of red light provides the energy necessary to drive the rotation around the double bond; the product P_{fr} is stable because of the high activation energy required to reverse the reaction. Alteration of the chromophore then causes subtle changes in the conformation of the apoprotein within the range of the chromophore domain (Song and Yamazaki 1987). The changes induced in the structure of the protein are important for the physiological activity of P_{fr}. Several lines of evidence

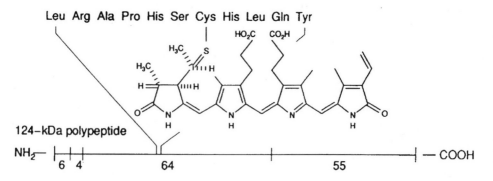

Figure 6.6 Structure of the 124 kDa *Avena* phytochrome molecule. (From Vierstra and Quail 1986)

suggest that the protein undergoes conformational changes. Phytochrome contains a higher proportion of hydrophobic amino acids than is typical of normal water-soluble proteins. It is thought that photoconversion of P_r to P_{fr} exposes more of the hydrophobic region of the molecule, which is consistent with the concept that P_{fr} functions in association with membranes. In addition, removal of the 6 kDa and 4 kDa fragments from the amino-terminal of the protein induces several changes in the molecule – a shift of approximately 8 nm in the absorption maximum of P_{fr}, a 10-fold increase in the rate of reversion to P_r and a decrease in the quantum yield for the photoconversion of P_r to P_{fr}. The amino acid side-chains of the P_r and P_{fr} forms are differentially reactive to various reagents. For example, the amino-terminal domain is more susceptible to proteolytic cleavage and more accessible to antibody binding when the pigment is in the P_r form than when it is in the P_{fr} form. On the other hand, the carboxyl domain of the protein is more susceptible to proteolysis when the pigment is in the P_{fr} form. The conformation changes induced by photoconversion must be slight, since they cannot be detected by methods generally used to distinguish between proteins – spectral assays, sedimentation coefficients, electrophoretic mobility. Immunological studies have also been used to show that P_r and P_{fr} differ conformationally. Results from such studies support the hypothesis that conformational changes occur in specific regions of the phytochrome protein as a result of photoconversion. Monoclonal antibodies have also been useful for comparisons of phytochromes from different species.

The study of phytochrome in green plants is hampered by interference from chlorophyll absorption and fluorescence in the red and far-red. Recently it has been found possible to circumvent this difficulty by using chemicals such as Norflurazan. Such herbicides inhibit carotenoid biosynthesis and as a result permit photo-bleaching of chlorophyll. The phytochrome in the resulting chlorophyll-free tissues can then be estimated spectrophotometrically *in vivo*. Findings from such studies have shown that phytochrome from light-grown *Avena* tissue is smaller (molecular mass 118 kDa) and has a shorter P_r absorption maximum (652 nm) compared with phytochrome from etiolated tissue (124 kDa; 666 nm). Furthermore, phytochrome from light-grown *Avena* seedlings can neither be immunoprecipitated nor recognized on immunoblots by antibodies raised against phytochrome from etiolated seedlings. The kinetic measurements for the photoconversions $P_r \leftrightarrow P_{fr}$ appear to be similar, but P_{fr} in light-grown tissue has a longer half-life. For example, the half-life of P_{fr} in seedlings of *Zea mays* treated with herbicide SAN 9789 (which inhibits chlorophyll biosynthesis in light) is about 8 h under continuous light, compared with 1.0–1.5 h in etiolated seedlings (Jabben and Deitzer 1979).

Phytochrome-mediated Responses

Phytochrome-mediated responses may be distinguished by the amount of light required to cause the response. The amount of light required is referred to as photon fluence and is measured in mole-photons per square metre (mol-photons m^{-2}). The photon fluence of light given to a plant is a function of the photon flux (mol-photons $m^{-2}s^{-1}$), designated N, and the irradiation time (t). Light responses can be divided into two categories – inductive responses and high-irradiance responses. Inductive responses usually require low levels of red light, are repeatedly R/FR reversible and show reciprocity between actinic intensity and duration of irradiance (Pratt 1979). Such responses are designated 'inductive' because P_{fr} continues to accumulate in darkness after the irradiation treatment has ceased. In some cases, reversibility is not shown because the amount of P_{fr} required to saturate the response is very low and sufficient P_{fr} is formed even on far-red exposure. High-irradiance responses (HIR), on the other hand, in etiolated plants, require long irradiation exposures (hours) to far-red (FR) or blue (B) and lack reciprocity between irradiance and duration of exposure.

For every phytochrome response there is a characteristic range of photon fluences over which the magnitude of the response is proportional to the photon fluence. Responses mediated by photon fluences of red light in the range 10^{-6} to $10^{-3}\,\mu mol\,m^{-2}$ (i.e. at least 100 times less than that required to induce a measurable conversion of P_r to P_{fr}) are classified as very-low-fluence responses (VLFR); low-fluence responses (LFR) require red light fluences in the range 1 to $10^{-3}\,\mu mol\,m^{-2}$; high-irradiance responses (HIR) require prolonged exposure to high irradiances of far-red or blue light greater than $10\,mmol\,m^{-2}$. HIRs are R/FR irreversible and may be more efficiently induced by far-red light (Terzaghi and Cashmore 1995).

Very-low-fluence responses

Red light in the range 10^{-6} to $10^{-3}\,\mu mol\,m^{-2}$ fluence range can affect mesocotyl and coleoptile growth in dark-grown oat seedlings. Such sensitive responses can be fully induced by low far-red fluences or even by a few seconds' exposure to the green 'safe light' that is universally used in photomorphogenetic work. Mandoli and Briggs (1981) germinated and grew oats in darkness for 4.3 days and then illuminated the seedlings with red, green and far-red light. All wavelengths applied inhibited elongation of the mesocotyl and promoted growth of the coleoptile; red light was by far the most effective in both responses. These results clearly underline the need to be aware of whether dark controls in photomorphogenesis work are cultured in 'true darkness' or under 'green safe-light' conditions. The amount of light needed to induce VLF responses converts less than 0.02% of the total phytochrome to P_{fr}. Since the far-red light that would normally reverse a red light effect converts 97% of the P_{fr} to P_r, about 3% of the phytochrome remains as P_{fr}, which is significantly more than is needed to induce VLF responses (Mandoli and Briggs 1981). Thus, far-red light cannot create VLF responses. It is the red-light action spectrum, rather than photoreversibility, that suggests that phytochrome is the photoreceptor for VLF responses.

Low-fluence responses

Other inducible responses cannot be initiated until photon fluence levels reach $1.0\,\mu mol\,m^{-2}$. These are referred to as low-fluence responses (LFR) and the fluence

Table 6.5 Selected photoreversible responses induced by phytochrome in a variety of plants

Group	Genus	Stage of developmen	P_{fr} effects
Angiospermae	*Lactuca* (lettuce)	Seed	Promotes germination
	Avena (oat)	Seedling (etiolated)	Promotes de-etiolation (e.g. leaf unrolling)
	Sinapis (mustard)	Seedling	Promotes formation of leaf primordia
	Pisum (pea)	Adult	Inhibits internode elongation
	Xanthium	Adult	Inhibits flowering (photo-periodic response)
Gymnospermae	*Pinus* (pine)	Seedling	Enhances rate of chlorophyll accumulation
Pteridophyta	*Onoclea* (sensitive fern)	Young gametophyte	Promotes growth
Bryophyta	*Polytrichum*	Germling	Promotes plastid replication
Chlorophyta	*Mougeotia*	Mature gametophyte	Promotes chloroplast orientation towards dim light

of red light required is in the range $1–10^3\,\mu\text{mol}\,\text{m}^{-2}$. For a LFR to occur, a high level of the physiologically active form P_{fr} is required, but only for a relatively short period of time. LFRs include most of the classical photoreversible responses, such as the promotion of lettuce seed germination, listed in Table 6.5.

Red/far-red photoreversibility with short-period exposures is characteristic of LFR and is the key test for phytochrome involvement in any physiological reaction. LFRs exhibit reciprocity between duration and fluence rate. The saturation of the response in the LFR range of red light corresponds well with the theoretical values (based on the absorption characteristics of P_r and P_{fr}) of the photon fluence required to reach photoequilibrium. LFRs are not dependent on fluence rate once equilibrium is established. In LFRs the level of response with far-red, either alone or as the last irradiation in sequence with red, is typically higher than in dark controls. In other words, far-red never reaches 100% photoreversibility; this is due to the overlap of the absorption spectra for P_r and P_{fr} (Fig. 6.4). The responses induced by LFRs are diverse and range in time from a few seconds to days and weeks. The very rapid responses constitute modulations of the response itself to R and FR, e.g. the orientation movement of the chloroplasts of *Mougeotia* (Serlin and Roux 1984). In contrast, slower responses such as the induction of seed germination, vegetative/floral transition and de-etiolation, result in an irreversible changed state.

High-irradiance responses

Since the 1950s it has become apparent that the action spectra for certain responses in dark-grown seedlings differ greatly depending on whether exposure to light is short (< 5 min) or long (several hours).

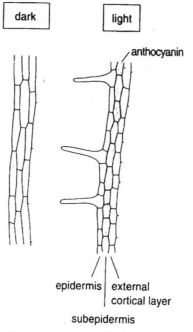

Figure 6.7   Outer layers of cells of mustard seedling hypocotyl: *left*, dark-grown; *right*, 24 h in far-red light. (From Mohr and Schopfer 1995)

On short exposures to red light, sub-epidermal cells of white mustard seedlings (*Sinapsis alba*) synthesize a purple pigment, anthocyanin, and exhibit a peak near 660 nm. Like other responses of etiolated seedlings, the initiation of anthocyanin synthesis is a classic phytochrome-dependent (P_{fr}) LFR. The red–far-red photo-reversibility, however, is confined to brief irradiations. When exposures of several hours are applied, the sub-epidermal cells of mustard hypocotyls synthesize large amounts of anthocyanin (Fig. 6.7); the action peak is shifted to the far-red (725 nm) region and a smaller peak in the blue region (450 nm), with reduced effectiveness in the red (Fig. 6.8). The effect of prolonged far-red irradiation is believed to be due to a low level of P_{fr} over time – long enough to avoid rapid degradation of the P_{fr}. Also, when etiolated lettuce seedlings are irradiated for long periods, hypocotyl elongation is inhibited and again the action spectrum shows a peak in the far-red region, with a lesser peak in the blue and UV-A regions and none in the red region where the P_r absorbs. Plant responses that have an absolute requirement for, or are strongly enhanced by long periods of high irradiance levels, show action spectra that are not typical of commonly observed phytochrome responses.

In the natural environment plants are exposed to long periods of sunlight. Under such conditions, photomorphogenic phenomena such as leaf expansion and stem elongation are more striking. Such light-dependent responses are called high-irradiance responses (HIR) (Shropshire 1972). The characteristics of HIR (Mancinelli 1980) include (a) irradiance dependency – full expression of the response requires long exposure to high levels of irradiance; (b) lack of reciprocity

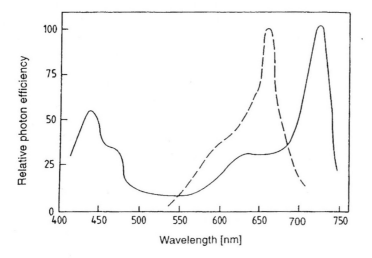

Figure 6.8   Typical steady-state action spectrum of photomorphogenesis in continuous light: the broken line represents the action spectrum for the same photoresponse under induction conditions. (From Mohr and Schopfer 1995)

Table 6.6   Some plant photomorphogenic responses caused by high irradiances (HIR)

Anthocyanin synthesis in various dicot seedlings
Inhibition of hypocotyl elongation in mustard, lettuce and other seedlings
Induction of flowering in *Hyoscyamus niger*
Opening of plumular hook in lettuce
Enlargement of cotyledons in mustard
Production of ethylene in sorghum

---

between irradiation and duration of light treatment – the magnitude of the response is a function of the fluence rate; (c) action maxima in the far-red and blue parts of the spectrum (Mancinelli and Rabino 1978); (d) in contrast to LFRs, HIRs are not fully R/FR photoreversible. Whereas most phytochrome responses are saturated at fluence levels of red light of the order of 10^{-3} mol m^{-2}, HIRs require at least 100 times more energy. However, Smith and Whitelam (1990) suggest that the term HIR is somewhat misleading, since the fluence level required to produce these responses is very much less than that provided by full sunlight. Some of the plant photomorphogenic responses that occur due to high irradiances are listed in Table 6.6.

HIRs have been implicated in a wide range of responses that also qualify as LFRs. For seedlings in general, LFRs and HIRs act synergistically to give the same response (e.g. de-etiolation), though they may be separated in time. In contrast, in seeds, HIR inhibits germination – it is antagonistic to the LFR-inductive reaction.

HIRs exhibit strikingly different action spectra depending on the species and growth conditions. For example, etiolated seedlings respond to blue, red and far-red light. After de-etiolation there is a shift from the FR-sensitive HIR to the R-sensitive HIR; on the basis of phytochrome alone it is difficult to explain the role of blue light in all HIRs. In a review of HIRs, Mancinelli (1980) concluded that, based on action spectra, at least three classes of HIRs exist. In one kind, there is a peak in a single spectral region, usually B/UV-A. Examples of this action spectrum include promotion of anthocyanin synthesis in dark-grown sorghum seedlings, unrolling of leaves of rice seedlings and coiling of tendrils in pea seedlings and phototropisms. In a second kind of action spectrum, peaks occur in two spectral regions, usually B/UV-A and R. These responses are seen in seedlings that have either been grown continuously in light or for some time in darkness followed by greening and de-etiolation in light. In a third type of action spectrum, three spectral regions show activity, namely B/UV-A, R and FR. This response is characteristic of many etiolated seedlings. In order to explain these differences, a possible conclusion is that there is an interaction in the B region between phytochrome and a B/UV-A-absorbing photoreceptor such as cryptochrome; in some cases, chlorophyll molecules may also act as photoreceptors in these responses (Mancinelli 1980). Mohr (1986) concluded that, for several responses in etiolated angiosperm seedlings, activation of cryptochrome (usually to bring about a HIR) is necessary for seedlings if they are to become competent to respond to red light acting through phytochrome. There appears to be interaction between the two photoreceptors. In other words, cryptochrome allows P_{fr} to be expressed fully.

Phytochrome Genes

Convincing evidence exists for the presence of immunochemically different pools of phytochrome in green tissue (Tokuhisha and Quail 1989; Wang et al. 1991). Determination and comparison of the N-terminal amino acid sequences provides evidence for the difference in primary structure of the two phytochrome species (Abe et al. 1989; Pratt et al. 1991). Consequently, the phytochrome dominating in etiolated tissue is designated as type 1 or light-labile phytochrome and that in green tissue as type 2 or light-stable phytochrome (Abe et al. 1989; Furuya 1989). Other studies have confirmed the coexistence of type 1 and type 2 phytochromes in both etiolated and green plants, at levels reasonably constant for type 2 and variable, depending on the light conditions, for type 1 (Tokuhisha and Quail 1989). Type 2 from green tissues is slightly smaller (118 kDa) than that from etiolated tissues (124 kDa) and has spectral properties similar but not identical to those of type 1. One important spectral difference between types 1 and 2 is that the P_r form of type 2 has an absorption maximum near 654 nm, as against 666 nm for type 1 P_r.

The presence of multiple phytochrome genes has been demonstrated in several plant species, including *Arabidopsis*, pea, tomato, potato, rice, oat (Pratt 1995). The phytochrome gene family of *Arabidopsis* is by far the best characterized with cDNA/genomic clones for five genes: PhyA, PhyB, PhyC, PhyD and PhyE (Clack *et al.* 1994; Cowl *et al.* 1994). The deduced amino acid sequences for these multiple molecular species exhibit 46–56% amino acid homology except for PhyB and PhyD, which share 80% amino acid identity (Clack *et al.* 1994). Consequently, these five genes are assigned to four subfamilies – PhyA, PhyB (which includes PhyD), PhyC and PhyE (Pratt 1995). Based on the sequence homology with previously determined type I phytochrome cDNA/genomic sequences, PhyA has been suggested to encode 'light-labile' or type I phytochrome, while PhyB, PhyC, PhyD and PhyE contribute to the stable pool of phytochrome. The deduced amino acid sequence of PhyB shows high-sequence homology with a partial amino acid sequence of pea type II phytochrome peptide fragments, suggesting further that PhyB may encode 'light-stable' phytochrome species (Tomozawa *et al.* 1990). As an initial step towards analysing the differential activities of multiple phytochrome species, their light-regulated, organ-specific and development-dependent expression was investigated. While PhyA is down-regulated by light, all other Phy genes are expressed in a light-independent manner. Phy genes do not differ greatly with respect to their spatial or temporal expression patterns at the transcriptional level (Clack *et al.* 1994); promoters from PhyA and PhyB genes also drive more or less similar organ-specific and stage-specific GUS expression in histochemical assays performed on transgenic plants (Adam *et al.* 1996; Sommers and Quail 1995). In contrast, the two members of the PhyB subfamily of tomato have been shown to be expressed in a distinctive organ-specific and mutually independent manner that may be important for defining their individual intracellular roles (Pratt 1995).

The structure–function relationships of the encoded polypeptides have been analysed through overexpression of intact and truncated versions of these genes in transgenic plants (Cherry and Vierstra 1994). Progress has been made in defining various functional domains in phytochrome apoproteins, particularly PhyA, which may mediate the dual functions of the photoreceptor, namely (a) photosensory, involving perception of light signals, and (b) regulatory, involving downstream transduction of the signals perceived (Boylan *et al.* 1994). However, despite delineation of several important regulatory sites on phytochrome apoproteins, no clue has emerged to a plausible biochemical mechanism of phytochrome action. An absence of any sequence similarity between phytochromes and other proteins currently available in the database raises the possibility of some novel, hitherto unidentified, mechanism of phytochrome action (Khurana *et al.* 1998).

Phytochrome Mutants

Analyses of photobiological responses of plants either overexpressing or lacking a particular phytochrome species have helped to determine (a) whether multiple phytochrome species govern independent sets of responses or they have overlapping roles, and (b) whether they act through common or distinct transduction pathways. Transgenic plants overexpressing cDNA clones for PhyA (Boylan and Quail 1989; Clough *et al.* 1995) or PhyB (Wagner *et al.* 1991) have been generated and studied for alteration in their photobiological responses. These plants contain increased levels of transgenically expressed phytochrome apoprotein and high levels of spectrophotometric activity, implying functional integrity of the overexpressed photoreceptor species. At the phenotype level, the host plants overexpressing PhyA exhibit light-dependent dwarfism, darker green leaves and reduced apical dominance (Boylan and Quail 1989; Keller *et al.* 1989). The assignment of individual phytochromes to specific responses in VLFR, LFR and HIR conditions has been achieved in some cases, largely by physiological and molecular analysis of the overexpression of individual phytochromes in mutant and wild-type *Arabidopsis* (McCormac *et al.* 1993; Quail *et al.* 1995; Quail 1997). For example, PhyA mediates hypocotyl growth inhibition in FR-HIR, whereas PhyB is responsible for the same response in R-HIR; PhyA exclusively mediates seed germination in VLFR conditions and PhyB is exclusively responsible for this response in LFR conditions (Shinomura *et al.* 1996). Other responses, however, involve more than one phytochrome: in some cases they appear to interact additively while in other cases it is antagonistically (Quail *et al.* 1995; Smith 1995). Hence, even though PhyA and PhyB have similar light-absorption profiles, they mediate responses in different light conditions. This is a consequence of additional properties such as their relative abundance and PhyA lability. In the natural environment, the ratio of R to FR light is likely to be the principal determinant of the relative capabilities of PhyA, PhyB and the other phytochromes to mediate physiological responses (Smith 1995).

In order to understand the molecular basis of light-regulated plant development, it is essential to identify all of the components involved in light-signal perception, its transduction and its final manifestation in the whole spectrum of photomorphogenic responses. Mutants impaired at any step in this information flow are aberrant in their photomorphogenic development. For example, some photoreceptor mutants may be deficient in virtually all molecular species of phytochrome due to a defect in the chromophore biosynthetic pathway while others may lack or have impaired individual molecular species due to a mutation in the apoprotein moiety. Detailed analysis of these mutants has reaffirmed the roles of PhyA and PhyB as inferred from overexpression studies. Photomorphogenic mutants have been raised for a wide variety of plant species, including monocots and dicots. The small weed *Arabidopsis thaliana* has been the plant of choice because of its small size, short life-cycle (less than four weeks), small genome size (100 Mb) and ease of mutagenesis (due to low repetitive DNA content, *ca.* 10%; Meyerowitz 1994). Photomorphogenic mutants characterized to date can be broadly classified into two phenotypic classes: (i) mutants showing dark-grown features when grown in light – mainly photoreceptor mutants, and (ii) mutants showing light-grown features when grown in darkness – mainly mutants impaired in signal transduction steps.

A mutation in any photoreceptor fragment is likely to make the plant blind to the quality of radiation sensed by the photoreceptor, thus resulting in lack of elicitation of corresponding photomorphogenic responses. The most dramatic effects of light can be visualized in young seedlings: stem growth inhibition, leaf differentiation and expansion, chloroplast development and chlorophyll accumulation. Stem growth inhibition is the most striking response mediated by phytochrome (and cryptochrome), making it the most promising screening parameter for identifying mutants impaired in photoreception or signal transduction steps. Photomorphogenic mutants have been selected that are impaired in the hypocotyl growth inhibition response to white light and show an etiolated apparance. They have been designated variously as: *hy* (long hypocotyl), *fre* (far-red elongated) and *fhy* (far-red elongated hypocotyl) in *Arabidopsis*; *lh* (long hypocotyl) in cucumber; *au* (*aurea*), *tri* (temporarily red-insensitive), *fri* (far-red-insensitive) and *yg-2* (yellow green-2) in tomato; *ein* (elongated internode) in *Brassica* (Kendrick *et al.* 1994; Koorneef and Kendrick 1994; Whitelam and Harberd 1994). The phenotypic aberrations exhibited by these mutants have been attributed to defects in chromophore biosynthesis, specific Phy genes and in positive regulatory loci encoding signal transduction components.

Mutations in *hy1*, *hy2* and *hy6* of *Arabidopsis* (Koorneef *et al.* 1980; Chory *et al.* 1989; Parks and Quail 1991), *au* and *yg-2* of tomato (Koorneef and Kendrick 1994; van Tuinen *et al.* 1996) have been associated with a defect in chromophore biosynthesis. These mutations exert pleiotropic effects on adult plant morphology, which include delayed seed germination, elongated appearance, increased apical dominance, yellow–green leaves, reduced Chl *a*/*b* ratio and reduced granal stacking (Koorneef *et al.* 1980, 1985; Chory *et al.* 1989).

Mutations in PhyA gene cause an inability to perceive and respond to continuous far-red light while retaining normal responses to continuous red or white light. Consequently, PhyA mutants grown under continuous far-red light show elongated hypocotyls, unopened cotyledonary hook and unexpanded cotyledons which are otherwise characteristic of dark-grown wild-type seedlings. This group is represented by *hy8* (Parks and Quail 1993), *fre1* (Nagatani *et al.* 1993) and *fhy2* (Whitelam *et al.* 1993) mutants of *Arabidopsis* and the *frei* mutants of tomato (van Tuinen *et al.* 1995). Physiological studies on wild-type plants (Smith and Whitelam 1990) and PhyA mutants (Nagatani *et al.* 1993; Parks and Quail 1993; Whitelam *et al.* 1993) clearly predict the role of PhyA to be the mediation of FR-HIR. As a result, PhyA mutant seeds fail to germinate under continuous far-red light (Johnson *et al.* 1994; Shinomura *et al.* 1994). Mutants of the PhyB group exhibit a normal hypocotyl inhibition response to FR irradiation but are impaired in red-light- as well as white-light-mediated hypocotyl growth inhibition. The PhyB *hy* mutants are exemplified by a number of mutants, such as *hy3* mutant of *Arabidopsis* (Koorneef *et al.* 1980), *lh* mutant of cucumber (Adamse *et al.* 1997), *tri* of tomato (van Tuinen *et al.* 1995). The PhyB apoprotein is severely reduced in these mutants while normal levels of PhyA can be detected. Thus the molecular aberration in these mutants is speculated to be the PhyB gene itself. The PhyB mutants of *Arabidopsis* (*hy3*) and *Brassica* (*ein*) and the putative PhyB mutant of cucumber (*lh*) are deficient in end-of-day far-red (EODFR) light-mediated stem-elongation responses and show attenuated shade-avoidance responses to low R/FR ratios (Nagatani *et al.* 1991; Whitelam and Smith 1991; Robson *et al.* 1993). In addition to mutants defective in photoreception, several mutants have been isolated that retain normal photoreception but still display aberrant photomorphogenic responses. Such mutatations can cause varying effects on photomorphogenic responses, depending on the relative position of the affected component in the signalling transduction cascade. Thus, mutations may be highly pleitropic in nature, affecting all major aspects of photomorphogenesis or specific aspects of it. Analysis of these mutants also suggests that they use common intermediates to integrate light signalling pathways with other regulatory pathways that influence the overall growth and development of plants.

Phytochrome and Changing Environments

The presence of a R/FR photoreversible pigment in plants implies that these wavelengths of light provide information that enables plants to adjust to a changing environment. The ratio of red to far-red light, designated ς varies with different environments. This ratio, which is a useful parameter for characterizing the light available to plants, is expressed as the ratio of the number of photons in the red band to that in the far-red band:

$$\varsigma = \frac{R}{FR} \quad \frac{\text{photon fluence at 655--665 nm}}{\text{photon fluence at 725--735 nm}}$$

The ratio of R to FR light differs substantially at different times of the day. At noon on a sunny day, the R:FR ratio for unfiltered daylight is typically in the range 1.05--1.25, whereas at twilight the R:FR ratio is 0.7--0.9. Similarly, the R:FR ratio for shade-light beneath a canopy differs greatly from that of unfiltered light: it is strongly shifted towards the FR and varies with the nature of the vegetation and the density of the canopy. These different ratios of R:FR light result in different ratios of $P_r{:}P_{fr}$ in the plant, with important consequences for plant development. For example, plants receiving the same total illumination, but a greater proportion of FR than R light, exhibit much greater internode elongation. Compared with direct daylight, there is relatively more FR at twilight, under 5 mm of soil and on a forest floor under the canopy of other plants. The green leaves of the canopy absorb the red and blue light because of their chlorophyll content but permit far-red light to pass through. Phytochrome can thus act as an indicator of the degree of shading of a plant by other plants, since a small change in ς would cause a relatively large change in the proportion of P_{fr} present. As shading increases, R/FR decreases and hence the ratio $P_r{:}P_{fr}$ decreases. When simulated natural radiation was used to vary the FR content, it was found that the higher the FR level (i.e. the lower the $P_{fr}{:}P_{total}$) the greater the stem elongation rate for 'sun plants' (plants that normally grow in an open habitat); in other words, simulated shading induced these plants to grow taller (Morgan and Smith 1976). However, this correlation did not hold for the woodland understorey plants that were tested. These 'shade plants' (which normally grow in a shaded environment), showed little or no extension growth when treated with higher R/FR levels. These findings suggest that there is a systematic relationship between phytochrome-controlled growth and species habitat and indicates the involvement of phytochrome in shade perception. In other words, when plants are grown for an extended period in light, development is modified by light quality, particularly by changes in the relative amounts of R and FR light brought about by leaf shading or reflectance of incident sunlight (Smith 1982). In addition to stem-extension growth these 'shade-avoidance' responses include enhanced apical dominance. Since the predominant phytochromes in light-grown tissues are stable (as P_{fr}, type 2), shade-avoidance responses are most likely regulated by the amount of P_{fr} rather than by the P_{fr}/P_{total} ratio. Some aspects of the shade avoidance response can be mimicked by giving plants a brief exposure to FR at the beginning of the daily dark period, in order to reduce the amount of P_{fr} present in darkness. This so-called 'end-of-day' treatment with FR is reversed by R, as in all inductive responses (Downs *et al.* 1957).

Light quality also plays a role in the germination of some seeds. Large seeds produced by some species contain enough reserve food material to sustain prolonged seedling growth in darkness (e.g. underground) and generally do not require light for germination. Small seeds, on the other hand, may remain dormant, even after imbibition if they are buried below the depth to which light penetrates. Even when such seeds are near the soil surface, their level of shading by the vegetation canopy (i.e. the R/FR they receive) may effect their germination; FR enrichment due to canopy density may even inhibit germination of certain small seeds.

Cryptochrome

The effects of blue light on plants were discovered in 1864 by Julius von Sachs, who measured the first crude action spectrum for phototropism using coloured glass and coloured solutions, demonstrating that wavelengths in the blue (and violet) region of the spectrum were among those that induced the response. The action spectra and possible photoreceptors for blue, violet and UV radiation have been reviewed by Senger and Lipson (1987) and Briggs and Huala (1999). A wide range of blue/UV-A responses have been known or suspected for a long time, and in 1977 Gressel coined the term *cryptochrome* to refer to the then unidentified photoreceptors. Although cryptochrome absorbs UV-A radiation (370 nm), the largest peak in its action spectrum is in the blue/violet region at around 450 nm with a shoulder at 480 nm (Fig. 6.9). Since more photons of blue and of violet light strike the plant compared with UV-photons, the responses are usually called *blue-light responses*. Action spectra and the fluence required to cause the response vary with the organism and the response.

Cryptochrome may act independently in photoresponses or sometimes the light-activated pigment may reinforce the effects of P_{fr} or of the UV-B photosensor. These three photosensors work together in photomorphogenesis:

$$\text{Phytochrome } (P_{fr}) \xrightarrow[\text{UV-B photoreceptor}]{\text{Cryptochrome}} \text{Photoresponse}$$

Phytochrome (P_{fr}) produces direct photoresponses. Light absorbed by crypto-chrome and the UV photoreceptor may determine the sensitivity of plants to P_{fr}. In this way a single effector (P_{fr}) can control gene expression and still be able to

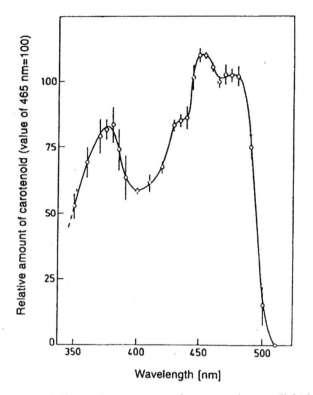

Figure 6.9 A representative action spectrum for cryptochrome (light-induced carotenoid synthesis in *Fusarium aquaeductum*). (From Rau 1976)

provide information on the entire spectrum of wavelengths that determine the extent of photomorphogenesis. The cooperation of the different photoreceptors in a photomorphogenic effect is seen in the opening of stomata in epidermal strips of *Commelina* (Fig. 8.10). Candidate chromophores for cryptochromes included carotenoids and flavins, or both. Flavins were rejected when it was found that the hypothesized differential flavoprotein-mediated destruction of auxin across a unilaterally illuminated coleoptile did not occur; carotenoids were rejected because maize coleoptiles devoid of detectable carotenoids responded to unilateral blue light with normal phototropic curvature. The cryptochrome-type action spectrum (Fig. 6.9) was used to support both the carotenoid and flavin concepts since it showed the spectral properties that were characteristic of each putative chromophore – fine structure in the blue typical of carotenoids, and a broad peak in the UV-A typical of flavins. When the issue could not be resolved biochemically, a genetic approach was used.

Cryptochrome Mutants

The long-hypocotyl mutant, *hy4*, of *Arabidopsis*, was originally isolated in a screen for mutants defective in inhibition of hypocotyl elongation in response to light

(Koorneef *et al.* 1980). The *hy4* mutants give an impaired response to blue light, a slightly inhibited response to UV light and no observed impairment to red or far-red light. Ahmad and Cashmore (1993) isolated the HY4 gene, by gene tagging, and cloned and sequenced it. The gene encodes a soluble protein of ~75 kDa with a considerable amino acid sequence similarity to prokaryotic DNA photolyases, a class of flavoproteins that catalyse the light-dependent repair of pyrimidine dimers in DNA damaged by UV light. However, subsequent work showed that the protein had no photolyase activity and contained a C-terminal extension not found in the photolyases. In addition, the 75 kDa protein is distinct from the DNA photolyases, in that a conserved tryptophan residue implicated in specific recognition of pyrimidine primers is not conserved in the protein and it also lacks DNA repair activity (Malhorta *et al.* 1995). HY4 lacks any membrane-spanning domains, implying that it is a soluble protein. It is expressed both in dark-grown *Arabidopsis* seedlings and in all organs of mature light-grown plants tested; protein levels are unaffected by white-light treatment (Lin *et al.* 1996a) and in this sense, the *Arabidopsis* chromoprotein is analogous to PHYB, which is light-stable (Quail 1991). In view of the photobiological, genetic and molecular properties of this protein, Ahmad and Cashmore (1993) concluded that the protein was the long-sought blue-light receptor and renamed the gene CRY1 (for cryptochrome). Further evidence that *cry1* may function as a photoreceptor results from the fact that several responses are elicited by green light as well as blue and UV-A light. Under redox conditions that might exist in a cell, *cry1* holoprotein forms a stable flavo-semiquinone with significant absorption in the green (Lin *et al.* 1995a). Overexpression of *cry1* in tobacco can give the seedlings an exaggerated increase in the sensitivity for inhibition of the hypocotyl, not just in the blue and UV-A but in green light as well.

Physiological analysis of *hy4* mutants suggests that CRY1 regulates disparate sets of responses which include high-fluence blue-light-mediated hypocotyl growth inhibition and anthocyanin biosynthesis, probably by a single primary mode of action (Ahmad *et al.* 1995). HY4 possibly controls multiple blue-light-mediated growth responses and is a member of a small multigene family controlling the entire range of blue-light responses in plants (Lin *et al.* 1995b).

A second gene, CRY2, encoding a CRY1-related protein with extensive similarity to *cry1* in the photolyase domain, was isolated from *Arabidopsis* (Lin *et al.* 1996b). Like *cry1*, *cry2* has a C-terminal extension that is not found in the photolyases. However, the *cry2* extension shows no similarity to that of *cry1*. Also like *cry1*, *cry2* plays a role in blue-light-induced shortening of the hypocotyl, cotyledon expansion and anthocyanin production. Transgenic *Arabidopsis* plants that overexpress either photoreceptor are hypersensitive to blue light. In addition, mutations in the CRY2 gene confer a late-flowering phenotype, observed under blue-plus-red light but not under blue light alone, apparently reflecting a repression of PHYB activity by CRY2 in wild-type plants (Cashmore 1997).

Many plant genes exhibit circadian rhythms in their expression. CRY1 mediates photoentrainment of circadian expression (Lin *et al.* 1995a). Under low-intensity blue light ($<3\,\mu\mathrm{mol\,m^{-2}\,s^{-1}}$) the period of gene expression in the *cry1* mutant is increased by about 4 hours, which indicates that CRY1 functions as a blue photoreceptor in rhythm entrainment. Surprisingly, the *cry2* mutant, which affects the sensitivity of *Arabidopsis* flowering to photoperiod does not affect rhythm entrainment.

Phototropin

In 1988, Gallagher *et al.* first reported the blue-light-activated phosphorylation of a plasma-membrane protein in etiolated pea seedlings. It is present in all plant species studied and ranges in molecular weight from 100 to 130 kDa (Reymond *et al.* 1992a). Phosphorylation occurs on multiple serine and threonine residues. The system also possesses a biochemical memory for a light pulse at least *in vitro* where light activation of phosphorylation can be detected after several minutes of darkness following the light treatment (Briggs and Huala 1999). Studies with *Arabidopsis* mutants (Liscum and Briggs 1995) provide direct genetic evidence that the phosphoprotein plays a central role in phototropism. The protein (120 kDa in *Arabidopsis*) exhibits severely reduced phosphorylation in the *nph1* mutant (Reymond *et al.* 1992b). After extensive biochemical, genetic and physiological characterization (Short and Briggs 1994) there is strong evidence that this protein is not only the photoreceptor, substrate and kinase for its own phosphorylation but a photoreceptor for phototropism as well. Direct evidence in support of this claim is provided by the cloning of the NPH1 gene (Huala *et al.* 1997) from *Arabidopsis*. The sequencing of the cDNA and genomic clones of NPH1 solved the kinase question: the *nph1* protein (996 amino acids) contains all 11 of the signature sequences for serine–threonine kinases (i.e. the conserved C-terminal motifs typical of serine–threonine protein kinases) (Hanks and Quinn 1991). Earlier, a 120 kDa phosphoprotein from maize and pea had been shown to be kinase autophos-phorylating the serine and threonine residues (Reymond *et al.* 1992a). It is likely that all of these are functionally similar proteins. The *nph1* itself is hydrophilic and lacks any membrane-spanning domains (Huala *et al.* 1997). The sequence-homology search also revealed two related domains designated LOV1 and LOV2 (for light, oxygen and voltage) in the N-terminal region. The redox status of these proteins is known to be regulated by light, oxygen and voltage and probably involves a flavin moiety (Huala *et al.* 1997). Christie *et al.* (1998) resolved the photoreceptor question by demonstrating that the *nph1* expressed in insect cells grown in the dark exhibited blue-light-activated phosphorylation with the same fluence dependence and kinetics as the native *Arabidopsis* protein; they designated the *nph1* holoprotein, *phototropin*.

nph1

Jarillo *et al.* (1998) reported an NPH1 homologue in *Arabidopsis*, designated NPL1 (NPHl-like), that encodes a protein slightly shorter than *nph1* (916 amino acids compared with 996 for *nph1*). It shows approximately 58% identity and 67%

similarity with *nph1* at the amino acid level and contains LOV1, LOV2 and kinase domains. The *npl1* protein has been shown to mediate the avoidance response of chloroplasts to high-intensity blue light in *Arabidopsis*. Like *nph1*, *npl1* binds noncovalently the chromophore flavin mononucleotide (FMN) within the two LOV domains (Saki *et al.* 2001). In addition, when expressed in insect cells, *npl1*, like *nph1*, undergoes light-dependent autophosphorylation, indicating that *npl1* also functions as a light-receptor kinase. Consistent with this is the fact that an *nph1.npl1* double mutant exhibits an impaired phototropic response under both low- and high-intensity blue light. Hence, *npl1* functions as a second photoreceptor under high-fluence-rate conditions and is, in part, functionally redundant to *nph1* (Saki *et al.* 2001).

Adiantum phy3

Nozue *et al.* (1998) identified a hybrid photoreceptor from the fern *Adiantum capillus-veneris* and designated it *phy3*. The N-terminal 566 amino acids of this protein show high homology to phytochrome. Moreover, recombinant protein, expressed in *E. coli* and reconstituted with a phycocyanobolin chromophore, shows the R- and FR-photoreversibility characteristic of phytochrome. However, downstream of a linking domain, the protein shows remarkable similarity to phototropin, containing two LOV domains and a serine–threonine domain. As with the LOV domains from *Arabidopsis* and oat *nph1*, those from *Adiantum* protein also bind FMN tightly (Christie *et al.* 1999). It will be interesting to learn if this complex three-chromophore photoreceptor undergoes light-activated phosphorylation and if so, can both blue and red light induce it?

UV-B Responses

Ultraviolet (UV) describes the spectral range of electromagnetic radiation (200–400 nm) which borders the visible range (light). UV radiation is divided into three ranges: UV-A (320–400 nm), UV-B (280–320 nm) and UV-C (200–280 nm). Less than 7% of the sun's radiation reaching the earth's surface is in the UV range and this is exclusively in the 295–400 nm range (Fig. 6.10).

All shorter wavelengths are filtered out by the ozone layer, at a height of 15–30 km. Plants have adapted to the natural UV radiation (UV-A plus the longer-wave UV-B) and important pigments such as chlorophylls, carotenoids and flavones absorb both light (visible spectrum) and long-wave UV. In contrast to this, the short-wave UV (UV-C plus the shorter-wave UV-B, 200–295 nm) cover a spectral range that is not part of natural radiation. All UV radiation of < 295 nm from artificial sources such as low-pressure mercury lamps (254 nm) usually have destructive effects on plant tissues. Short-wave UV is absorbed by functionally important aromatic compounds. The aromatic amino acids phenylalanine, tyrosine and tryptophan, and to a lesser extent, the sulphur-containing cysteine, are responsible for UV absorption by proteins at 280 nm. Purines and pyrimidines

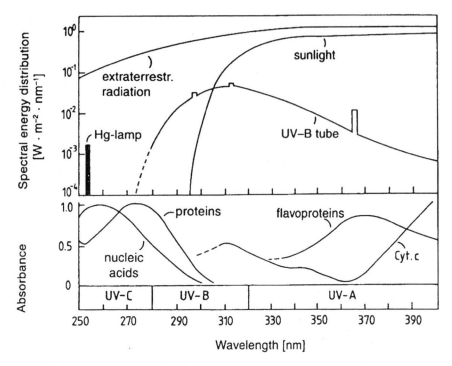

Figure 6.10   Emission spectra of UV sources and absorption spectra of some UV-absorbing molecules. (From Caldwell 1981)

absorb strongly at 260 nm and peak at this wavelength in the absorption spectrum of nucleic acids.

Absorption of UV radiation in a photochemical reaction can break covalent linkages which may lead to discrepancies in the transcription and replication of DNA with consequent occurrence of gene mutations accompanied by failure of certain vital proteins. This process is exploited for killing bacteria and other microorganisms ('sterilization') by short-wave UV (254 nm) radiation. If the lethal UV radiation at 254 nm is quickly followed by irradiation with UV-A (320–400 nm) or blue light (400–520 nm), inactivation is reversed. This remarkable phenomenon, which is called *photoreactivation*, is based on the existence of the enzyme DNA-photolyase, which splits thymine dimers in DNA. The reaction is sensitized by radiation in the range 300–500 nm (flavin as a photoreceptor) and thus recreates the original, intact structure of the macromolecule. This enzyme-repair mechanism has been found not only in bacteria but also in eukaryotes, including higher plants. Light induces the formation of the photolyase and phytochrome has been shown to be the photoreceptor; the enzyme is subject to control by blue light. Thus, plants are effectively protected by the blue/UV-A part of the daylight spectrum from lasting DNA damage resulting from short-wave UV radiation.

The positive effect of UV radiation on anthocyanin accumulation and the fact that sunlight filtered through glass (which absorbs UV rays) is less efficient than unfiltered light has been known for a long time. Wellmann (1983) showed that flavonoid biosynthesis in parsley (*Petroselinum hortense*) cell suspensions and seedlings was induced by UV-B (280–320 nm) radiation, with maximum effectiveness at 290–300 nm. By 1986, eleven species of higher plants had been listed (Beggs *et al.* 1986) in which UV-B radiation induced anthocyanin and flavonoid biosynthesis in coleoptiles, hypocotyls, seedling roots and cell cultures. After pretreatment with UV-B, phytochrome (or more correctly the level of P_{fr}) regulates flavonoid synthesis; without pretreatment, neither R/FR nor blue/UV-A is effective. Flavonoid biosynthesis in parsley apparently results from the cooperation of three pigments – phytochrome, cryptochrome and a UV-B receptor. Flavonoids, in particular, are found to accumulate in a light intensity- and wavelength-dependent manner and their level can be correlated with the extent of protection against UV radiation (Lois 1994; Lois and Buchanan 1994).

In addition to the many positive effects on photomorphogenesis, long-wave UV (295–400 nm) can lead to UV stress in plants. This form of radiation stress occurs in plants growing at high altitudes. The UV of sunlight can cause photoinhibition of the photosynthetic apparatus resulting in photobleaching and death of leaves. Strong sunlight can also cause DNA damage as nucleic acids can absorb wavelengths up to 310 nm. Also, plants grown in darkness or in UV-free (dim) white light react very sensitively to long-wave UV radiation. This forms the basis for the 'transplanting shock' suffered by young plants grown under UV-impermeable glass and then transplanted into the open; it is expressed as transient wilting, bleaching and poor development. Plants usually recover after some time and become UV-resistant. This acquired UV-resistance is due, in part, to photoreactivation and the induction of flavonoids and cuticular waxes which provide a protective filter for the assimilatory parenchyma cells beneath the epidermis.

UV-B Photoreceptors

The existence of specific plant responses to UV-B radiation raises the question of the nature of the photoreceptor. Various cellular components such as DNA, phytochromes, flavins and pterins, aromatic residues in proteins and growth regulators such as auxin, all of which have strong absorption in the UV-B region, have been considered as potential photoreceptors for UV-B-induced responses (Ballaré *et al.* 1995a). A relatively low UV-B irradiance (*ca.* 0.5% of the total photons) was shown to rapidly and effectively inhibit hypocotyl elongation in etiolated tomato seedlings of both the wild-type and a *phy*-deficient mutant, *aurea* (Ballaré *et al.* 1995b). This would appear to eliminate phytochrome as photoreceptor for this UV-B response. An action spectrum peak at 300 nm argues against DNA or aromatic residues in proteins as candidates for UV-B chromophores. The ability of potassium iodide (KI) to cancel UV-B-mediated hypocotyl growth inhibition at low concentrations (10^{-4} M) correlates with its effective quenching of the triplet excited state of flavins but not of photoexcited pterins or the singlet state of flavins (Ballaré *et al.* 1995a). This and other experimental evidence suggests that a flavin may be the chromophore for the UV-B photoreceptor in tomato seedlings. The finding that exogenously supplied flavin had an inductive effect on chalcone synthase and flavonoid biosynthesis in parsley suspension cells after treatment with UV-B light (Ensminger and Schäfer 1992) supports this hypothesis.

A genetic approach to identifying the photoreceptor of UV-B responses is still in its infancy. Mutants that are hypersensitive to UV-B treatment have been isolated and the lesions have been associated with reduced biosynthesis of photoprotective compounds such as flavonoids and phenolics (Shirley 1996; Bharti and Khurana 1997). However, mutants deficient in specific UV-B-induced photomorphogenic responses such as hypocotyl growth suppression and cotyledon curling, which would be more useful for the identification of a UV-B photoreceptor (on the basis of phytochrome and blue-light photoreceptor mutants) have yet to be found.

Cross-talk between Light and Plant-growth Regulators

Plant-growth regulators play an important role in plant growth and development and so their involvement in light-mediated morphogenic responses is to be expected. Analysis of mutants that define components responsible for coordinating/integrating the overlapping signals may articulate the cross-talk that occurs between light and plant growth regulators.

Auxins

The stem elongation due to end-of-day far-red (EODFR) treatment in peas is accompanied by a red-reversible 40% increase in epidermal IAA levels (Behringer et al. 1992). This is corroborated by the presence of abnormally high epidermal IAA levels in the PHYB-deficient lv mutant of pea, which otherwise is typical of FR-light-treated wild-type plants. Thus, it appears that phytochrome regulation of stem elongation may occur, in part, through modulation of epidermal IAA levels. Phytochrome may regulate stem elongation rates by depleting auxin within the epidermis, which, in turn, could halt the growth of the entire stem. Work on mutants of tobacco suggests that the light-stable form of phytochrome (P_r) may play a major role in the control of IAA levels (Kraepiel et al. 1995). The antagonistic interaction of auxin and light has also been suggested by an elongated appearance of light-grown transgenic Arabidopsis plants constitutively expressing tryptophan monooxygenase (iaaM) leading to four-fold higher levels of endogenous auxin (Romano et al. 1995).

Cytokinins

Alterations in plant growth substance metabolism or responsiveness can also alter the morphology of dark-grown seedlings. The Arabidopsis amp (altered meristem programme) mutant has a short hypocotyl, produces leaves in the dark and accumulates six times as much cytokinin as wild-type plants. The amp phenotype can be mimicked in wild-type seedlings grown in the dark by the addition of exogenous cytokinins (Chaudhury *et al.* 1993; Chin-Atkins *et al.* 1996). The phenotype of *det1* mutants provides an example of how plant growth substances might act on phytochrome-mediated signal pathways. Applied cytokinin mimics the phenotypes of *Arabidopsis det* (de-etiolation) mutants causing a de-etiolated morphology, development of chloroplasts, and expression of light-regulated genes in dark-grown wild-type seedlings. Thus, in *Arabidopsis*, increased cytokinins alone can override the light requirement. It appears that cytokinins and light may act independently or sequentially through common signal transduction intermediates such as DET1 and DET2 to influence downstream light-regulated responses. The effects of cytokinin and light on hypocotyl elongation and apical hook opening have, in fact, been shown to be independent and additive (Su and Howell 1995).

Gibberellins

Gibberellins stimulate seed germination and stem elongation and are required for flowering under non-inductive short days in *Arabidopsis*. These responses are constituents of the phytochrome-mediated photomorphogenic development induced by light. In several plants, alterations in gibberellin metabolism or response can result in phenotypes that resemble the elongation and flowering phenotypes of *phyb* mutants. Increased levels of gibberellic acid (GA) have been found in the $ma_3{}^R$ mutant of *Sorghum* (Foster and Morgan 1995) and the *ein* mutant of *Brassica* (Rood *et al.* 1990) and increased responsiveness to GA in the *lv* mutant of pea (Reid and Ross 1988), the *lh* of cucumber (Lopez-Juez *et al.* 1995) and *hy3* mutant of *Arabidopsis* (Reed *et al.* 1996), all of which are deficient in phyB. Far-red irradiation of wild-type cucumber seedlings (which causes depletion of P_{fr}) results in increased GA responsiveness (Lopez-Juez *et al.* 1995) and phytochrome reduces the effectiveness of gibberellins to induce cell elongation in rice seedlings (Toyomasu *et al.* 1994). Phytochrome A overexpression in transgenic tobacco leads to a dwarf phenotype and correlates with elevated levels of P_{fr} and a reduction in endogenous GA levels. Accordingly, exogenously applied GA can eliminate the dwarfism caused by *phya* overexpression (Jordan *et al.* 1995). Analysis of GA- and *phy*-deficient double mutants have shown (Peng and Harberd 1997) that the interaction between the two systems depends on the developmental stage of the plant and that the GA-deficiency suppresses the elongated phenotype characteristic of the *phy*-deficient mutants.

Abscisic acid

Light-regulated physiological processes such as seed germination, stomatal movement and senescence are also known to be regulated by abscisic acid (ABA). It is likely that light may affect ABA levels or tissue sensitivity to ABA in the control of these processes. Kraepiel *et al.* (1994) reported increased ABA levels in a chromophore-deficient tobacco mutant, *pew1*, which is deficient in all phytochromes and exhibits a phenotype expected from an ABA overproducer or hypersensitive mutant. They concluded that phytochrome-mediated light signals caused ABA breakdown rather than inhibiting its biosynthesis. Increases in ABA levels on dark-incubation of light-grown *Arabidopsis* and down-regulation of ABA levels in *Lemna*, further suggest that phytochrome effects may be mediated by changes in ABA levels (Weatherwax *et al.* 1996). The expression pattern of a novel *Arabidopsis* intrinsic membrane-protein gene, which is specifically induced by blue light and ABA, also supports the concept of common transduction elements in ABA and blue-light signalling pathways (Kaldenhoff *et al.* 1993).

Ethylene

Certain light-regulated responses, such as inhibition of stem growth, opening of the plumular hook, promotion of seed germination and flowering can also be induced by the plant growth substance ethylene (Ecker 1995; Fluhr and Mattoo 1996). Thus it is possible that the light- and ethylene-signalling pathways may converge to mediate these responses. An ethylene-response mutant *hls1* (hookless 1) in *Arabidopsis* whose expression is up-regulated by ethylene, results in constitutive hook curvature. Experimental evidence suggests that the HLS1 gene controls differential cell growth by regulating auxin activity (Lehman *et al.* 1996). Thus, possibly ethylene and auxin together are needed to maintain an apical hook in the dark. Ethylene has also been found to stimulate hypocotyl growth in *Arabidopsis* seedlings grown in the light by promoting cell expansion, which is opposite to its inhibitory effect on this response in the dark (Smalle *et al.* 1997).

Brassinosteroids

A relatively new class of plant growth regulators, brassinosteroids, was first isolated from pollen extracts of *Brassica napus* (Mitchell *et al.* 1970). Studies on a subset of *Arabidopsis* de-etiolated mutants with a dwarf structure in the light has generated interest in this class of plant steroids and implicated brassinolide particularly in the light development of plants (Li *et al.* 1996; Szekeres *et al.* 1996). In the dark, these mutants are defective in brassinosteroid biosynthesis, show pleiotropic phenotypic aberrations including dwarfism, de-etiolation, anthocyanin accumulation and a 10–20-fold de-repression of several light-responsive genes. In the light, these mutants are smaller and darker green than wild-type plants, show reduced cell size in several tissues, have reduced apical dominance and male fertility and have altered photoperiodic responses (Chory *et al.* 1991). The mutants defined by these traits include *det2* (de-etiolation), *cpd* (constitutive photomorphogenesis and dwarfism), several *dwf* (dwarfism) and *cbb* (cabbage) lines. The DET2 gene has been cloned and found to encode a steroid 5α-reductase that catalyses the conversion of the brassinolide precursor campesterol to campestanol. Addition of exogenous brassinolide to *det2* mutants in the dark has been shown to rescue their short-hypocotyl phenotype and to suppress their dwarf phenotypes in light-grown plants (Li *et al.* 1996). The CPD gene has also been cloned and shown to encode a cytochrome P450 mono-oxygenase that has homology with steroid hydroxylases and which is also involved in brassinosteroid biosynthesis. Mutants in this gene show de-etiolation and de-repression of certain photosynthetic genes in the dark. Further analysis of these mutants is likely to reveal the nature of molecular cross-talk between light and brassinosteroid signalling pathways (Russell 1996).

Summary

Light acts as a source of information which influences the structural development of plants in the process of photomorphogenesis. To be effective, the different wavelengths of light must first be absorbed by a photoreceptor which reads the information and processes the light stimulus through a signal transduction pathway. Three main photoreceptors occur in plants – phytochrome, crypto-chrome and UV-B photoreceptors.

Phytochrome is the pigment involved in most photomorphogenic responses. Phytochrome is a dimeric protein with a molecular mass of 124 kDa. The attached chromophore is an open-chain tetrapyrrole, similar in structure to phycocyanin. Phototransformations of phytochrome involve conformational changes in both the chromophore and apoprotein. Phytochrome exists in two interconvertible forms – the red-absorbing (P_r) and the far-red-absorbing (P_{fr}) forms. Absorption of red light by P_r converts it to P_{fr} while absorption of far-red light by P_{fr} converts it to P_r. The absorption spectra overlap in the red part of the spectrum, leading to an equilibrium between the two forms. P_{fr} is the physiologically active form; a number of intermediate, short-lived forms also exist.

Both the synthesis and the degradation of phytochrome play an important role in the biological effect of the pigment. Both forms are degraded by proteolytic enzymes and are broken down in the normal turnover process in the cell; the mechanism of breakdown is thought to occur through ubiquitination. Two distinct types of phytochrome occur – Type 1 and Type 2. Type 1 predominates in etiolated tissues, whereas in green tissues the amounts of both types are about the same.

Each phytochrome response has a characteristic range of photon fluences and the magnitude of the response is proportional to the particular photon fluence. The responses fall into three main categories – very-low-fluence responses (VLFR), low-fluence responses (LFR) and high-irradiance responses (HIR). Phytochrome also plays an important role in shade detection and in the regulation of circadian rhythms in plants. In addition, phytochrome regulates the transcription of a number of genes. Red light activates the genes which encode the small subunit of RubisCO and the chlorophyll a/b protein of the light-harvesting complex. Red light may also repress the transcription of certain genes, including the gene for phytochrome itself.

Several different blue/UV-A photosensors are involved in light responses in plants; none have been clearly identified and hence the name cryptochrome. Cryptochrome absorbs blue/UV-A light, the largest peak in its action spectrum being in the blue/violet region (450 nm). Candidate chromophores for crypto-chromes are carotenoids, flavins and pterins. A genetic approach has isolated the

HY4 gene that encodes a 75 kDa chromoprotein. The gene has been renamed CRY1 (cryptochrome) and the *cry1* protein the photoreceptor. A second gene, CRY2, has also been isolated. Genetic studies revealed the NPH1 (non-phototropic hypocotyl) gene that encodes a 100–130 kDa protein *nph1* protein (phototropin), the blue-light photoreceptor for phototropism.

Experimental evidence suggests that a flavin may be the chromophore for the UV-B photoreceptor in plants.

Light signals and plant growth regulators are known to control an overlapping set of processes in developing seedlings. Mutant plants are employed to study the cross-talk between light and the well-characterized plant-growth regulators – auxins, cytokinins, gibberellins, ethylene, abscisic acid and brassinosteroids.

References

Abe, H., Takio, K., Titani, K. and Furuya, M. (1989). Amino-terminal amino acid sequences of pea phytochrome II fragments obtained by limited proteolysis. *Plant Cell Physiol.* **30**: 1089–1097.

Adam, E., Kozma-Bognar, L., Kolar, C., Schafer, E. and Nagy, F. (1996). The tissue-specific expression of a tobacco phytochrome B gene. *Plant Physiol.* **110**: 1081–1088.

Adamse, P., Jaspers, P.A.P.M., Kendrick, R.E. and Koorneef, M. (1997). Photomorphogenetic responses of a long-hypocotyl mutant of *Cucumis sativa* L. *J. Plant Physiol.* **127**: 481–491.

Ahmad, M. and Cashmore, A.R. (1993). HY4 gene of *A. thaliana* encodes a protein with characteristics of a blue-light photoreceptor. *Nature* **366**: 162–166.

Ahmad, M., Lin, C. and Cashmore, A.R. (1995). Mutations throughout an *Arabidopsis* blue-light photoreceptor impair blue-light-responsive anthocyanin accumulation and inhibition of hypocotyl elongation. *Planta J.* **8**: 653–658.

Ballaré, C.L., Barnes, P.W. and Flint, S.D. (1995a). Inhibition of hypocotyl elongation by ultraviolet-B radiation in de-etiolating tomato seedlings. I. The photoreceptor. *Physiol. Plant.* **93**: 584–592.

Ballaré, C.L., Barnes, P.W., Flint, S.D. and Price, S. (1995b). Inhibition of hypocotyl elongation by ultraviolet-B radiation in de-etiolating tomato seedlings. II. Time-course comparison with flavonoid responses and adaptive significance. *Physiol. Plant.* **93**: 593–601.

Beggs, C.J., Wellmann, E. and Grisebach, H. (1986). Photocontrol of flavonoid biosynthesis. In: *Photomorphogenesis in Plants*, R.E. Kendrick and G.H. Kronenberg (eds), pp. 467–499. Martinus Nijhoff, Dordrecht.

Behringer, F.J., Davies, P.J. and Reid, J.B. (1992). Phytochrome regulation of stem growth and indole-3-acetic acid levels in the *lv* and Lv genotypes of *Pisum*. *Photochem. Photobiol.* **56**: 677–684.

Bharti, A.K and Khurana, J.P. (1997). Mutants of *Arabidopsis* as tools to understand the regulation of phenylpropanoid pathway and UV-B protection mechanisms. *Photochem. Photobiol.* **65**: 765–776.

Borthwick, H.A. and Parker, M.W. (1952). Light in relation to flowering and vegetative development. *Proc. 13th Int. Hort. Congress*, The Netherlands, p. 801.

Borthwick, H.A., Hendricks, S.B. and Parker, M.W. (1948). Action spectrum for the photoperiodic control of floral initiation of a long-day plant, Wintex barley (*Hordeum vulgare*). *Bot. Gaz.* **110**: 103–118.

Borthwick, H.A., Hendricks, S.B. and Parker, M.W. (1951). Action spectrum for inhibition of stem growth in dark-grown seedlings of albino and non-albino barley (*Hordeum vulgare*). *Bot. Gaz.* **113**: 95–105.

Borthwick, H.A., Hendricks, S.B., Toole, E.H. and Toole, V.K. (1954). Action of light on lettuce seed germination. *Bot. Gaz.* **115**: 205–225.

Boylan, M. and Quail, P.H. (1989). Oat phytochrome is biologically active in transgenic tomatoes. *Plant Cell* **1**: 765–773.

Boylan, M., Douglas, N. and Quail, P.H. (1994). Dominant negative suppression of *Arabidopsis* photoresponses by mutant phytochrome A sequences identifies spatially discrete regulatory domains in the photoreceptor. *Plant Cell* **6**: 449–460.

Briggs, W.R. and Huala, E. (1999). Blue-light photoreceptors in higher plants. *Ann. Rev. Cell Dev. Biol.* **15**: 33–62.

Butler, W.L., Norris, K.H., Siegelman, H.W. and Hendricks, S.B. (1959). Detection, assay and preliminary purification of the pigment controlling photoresponsive development of plants. *Proc. Natl. Acad. Sci. USA* **25**: 1703–1708.

Caldwell, M.M. (1981). Plant response to ultraviolet radiation. *Encyclopedia of Plant Physiology* NS, **12A**: 169–197. Springer, Berlin, Heidelberg and New York.

Cashmore, A.R. (1997). The cryptochrome family of photoreceptors. *Plant, Cell and Environment* **20**: 764–767.

Chaudhury, A.M., Letham, S., Craig, S. and Dennis, E.S. (1993). *amp1*: a mutant with high cytokinin levels and altered embryonic pattern, faster vegetative growth, constitutive photomorphogenesis and precocious flowering. *Plant J.* **4**: 907–916.

Cherry, J.R. and Vierstra, R.D. (1994). The use of transgenic plants to examine phytochrome structure/function. In: *Photomorphogenesis in Plants*, R.E. Kendrick and G.H.M. Kronenberg (eds), pp. 271–297. Kluwer Academic Publishers, Dordrecht.

Chin-Atkins, A.N., Craig, S., Hocart, C.H., Dennis, E.S. and Chaudhury, A.M. (1996). Increased endogenous cytokinin in the *Arabidopsis amp1* mutant corresponds with de-etiolation responses. *Planta* **198**: 549–556.

Chory, J., Peto, C.A., Ashbaugh, M., Saganich, R., Pratt, L. and Ausubel, F. (1989). Different roles for phytochrome in etiolated and green plants deduced from characterization of *Arabidopsis thaliana* mutants. *Plant Cell* **1**: 867–880.

Chory, J., Nagpal, P. and Peto, C.A. (1991). Phenotypic and genetic analysis of *det2*, a new mutant that affects light-regulated seedling development in *Arabidopsis*. *Plant Cell* **3**: 445–459.

Christie, J.M., Reymond, P., Powell, G.P., Bernasconi, P. and Raibekas, A.A. (1998). *Arabidopsis* NPH1: a flavoprotein with the properties of the photoreceptor for phototropism. *Science* **282**: 1698–1701.

Christie, J.M., Salomon, M., Nozue, K., Wada, M. and Briggs, W.R. (1999). LOV (light, oxygen, or voltage) domains of the blue-light photoreceptor phototropin (*nph1*): binding sites for the chromophore flavin mononucleotide. *Proc. Natl. Acad. Sci. USA* **96**: 8779–8783.

Clack, T., Mathews, S. and Sharrock, R.A. (1994). The phytochrome apoprotein family in *Arabidopsis* is encoded by five genes: the sequences and expression of *PHYD* and *PHYE*. *Plant Mol. Biol.* **25**: 413–427.

Clough, R.C., Casal, J.J., Jordon, E.T., Christou, P. and Vierstra, R.D. (1995). Expression of functional oat phytochrome A in transgenic rice. *Plant Physiol.* **109**: 1039–1045.

Cordonnier, M.-M. (1989). Yearly review: monoclonal antibodies: molecular probes for the study of phytochrome. *Photochem. Photobiol.* **49**: 821–831.

Cowl, J.S., Hartley, N., Xie, D.-X., Whitelam, G.C., Murphy, G.P. and Harberd, N.P. (1994). The *PHYC* gene of *Arabidopsis*: absence of the third intron found in *PHYA* and *PHYB*. *Plant Physiol.* **106**: 813–814.

Downs, R.J. (1956). Photoreversibility of flower initiation. *Plant Physiol.* **31**: 279–282.

Downs, R.J., Hendricks, S.B. and Borthwick, H.A. (1957). Photoreversible control of extension of pinto beans and other plants under normal conditions of growth. *Bot. Gaz.* **118**: 199–208.

Ecker, J.P. (1995). The ethylene signal transduction pathway in plants. *Science* **268**: 667–675.

Ensminger, P.A. and Schäfer, E. (1992). Blue and ultra-violet-B light photoreceptors in parsley cells. *Photochem. Photobiol.* **55**: 437–447.

Flint, L.H. and McAlister, E.D. (1935). Wavelengths of radiation in the visible spectrum inhibiting the germination of light-sensitive lettuce seeds. *Smithson. Misc. Collns.* **94** : No. 5, 1–11.

Fluhr, R. and Mattoo, A.K. (1996). Ethylene – biosynthesis and perception. *Crit. Rev. Plant Sci.* **15**: 479–523.

Foster, K.R. and Morgan, P.W. (1995). Genetic regulation of development in *Sorghum bicolor*. IX. The ma_3^R allele disrupts diurnal control of gibberellin biosynthesis. *Plant Physiol.* **108**: 337–343.

Furuya, M. (1989). Molecular properties and biogenesis of phytochrome I and II. *Adv. Biophys.* **25**: 133–167.

Gallagher, S., Short, T.W., Pratt, L.H., Ray, P.M. and Briggs,W.R. (1988). Light-induced changes in two proteins found associated with plasma membrane fractions from pea stem sections. *Proc. Natl. Acad. Sci. USA* **85**: 8003–8007.

Garner, W.W. and Allard, H.A. (1920). Effect of the relative length of day and night and other factors of the environment on growth and reproduction in plants. *J. Agricultural Research* **18**: 553–606.

Gressel, J. (1977). Blue light photoreception. *Photochem. Photobiol.* **30**: 749–754.

Hamner, K.C. and Bonner, J. (1938). Photoperiodism in relation to hormones as factors in floral initiation and development. *Bot. Gaz.* **100**: 388–431.

Hanks, S.K. and Quinn, A.M. (1991). Protein kinase catalytic domain sequence database: identification of conserved features of primary structure and classification of family members. *Methods Enzymol.* **200**: 38–62.

Hershey, H.P., Barker, R.F., Idler, K.P., Lissemore, J.L. and Quail, P.H. (1985). Analysis of cloned cDNA and genomic sequences for phytochrome: complete amino acid sequences for two gene products in etiolated *Avena*. *Nucl. Acids Res.* **3**: 8543–8559.

Huala, E., Oeller, P.W., Liscum, E., Han, I.-S., Larsen, E. and Briggs, W.R. (1997). *Arabidopsis* NPH1: a protein kinase with a putative redox-sensing domain. *Science* **278**: 2120–2123.

Jabben, M. and Deitzer, G.F. (1979). Effects of the herbicide San 9789 on photomorphogenic responses. *Plant Physiol.* **63**: 481–485.

Jarillo, J.A., Ahmad, M. and Cashmore, A.R. (1998). NPL1 (accession no. AF053941): a second member of the NPH serine/threonine kinase family of *Arabidopsis* (PGR 98-100). *Plant Physiol.* **117**: 719.

Johnson, E., Bradley, M., Harberd, N.P. and Whitelam, G.C. (1994). Photoresponses of light-grown phyA mutants of *Arabidopsis*. Phytochrome A is required for the perception of daylength extensions. *Plant Physiol.* **105**: 141–149.

Jones, A.M., Vierstra, R.D., Daniels, S.M. and Quail, P.H. (1985). The role of separate molecular domains in the structure of phytochrome from etiolated *Avena sativa*. *Planta* **164**: 501–506.

Jordan, B.R., Partis, M.D. and Thomas, B. (1986). The biology and molecular biology of phytochrome. In: *Oxford Surveys of Plant Molecular and Cell Biology 3*, B.J. Miflin (ed.), pp. 315–362. Oxford University Press, Oxford.

Jordan, E.T., Hatfield, P.M., Hondred, D., Talon, M., Zeevaart, J.A.D. and Vierstra, R.D. (1995). Phytochrome A overexpression in transgenic tobacco. Correlation of dwarf phenotype with high concentrations of phytochrome in vascular tissue and attenuated gibberellin levels. *Plant Physiol.* **107**: 797–805.

Kaldenhoff, R., Kölling, A. and Richter, G. (1993). A novel blue light- and abscicic acid-inducible gene of *Arabidopsis thaliana* encoding an intrinsic membrane protein. *Plant Mol. Biol.* **23**: 1187–1198.

Keller, J.M., Shanklin, J., Vierstra, R.D. and Hershey, H.P. (1989). Expression of a functional monocotyledonous phytochrome in transgenic tobacco. *EMBO J.* **8**: 1005–1012.

Kendrick, R.E and Kronenberg, G.H.M. (1994). *Photomorphogenesis in Plants.* Kluwer Academic Publishers, Dordrecht.

Kendrick. R.E. and Spruit, C.J.P. (1977). Phototransformations of phytochrome. *Photochem. Photobiol.* **26**: 201–214.

Kendrick, R.E., Kerckhoffs, L.H.J., Pundsnes, A.S., van Tuinen, A., Koorneef, M., Nagatani, A., Terry, M.J., Tretyn, A., Cordonnier-Pratt, M.-M., Hauser, B. and Pratt, L.H. (1994). Photomorphogenic mutants in tomato. *Euphytica* **79**: 227–234.

Khurana, J.P., Kochlar, A. and Tyagi, A.K. (1998). Photosensory perception and signal transduction in higher plants – molecular genetic analysis. *Crit. Revs. in Plant Scis.* **17**: 465–539.

Kohen, E., Santus, R. and Hirschberg, J.G. (1995). Photoregulatory mechanisms, *Photobiology*, pp. 227–257. Academic Press, London.

Koorneef, M. and Kendrick, R.E. (1994). Photomorphogenic mutants of higher plants. In: *Photomorphogenesis in Plants*, R.E. Kendrick and G.H.M. Kronenberg (eds), pp. 601–628. Kluwer Academic Publishers, Dordrecht.

Koorneef, M., Rolff, E. and Spruit, C.J.P. (1980) Genetic control of light-inhibited hypocotyl elongation in *Arabidopsis thaliana* (L.) Heynh. *Z. Pflanzenphysiol.* **100**: 147–160.

Koorneef, M., Cone, J.W., Dekens, R.G., O'Herne-Robers, E.G., Spruit, C.P.J. and Kendrick, R.E. (1985). Photomorphogenic responses of long hypocotyl mutants of tomato. *J. Plant Physiol.* **120**: 153–165.

Kraepiel, Y., Rousslin, P., Sotta, B., Kerhoas, L., Einhorn, J., Caboche, M. and Miginiac, E. (1994). Analysis of phytochrome- and ABA-deficient mutants suggests that ABA degradation is controlled by light in *Nicotiana plumbaginifolia. Plant J.* **6**: 665–672.

Kraepiel, Y., Marrec, K., Sotta, B., Caboche, M. and Miginiac, E. (1995). *In vitro* morphogenic characteristics of phytochrome mutants in *Nicotiana plumbaginifolia* are modified and correlated to high indole-3-acetic acid levels. *Planta* **197**: 142–146.

Lehman, A., Black, R. and Ecker, J.R. (1996). *Hookless1*, an ethylene response gene, is required for differential cell elongation in the *Arabidopsis* hypocotyl. *Cell* **85**: 183–194.

Li, J., Nagpal, P., Vitart, V., McMorris, T.C. and Chory, J. (1996). A role for brassinosteroids in light-dependent development of *Arabidopsis. Science* **272**: 398–401.

Lin, C., Ahmad, M., Gordon, D. and Cashmore, A.R. (1995a). Expression of an *Arabidopsis* cryptochrome gene in transgenic tobacco results in hypersensitivity to blue, UV-A and green light. *Proc. Natl. Acad. Sci. USA* **92**: 8423–8427.

Lin, C., Robertson, D.E., Ahmad, M., Raibekas, A.A. and Jorns, M.S. (1995b). Association of flavin adenine dinucleotide with the *Arabidopsis* blue light receptor CRY1. *Science* **269**: 968–970.

Lin, C., Ahmad, M. and Cashmore, A.R. (1996a). *Arabidopsis* cryptochrome 1 is a soluble protein mediating blue light-dependent regulation of plant growth and development. *Plant J.* **10**: 893–902.

Lin, C., Ahmad, M., Chan, J. and Cashmore, A.R. (1996b). CRY2: a second member of the *Arabidopsis* cytochrome gene family. *Plant Physiol.* **110**: 1047.

Liscum, E. and Briggs, W.R. (1995). Mutations in the NPH1 locus of *Arabidopsis* disrupt the perception of phototropic stimuli. *Plant Cell* **7**: 473–485.

Lois, R. (1994). Accumulation of UV-absorbing flavonoids induced by UV-B radiation in *Arabidopsis thaliana* L. I. Mechanisms of UV-resistance in *Arabidopsis. Planta* **194**: 498–503.

Lois, R. and Buchanan, B.B. (1994). Severe sensitivity to ultraviolet radiation in an *Arabidopsis* mutant deficient in flavonoid accumulation. II. Mechanisms of UV-resistance in *Arabidopsis. Planta* **194**: 504–509.

Lopez-Juez, E., Kobayashi, M., Sakurai, A., Kamiya, Y. and Kendrick, R.E. (1995). Phytochrome, gibberellins, and hypocotyl growth. A study using cucumber (*Cucumis sativa*) long-hypocotyl mutant. *Plant Physiol.* **107**: 131–140.

Lumsden, P.J., Yamamoto, K.T., Nagatani, A. and Furuya, M. (1985). Effect of monoclonal antibodies on the *in vitro* P_{fr} dark-reversion of pea phytochrome. *Plant Cell Physiol.* **26**: 1313–1322.

Malhorta, K., Kim, S.-T., Batschauer, A., Dawut, L. and Sancar, A. (1995). Putative blue-light photoreceptors from *Arabidopsis thaliana* and *Sinapsis alba* with a high degree of sequence homology to DNA photolyase contain the two photolyase cofactors but lack DNA repair activity. *Biochemistry* **34**: 6892–6899.

Mancinelli, A. (1980). The photoreceptors of the high irradiance responses of plant photomorphogenesis. *Photochem. Photobiol.* **32**: 853–857.

Mancinelli, A. and Rabino, I. (1978). The high-irradiance response of plant photomorphogenesis. *Bot. Rev.* **44**: 129–180.

Mandoli, D.F. and Briggs, W.R. (1981). Phytochrome control of two low-irradiance responses in etiolated oat seedlings. *Plant Physiol.* **67**: 733–739.

McCormac, A.C., Wagner, D., Boylan, M.T., Quail, P.H., Smith, H. and Whitelam, G.C. (1993). Photoresponses of transgenic *Arabidopsis* seedlings expressing introduced phytochrome B-encoded cDNAs: evidence that phytochrome A and phytochrome B have distinct photoregulatory functions. *Plant J.* **4**: 19–27.

Meyerowitz, E.M. (1994). Structure and organization of the *Arabidopsis thaliana* nuclear genome. In: *Arabidopsis*, E.M. Meyerowitz and C.R. Sommerville (eds), pp. 21–36. Cold Spring Harbor Laboratory Press, Cold Spring Harbor, NY.

Mitchell, J.W., Mandava, N., Worley, J.F., Plimmer, J.R. and Smith, M.V. (1970). Brassins – a new family of plant hormones from rape pollen. *Nature* **225**: 1065–1066.

Mohr, H. (1986). Coaction between pigment systems. In: *Photomorphogenesis in Plants*, R.E. Kendrick and G.H.M. Kronenberg (eds), pp. 547–564. Martinus Nijhoff, Boston.

Mohr, H. and Schopfer, P. (1995). Photomorphogenesis, *Plant Physiology*, pp. 345–373. Springer-Verlag, Berlin.

Morgan, D.C. and Smith, H. (1976). Linear relationship between phytochrome equilibrium and growth in plants under simulated natural radiation. *Nature* **262**: 210–212.

Mumford, F.E. and Jenner, E.L. (1966). Purification and characterization of phytochrome from oat seedlings. *Biochem.* **5**: 3657–3662.

Nagatani, A., Chory, J. and Furuya, M. (1991). Phytochrome B is not detectable in the *hy3* mutant of *Arabidopsis*, which is deficient in responding to end-of-day red light treatments. *Plant Cell Physiol.* **32**: 1119–1122.

Nagatani, A., Reed, J.W. and Chory, J. (1993). Isolation and initial characterization of *Arabidopsis* mutants that are deficient in phytochrome A. *Plant Physiol.* **102**: 269–277.

Nozue, K., Kanegae, T., Imaizumi, T., Fukada, S., Okamoto, H., Yeh, K.-C., Lagarias, J.C. and Wada, M. (1998). A phytochrome from the fern *Adiantum* with features of the putative photoreceptor NPH1. *Proc. Natl. Acad. Sci. USA* **95**: 15826–15830.

Parker, M.W., Hendricks, S.B., Borthwick, H.A. and Went, F.W. (1949). Spectral sensitivities of leaf and stem growth in pea seedlings and their similarity to action for photoperiodism. *Am. J. Bot.* **36**: 194–204.

Parker, M.W., Hendricks, S.B. and Borthwick, H.A. (1950). Action spectrum for the photoperiodic control of floral initiation of the long-day plant *Hyoscyamus niger*. *Bot. Gaz.* **111**: 242–252.

Parks, B.M and Quail, P.H. (1991). Phytochrome-deficient *hy1* and *hy2* long hypocotyl mutants of *Arabidopsis* are defective in phytochrome chromophore biosynthesis. *Plant Cell* **3**: 1177–1186.

Parks, B.M. and Quail, P.H (1993). *hy8*, a new class of *Arabidopsis* long hypocotyl mutants deficient in functional phytochrome A. *Plant Cell* **5**: 39–48.

Peng, J. and Harberd, N.P. (1997). Gibberellin deficiency and response mutations suppress the stem elongation phenotype of phytochrome-deficient mutants of *Arabidopsis*. *Plant. Physiol.* **113**: 1051–1058.

Pratt, L.H. (1979). Phytochrome: function and properties. In: *Photochemical and Photobiological Reviews*, K.C. Smith (ed.), **4**: 59–124. Plenum Press, New York.

Pratt, L.H. (1995). Phytochromes: differential properties, expression patterns and molecular evolution. *Photochem. Photobiol.* **61**: 10–21.

Pratt, L.H., Cordonnier, M.-M., Wang, Y.-C., Stewart, S.J. and Moyer, M. (1991). Evidence for three phytochromes in *Avena*. In *Phytochrome properties and Biological Action*, B. Thomas and C.B. Johnson (eds), pp. 39–55. Springer-Verlag, Berlin.

Quail, P.H. (1991). Phytochrome: a light-activated molecular switch that regulates plant gene expression. *Ann. Rev. Genet.* **25**: 389–409.

Quail, P.H. (1997). An emerging molecular map of the phytochromes. *Plant Cell Environ.* **20**: 657–665.

Quail, P.H., Boylan, M.T., Parks, B.M., Short, B.W., Xu, Y. and Wagner, D. (1995). Phytochromes: photosensory perception and signal transduction. *Science* **268**: 675–680.

Rau, W. (1976). Photoregulation of carotenoid biosynthesis in plants. *Pure Appl. Chem.* **47**: 237–243.

Reed, J.W., Foster, K.R., Morgan, P.W. and Chory, J. (1996). Phytochrome B affects responsiveness to gibberellins in *Arabidopsis*. *Plant Physiol.* **112**: 337–342.

Reid, J.B. and Ross, J.J. (1988). Internode length in *Pisum*. A new gene, *lv*, conferring an enhanced response to gibberellin A_1. *Physiol. Plant.* **72**: 595–604.

Reymond, P., Short, T.W. and Briggs, W.R. (1992a). Blue light activates a specific protein kinase in higher plants. *Plant Physiol.* **100**: 655–661.

Reymond, P., Short, T.W., Briggs, W.R. and Poff, K.L. (1992b). Light-induced phosphorylation of a membrane protein plays an early role in signal transduction for phototropism in *Arabidopsis thaliana*. *Proc. Natl. Acad Sci. USA* **89**: 4718–4721.

Robson, P.R.H., Whitelam, G.C. and Smith, H. (1993). Selected components of the shade-avoidance syndrome are displayed in a normal manner in mutants of *Arabidopsis thaliana* and *Brassica rapa* deficient in phytochrome B. *Plant Physiol.* **102**: 1179–1184.

Romano, C.P., Robson, P.R.H., Smith, H., Estelle, M. and Klee, H. (1995). Transgene-mediated auxin overproduction in *Arabidopsis*: hypocotyl elongation phenotype and interactions with the *hy6-1* hypocotyl elongation and *axr1* auxin-resistant mutants. *Plant Mol. Biol.* **27**: 1071–1083.

Rood, S.B., Williams, P.H., Pearce, D., Murofushi, N., Mander, L.N. and Phsaris, R.P. (1990). A mutant gene that increases gibberellin production in *Brassica*. *Plant Physiol.* **93**: 1168–1174.

Rüdiger, W. (1986). The chromophore. In: *Photomorphogenesis in Plants*, R.E. Kendrick and G.H.M. Kronenberg (eds), pp. 17–33. Martinus Nijhoff, Dordrecht.

Russell, D.W. (1996). Green light for steroid hormones. *Science* **272**: 370–371.

Sage, L.C. (1992). *Pigment of the Imagination. A History of Phytochrome Research*. Academic Press, San Diego, CA.

Saki, T., Kagawa, T., Kasahara, M., Swartz, T.E. Christie, J.M., Briggs, J.R., Wada, M. and Okada, K. (2001). *Arabidopsis nph1* and *npl1*: blue light receptors that mediate both phototropism and chloroplast relocation. *Proc. Natl. Acad. Sci. USA* **98**: 6969–6974.

Senger, H. and Lipson, D. (1987). Problems and prospects of blue and ultraviolet light effects. In: *Phytochrome and Photoregulation in Plants*, M. Furuya (ed.), pp. 315–331. Academic Press, New York.

Serlin, B.S. and Roux, S.J. (1984). Modulation of chloroplast movement in the green alga *Mougeotia* by the Ca^{2+} ionophore A23187 and by calmodulin antagonists. *Proc. Natl. Acad. Sci. USA* **81**: 6368–6372.

Shanklin, J., Jabben, M. and Vierstra, R.D. (1987). Red-light induced formation of ubiquitin-phytochrome conjugates: identification of possible intermediates of phytochrome degradation. *Proc. Natl. Acad. Sci. USA* **84**: 359–363.

Sharrock, R.A. and Quail, P.H. (1989). Novel phytochrome sequences in *Arabidopsis thaliana*: structure, evolution and differential expression of a plant regulatory photoreceptor family. *Genes Dev.* **3**: 1745–1757.

Shimazaki, Y. and Pratt, L.H. (1985). Immunochemical detection with rabbit polyclonal and mouse monoclonal antibodies of different pools of phytochrome from etiolated and green Avena shoots. *Planta* **164**: 333–344.

Shimazaki, Y., Cordonnier, M. and Pratt, L.H. (1983). Phytochrome quantitation in crude extracts of *Avena* by enzyme-linked immunosorbent assay with monoclonal antibodies. *Planta* **159**: 534–544.

Shinomura, T., Nagatani, A., Chory, J. and Furuya, M. (1994). The induction of seed germination in *Arabidopsis thaliana* is regulated principally by phytochrome B and secondarily by phytochrome A. *Plant Physiol.* **104**: 363–371.

Shinomura, T., Nagatani, A., Hanzawa, H., Kubota, M., Watanabe, M. and Furuya, M. (1996). Action spectra of phytochrome A- and B-specific photoinduction of seed germination in *Arabidopsis thaliana*. *Proc. Natl. Acad. Sci. USA* **93**: 8129–8133.

Shirley, B.W. (1996). Flavonoid biosynthesis: 'new' functions for an old pathway. *Trends Plant Sci.* **1**: 377–382.

Short, T.W. and Briggs, W.R. (1994). The transduction of blue light signals in higher plants. *Ann. Rev. Plant Physiol. Plant Mol. Biol.* **45**: 143–171.

Shropshire, W., Jr. (1972). Phytochrome-photochromic sensor. In: *Photophysiology*, A.C. Giese (ed.). **7**: 34–72.

Siegelman, H.W. and Firer, E.M. (1964). Purification of phytochrome from oat seedlings. *Biochem.* **3**: 418–423.

Smalle, J., Haegman, M., Kurepa, J., van Montagu, M. and van der Straeten, D. (1997). Ethylene can stimulate *Arabidopsis* hypocotyl elongation in the light. *Proc. Natl. Acad. Sci. USA* **94**: 2756–2761.

Smith, H. (1982). Light quality, photoreception and plant strategy. *Ann. Rev. Plant Physiol.* **33**: 481–518.

Smith, H. (1995). Physiological and ecological function within the phytochrome family. *Ann. Rev. Plant Physiol. Plant Mol. Biol.* **46**: 289–315.

Smith, H. and Whitelam, G.C. (1990). Phytochrome, a family of photo-receptors with multiple physiological roles. *Plant, Cell and Environment* **13**: 695–707.

Sommers, D.E. and Quail, P.H. (1995). Temporal and spatial expression patterns of *PHYA* and *PHYB* genes in *Arabidopsis*. *Plant J.* **7**: 413–427.

Song, P.-S. and Yamazaki, I. (1987). Structure–function relationship of the phytochrome chromophore. In: *Phytochrome and Photoregulation in Plants*, M. Furuya (ed.), pp. 139–156. Academic Press, New York.

Su, W. and Howell, S.H. (1995). The effects of cytokinin and light on hypocotyl elongation in *Arabidopsis* seedlings are independent and additive. *Plant Physiol.* **108**: 1423–1430.

Szekeres, M., Németh, K., Koncz-Kálmán, Z., Mathur, J., Kauschmann, A., Altmann, T., Rédei, J.P., Nagy, F., Schell, J. and Koncz, C. (1996). Brassinosteroids rescue the deficiency of CYP90, a cytochrome P450, controlling cell elongation and de-etiolation in *Arabidopsis*. *Cell* **85**: 171–182.

Taiz, L. and Zeiger, E. (1991). Auxins: growth and tropisms, *Plant Physiology*, pp. 398–425. Benjamin/Cummings.

Terzaghi, W.B. and Cashmore, A.R. (1995). Light-regulated transcription. *Ann. Rev. Plant Physiol. Plant Mol. Biol.* **46**: 445–474.

Thomas, B., Crook, N.E. and Penn, S.E. (1984). An enzyme-linked immunosorbent assay for phytochrome. *Physiol. Plant.* **60**: 409–415.

Tokuhisha, J.G. and Quail, P.H. (1989). Phytochrome in green tissue: partial purification and characterization of the 118-kilodalton phytochrome species from light-grown *Avena sativa* L. *Photochem. Photobiol.* **52**: 143–152.

Tokuhisha, J.G., Daniels, S.M. and Quail, P.H. (1985). Phytochrome in green tissue: spectral and immunochemical evidence for two distinct molecular species of phytochrome in light-grown *Avena sativa*. *Planta* **164**: 321–332.

Tomozawa, K., Nagatani, A. and Furuya, M. (1990). Phytochrome genes: studies using the tools of molecular biology and photomorphogenetic mutants. *Photochem. Photobiol.* **52**: 265–275.

Toyomasu, T., Yamane, H., Murofushi, N. and Nick, P. (1994). Phytochrome inhibits the effectiveness of gibberellins to induce cell elongation in rice. *Planta* **194**: 256–263.

van Tuinen, A., Kerckhoffs, L.H.J., Nagatani, A., Kendrick, R.E. and Koorneef, M. (1995). Far-red light-insensitive mutant of tomato lacks a light-stable B-like phytochrome. *Plant Physiol.* **108**: 939–947.

van Tuinen, A., Hanhart, C.J., Kerckhoffs, L.H.J., Nagatani, A., Boylan, M.T., Quail, P.H., Kendrick, R.E. and Koorneef, M. (1996). Analysis of phytochrome-deficient *yellow-green-2* and *aurea* mutants of tomato. *Plant J.* **9**: 173–182.

Vierstra, R.D. (1993). Illuminating phytochrome functions. There is light at the end of the tunnel. *Plant Physiol.* **103**: 679–684.

Vierstra, R.D. and Quail, P.H. (1983). Purification and initial characterization of 124-kilodalton phytochrome from *Avena*. *Biochemistry* **22**: 2498–2505.

Vierstra, R.D. and Quail, P.H. (1986). Phytochrome: the protein. In: *Photomorphogenesis in Plants*, R.E. Kendrick and G.H.M. Kronenberg (eds), pp. 35–60. Martinus Nijhoff, Boston.

Vierstra, R.D., Cordonnier, M.-M., Pratt, L.H. and Quail, P.H. (1984). Native phytochrome: immunoblot analysis of relative molecular mass and *in vitro*: proteolytic degradation for several plant species. *Planta* **160**: 521–528.

Wagner, D., Tepperman, J.M. and Quail, P.H. (1991). Overexpression of phytochrome B induces a short hypocotyl phenotype in transgenic *Arabidopsis*. *Plant Cell* **3**: 1275–1288.

Wang, Y.C., Stewart, S.J., Cordonnier, M.-M. and Pratt, L.H. (1991). *Avena sativa* contains three phytochromes, only one of which is abundant in etiolated tissue. *Planta* **184**: 96–104.

Weatherwax, S.C., Ong, M.S., Degenhardt, J., Bray, E.A. and Tobin, E.M. (1996). The interaction of light and abscisic acid in the regulation of plant gene expression. *Plant Physiol.* **111**: 363–370.

Wellmann, E. (1983). UV radiation in photomorphogenesis. In: W. Shropshire, Jr. and H. Mohr (eds), *Encyclopedia of Plant Physiology*, New Series, **16B**: 745–756, Springer-Verlag, Berlin.

Whitelam, G.C. and Harberd, N.P. (1994). Action and function of phytochrome family members revealed through the study of mutant and transgenic plants. *Plant Cell Environ.* **17**: 615–625.

Whitelam, G.C. and Smith, H. (1991). Retention of phytochrome-mediated shade avoidance responses in phytochrome-deficient mutants of *Arabidopsis*, cucumber and tomato. *J. Plant Physiol.* **139**: 119–125.

Whitelam, G.C., Johnson, E., Peng, J., Carol, P., Anderson, M.L., Cowl, J.S. and Harberd, N.P. (1993). Phytochrome A null mutants of *Arabidopsis* display a wild-type phenotype in white light. *Plant Cell* **5**: 757–768.

Withrow, R.B., Klein, W.H. and Elstad, V. (1957). Action spectra of photomorphogenic induction and its photoinactivation. *Plant Physiol.* **32**: 453–462.

7
Photoperiodism

Photoperiodism in Plants – Early Work

For many years it was thought that flowering was a consequence of the accumulation of the products of photosynthesis synthesized during long days. Julien Tournois, a French scientist, was probably the first to realize the role of day length in flowering. He observed that hemp (*Cannibis sativa*) planted in early spring flowered vigorously in summer but remained vegetative if planted in late spring or early summer. He wrote (1914) that 'Precocious flowering in young plants of hemp and hops occurs when, from germination, they are exposed to very short periods of daily illumination'. Also, 'Precocious flowering is not so much caused by shortening of the days as by the lengthening of the nights'. He died at the front in the First World War. About the same time, Georg Klebs, in Heidelberg, Germany, probably also discovered the role of day length in flowering. Klebs (1918) showed that the house leek (*Sempervivum funkii*), treated with artificial light in the glasshouse, could be induced to flower in winter, although the normal time is in June. Klebs concluded that, in this case, light acted as a catalyst rather than providing nutrition from the additional photosynthesis.

The work of W.W. Garner and H.A. Allard, at the United States Department of Agriculture, Beltsville, Maryland, in the 1920s, resulted in the first clearly stated hypothesis on photoperiodism. They noted that soybeans, planted at various times during the spring, all flowered about the same time in late summer/early autumn, regardless of how long they had been growing (Fig. 7.1). They observed that Biloxi soybean (*Glycine max*) plants flowered in September/October, even if they were germinated over a three-month period, May–July (Table 7.1). They found that a variation of 59 days in the date of germination (May 2 to June 30) resulted in a difference of only 11 days (September 4–15) to the first open flowers. The number of days from germination to blossoming and the final height of the plants decreased as the season continued. They concluded that in soybean plants there is a seasonal timing mechanism. Garner and Allard also planted a commercial strain of tobacco (*Nicotiana tabacum*) Maryland Narrowleaf in spring and it flowered in summer. Sometimes a mutant appeared in the crop; it grew taller and produced large leaves

Photobiology of Higher Plants. By M. S. Mc Donald.
© 2003 John Wiley & Sons, Ltd: ISBN 0 470 85522 3; ISBN 0 470 85523 1 (PB)

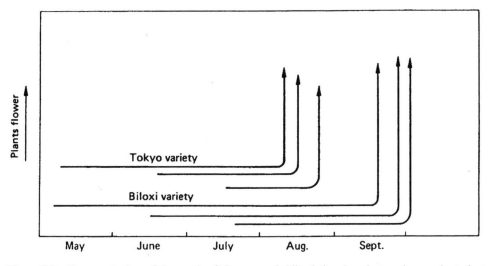

Figure 7.1 Representation of the work of Garner and Allard showing that soybeans planted at different times all came into flower at about the same time in late summer. (From Bidwell 1974)

Table 7.1 Growth and flowering of Biloxi soybean as a function of germination date. (From Garner and Allard 1920)

Germination date	Date of first open blossoms	Maximum height (inches)	Days to blossoming
May 2	September 4	52	125
June 2	September 4	52	94
June 16	September 11	48	92
June 30	September 15	48	77
July 15	September 22	44	69
August 2	September 29	28	58
August 16	October 16	20	61

and was given the name Maryland Mammoth (Fig. 7.2). The mutant had obvious commercial value for tobacco production and there was great interest in obtaining seed for further production. The Maryland Mammoth plants, however, did not flower during the summer or early autumn and the plants were killed by the cold weather of winter. When the plants were lifted from the field and grown in the glasshouse under the relatively shorter photoperiods of winter, they flowered profusely in mid-December and set seed. From seed produced in this way, it was found that Maryland Mammoth was true-breeding and the strain was used for commercial production. It was necessary, however, to transfer field-grown plants into the glasshouse and grow them during the winter to produce seeds. Like Tournois and Klebs, Garner and Allard ruled out other factors such as

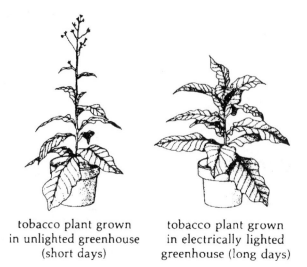

tobacco plant grown
in unlighted greenhouse
(short days)

tobacco plant grown
in electrically lighted
greenhouse (long days)

Figure 7.2 Maryland Mammoth tobacco plants in which photoperiodism was first observed

temperature, nutrition and light intensity and concluded that flowering was controlled by the relative lengths of day and night. This conclusion was supported by the finding that the tobacco could be maintained in the vegetative state during the winter months by lengthening the days with artificial light. Garner and Allard termed the response of Maryland Mammoth to daylength, *photoperiodism* (Fig. 7.2).

Terminology

The Maryland Mammoth mutant was called a 'short-day plant' because it flowered under short-day conditions. Plants vary greatly in their response to daylength. Some plants require long photoperiods to induce flowering. Other plants flower under both short- and long-day conditions, while still others respond to photoperiods somewhere between short and long daylengths. The terms proposed are all based on a 24-hour cycle of light and darkness. The classification of plants according to their photoperiodic response is usually made on the basis of flowering, rather than on any other criterion. Representatives of plants showing the main response types are shown in Table 7.2.

The two main categories are short-day plants (SDPs), in which flowering occurs only on short days (*qualititative* SDPs) or is accelerated by short days (*quantitative* SDPs), and long-day plants (LDPs), in which flowering occurs only on long days (*qualitative* LDPs) or is accelerated by long days (*quantitative* LDPs). A third category, day-neutral plants (DNPs), flower after a period of vegetative growth, regardless of the photoperiod. Some plants have more specialized daylength requirements closely tied to seasons. Some relatively rare plants require a long

Table 7.2 Representative plants of the principal photoperiodic response types

Short-day Plants	
Chenopodium rubrum	red goosefoot
Chrysanthemum spp.	chrysanthemum
Euphorbia pulcherrima	poinsettia
Glycine max	soybean
Nicotiana tabacum	tobacco
Pharbitis nil	Japanese morning glory
Xanthium strumarium	cocklebur
Long-day Plants	
Beta vulgaris	beet
Hyoscyamus niger	black henbane
Lolium spp.	rye grass
Raphinus sativus	radish
Secale cerale	rye
Sinapsis alba	white mustard
Spinacea oleracea	spinach
Triticum aestivum	wheat
Day-neutral Plants	
Cucumis sativus	cucumber
Helianthus abbus	sunflower
Phaseolus vulgaris	bean
Pisum sativum	pea
Zea mays	maize

photoperiod followed by a short photoperiod in order to flower; these long–short-day plants flower only in autumn during the shortening days that follow the long summer days. Other plants require a short photoperiod followed by a long photoperiod; these short–long-day plants flower only during the lengthening days of early summer following the short days of spring. A few species (e.g. sugarcane) are intermediate-daylength plants and flower only in daylengths that are intermediate between short and long days. Another category of plant are amphophotoperiodic, that is, they flower either on short days or long days but not on intermediate days. A few plants require special daylengths only early in their development and later become day-neutral. A small number of plants show interrelations with other environmental factors such as temperature: the poinsettia (*Euphorbia pulcherrima*) and morning glory (*Ipomoea purpurea*) are short-day plants at high temperatures and long-day plants at low temperatures.

The essential distinction to be made between short-day and long-day responses is that flowering in SDPs occurs only when the daylength is less than a specific, critical length (daylength in excess of this period will keep the plant vegetative), while promotion of flowering in LDPs occurs only after a critical daylength is exceeded.

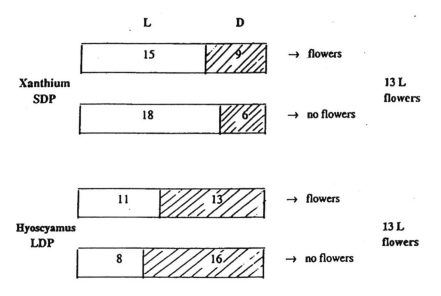

Figure 7.3 *Xanthium*, a short-day plant and *Hyoscyamus*, a long-day plant, both flower if subjected to the same photoperiod of 13 hours

This classification does not mean to imply that all short-day plants flower at photoperiods that are shorter than photoperiods that induce flowering in long-day plants. For example, *Xanthium* is a short-day plant with a critical daylength of 15 hours and flowers if this critical maximum is not exceeded. *Hyoscyamus* is a long-day plant with a critical daylength of 11 hours and flowers when this critical minimum is exceeded. Both *Xanthium* (SDP) and *Hyoscyamus* (LDP) will flower when subjected to a 13-hour photoperiod (Fig. 7.3). An important distinction here is not the absolute number of hours of light received but when a plant will flower – before or after a critical length of time. Long-day plants can effectively measure the lengthening days of spring into early summer and delay flowering until the critical daylength is reached; many cultivars of wheat behave in this way. In short-day plants flowering often occurs in autumn when the days shorten below a critical length, as in many cultivars of *Chrysanthemum*. However, since daylength cannot distinguish between spring and autumn, plants have to use a number of strategies to obviate this ambiguity. One is the coupling of a temperature requirement to a photoperiodic response. For example, in some short-day cultivars of strawberry (*Fragaria*), flowers are initiated when the critical daylength is reached in the autumn but do not emerge until spring. The low temperatures of winter and early spring prevent flower production, even though daylength conditions are appropriate.

Photoperiodism in Other Organisms

With the realization that flowering in plants is determined by the relative lengths of day and night, it was soon afterwards established that other organisms respond to

Table 7.3 Some reactions of plants and animals to photoperiod

	Seasonal responses	
Plant/Animal	Long nights and short days	Short nights and long days
Alga (*Ulothrix*)	Zoospore production	Vegetative growth
Fern (*Salvinia*)	Sporocarp production	Vegetative growth
Seed Plants		
Wheat	Vegetative growth	Flower formation
Chrysanthemum	Flower formation	Vegetative growth
Maple	Bud dormancy	Vegetative growth
Potato	Tuber formation	Vegetative growth
Onion	Vegetative growth	Bulb formation
Lower Animals		
Red spider mites	Laying of dormant eggs	Laying of non-dormant eggs
Pulmonate snails	Non-reproductive growth	Laying of eggs
Vertebrates		
Brook trout	Laying of eggs	Non-reproductive growth
Pheasant	Non-reproductive growth	Laying of eggs
Sheep	Mating	Bearing of young
Horse	Non-reproductive growth	Mating/bearing of young

other photoperiodic regimes (Table 7.3). Long nights and short days promote reproductive activity in *Ulothrix*, an alga, and in *Salvina*, a fern, while short nights and long days promote vegetative growth. Both invertebrate and vertebrate animals also display patterns of reproductive activity controlled by daylength. Red spider mites and pulmonate snails, for example, engage in reproductive activity under conditions of short nights and long days. Similarly, the reproductive behaviour of a number of vertebrates is controlled by daylength. While the basic mechanisms responsible for photoperiodic responses are different in plants and animals, there is probably one common time-measuring system (biological clock) involved in both.

Importance of the Dark Period

Under normal conditions plants are subjected to a 24-hour cycle of light and darkness. Workers soon found that a more analytical view of photoperiodism could be obtained by using cycles such as 8 hours' light followed by 8 hours' darkness or of 16 hours' light followed by 16 hours' darkness (Fig. 7.4). By subjecting plants to cycles other than 24 hours it was shown that flowering in plants is more a response to darkness than to light. In other words, short-day plants flower when a certain

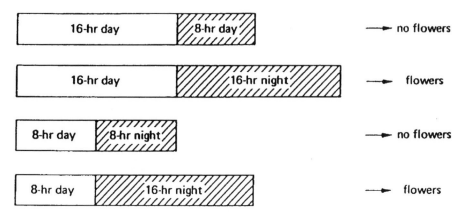

Figure 7.4   The effects of long and short days and nights on flowering in cocklebur (*Xanthium strumarium*)

critical period of darkness is exceeded and long-day plants flower when the duration of darkness is less than a critical value.

Hamner and Bonner (1938) at USDA, showed that cocklebur (*Xanthium strumarium*), a short-day plant, would flower when the dark period in the daily cycle was greater than 9 hours, regardless of the daylength. Hamner and his associates showed that a brief interruption of the dark period with a flash of light would nullify the long night; the effect works equally with short-day and long-day plants (Fig. 7.5).

Hamner and Bonner (1938) showed that the short-day plant *Xanthium* can be kept from flowering, even though on the correct photoinductive cycle, by

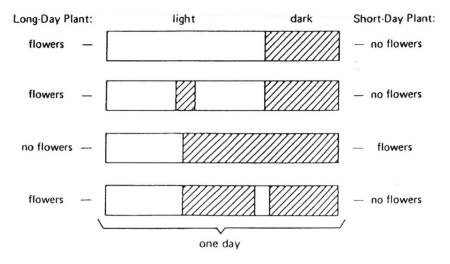

Figure 7.5   The effect of a dark interruption of a long day and of a light interruption of a long night on flowering in plants

interrupting the dark period with a brief flash of light; interrupting the light period with a brief period of darkness had no effect. In other words, the long dark period necessary for flowering in *Xanthium* was broken into two short dark periods, thus keeping the plant in a vegetative state. The light interruption of the dark period is referred to as the 'night-break'. The light intensity need not be great, several foot-candles being sufficient for some plants. Certain short-day plants avoid the effects of moonlight by folding their leaves at night. Certain other short-day plants require a critical period of low-intensity light (comparable to moonlight) instead of a dark period. Thus, the initiation of floral primordia in short-day plants is promoted by uninterrupted long, dark periods. Flower initiation in long-day plants, on the other hand, is inhibited by uninterrupted long, dark periods. If the long dark period is broken up (night break) by a brief flash of dim light, long-day plants will initiate floral primordia. In fact, many long-day plants flower under continuous light and do not require any dark period. Commercial flower growers take advantage of the night break phenomenon to regulate the time of flowering in certain plants.

Importance of the Photoperiod

The concept that the dark period is the critical part of the photoinductive cycle was given considerable support by the findings of Hamner in 1940. Working with Biloxi soybean, a short-day plant, Hamner showed that flowering cannot be induced unless the plant first receives more than 10 hours' darkness, while the length of the photoperiod does not matter (Fig. 7.6). While the length of the photoperiod has no effect on flower initiation, it has a quantitative effect: an increase in the length of the

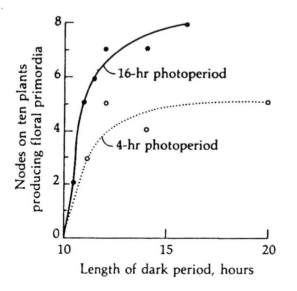

Figure 7.6 Relationship between length of dark period and initiation of floral primordia in Biloxi soybean. (From Hamner 1940). Reproduced with permission of The University of Chicago Press

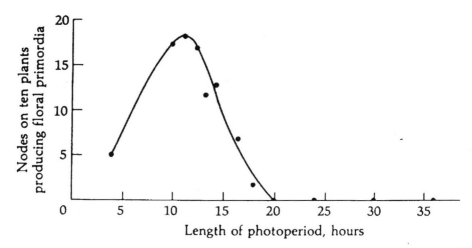

Figure 7.7 Relationship between light duration and initiation of floral primordia by Biloxi soybean. (From Hamner 1940). Reproduced with permission of The University of Chicago Press

photoperiod (to 16 hours rather than 4 hours) increases the number of floral primordia. In other words, the length of the dark period determines actual initiation of floral primordia; the length of the light period determines the number of floral primordia initiated.

The optimal response for the Biloxi soybean was found to be a photocycle consisting of 16 hours' darkness and 11 hours' light (Fig. 7.7); photoperiods greater or less than 11 hours produce fewer primordia.

Intensity of Light

Light intensity could have a direct effect by controlling the flow of sugars to the meristems capable of initiating floral primordia. In addition, the effectiveness of the photoperiod diminishes in the absence of CO_2. The enhancing effect of externally applied sugar and CO_2 was partially successful in bringing about flowering in the dark by supplying plants with sugar solutions (Takimoto 1960). This suggests that substrates provided by photosynthesis have some effect on the ability of the plant to produce flowers. Light intensity could have a direct effect, possibly in the synthesis of some factor, e.g. a hormone, necessary for flower formation. Hamner (1940) studied the quantitative effect of light duration and intensity on flower initiation in Biloxi soybean on a photoinductive cycle. He found that, at light intensities below 100 foot-candles, no flowers were produced. Increase in light intensity above this level increased the number of flowers produced and the longer photoperiod produced the greater number of flowers (Fig. 7.8).

Photoperiodic Floral Induction

Despite the many descriptive studies of photoperiodism, the actual mechanism of flower initiation and subsequent flower development is still not clearly understood.

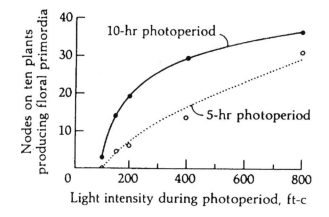

Figure 7.8 Quantitative effect of light duration and intensity on floral induction by Biloxi soybean. (From Hamner 1940). Reproduced with permission of The University of Chicago Press

It is difficult to pinpoint the precise time at which the pattern of growth changes from the vegetative to the reproductive. By the time that floral primordia are macroscopically visible, molecular, subcellular and cellular changes have already taken place. In addition, the flowering response to photoperiod is quite diverse. Many plants require more or less continuous exposure to the appropriate photoperiod in order to flower successfully. Such plants are said to be *induced* and the appropriate photoperiod is referred to as the *inductive cycle*. After a sufficient number of favourable cycles, flowering will often occur even when the plants are returned to non-inductive cycles; this is true for both SDPs and LDPs. The number of photoinductive cycles needed to induce flowering differs widely among different plant species. In one extreme, for example *Xanthium pennsylvanicum*, a single photoinductive cycle is sufficient to initiate floral primordia. In such plants, and in plants requiring only a few photoinductive cycles, no morphological changes can be observed in the shoot apex at the end of the induction period. Photoperiodism has been studied in far greater detail in *Xanthium* than in any species because only a single short-day cycle will irreversibly lead to flowering even if the plant is returned to a non-inductive cycle. Other short-day plants such as Japanese morning glory (*Pharbitis nil*), duckweed (*Lemna purpurilla*) and pigweed (*Chenopodium rubrum*) also require only one photoinductive cycle for the initiation of floral primordia. In fact, seedlings of *Pharbitis nil* may be induced to flower by exposing the cotyledons to an appropriate photoperiod, rather than waiting for fully formed leaves to be produced as with *Xanthium*. At the other extreme, photoinductive cycles must be continued until the apex has become recognizably floral (Battey and Lyndon 1990). In the short-day plant *Impatiens balsamina*, for example, returning the plant to a non-inductive cycle causes the tip of the placenta to resume vegetative growth, even after the flower has formed fertile anthers and an ovary with abortive ovules. *Salvia occidentalis*, a short-day plant, requires at least

17 photoinductive cycles to flower; *Plantago lanceolata*, a long-day plant, requires 25 photoinductive cycles for maximum floral response. Other long-day plants widely used in photoperiodism studies are black henbane (*Hyoscyamus niger*), which requires at least three photoinductive cycles and rye grass (*Lolium temulentum*), which has been shown to flower following a single photoinductive cycle.

The fact that plants can be induced to flower after the appropriate number of nights (long or short) suggests that the formation of flowers by a plant is an all-or-nothing affair with respect to photoperiodism. Once a plant has received a minimum number of photoinductive cycles, it will flower, even if it is returned to a non-inductive regime. Consider, for example, *Xanthium strumarium*, a short-day plant with a critical daylength of 15.5 hours. On a light regime of 16 h light/8 h darkness, *Xanthium* produces leaves and remains vegetative indefinitely. If it receives a single photoinductive cycle of 15 h light/9 h darkness and is then returned to the non-inductive 16 h light/8 h darkness, microscopic evidence of flower initiation is visible within 2–3 days. Some short- and long-day plants exhibit fractional induction – the summation of inductive cycles despite the intercalation of non-inductive cycles. The long-day plant *Plantago lanceolata* needs 25 photo-inductive cycles for 100% inflorescence production. If the plant is given 10 photoinductive cycles and then subjected to a number of non-inductive cycles it will not flower. However, if the plant is then returned to a photoinductive regime, only 15 additional cycles are needed to produce 100% inflorescence. The plant is able to sum the photoinductive cycles, despite the interruption of non-inductive cycles. Thus, it appears that after an adequate number of favourable cycles the change produced is relatively stable and can be maintained for some time (the exact duration depending on the plant); while fractional induction is insufficient to cause flowering it can be added to on exposure to subsequent favourable cycles.

The results from fractional induction treatments lend support to the concept that induction is an all-or-nothing process. There is experimental evidence to support the concept that there are degrees of induction. For example, *Xanthium* plants given only a single photoinductive short-day cycle progress only slowly towards flowering over several months; when given multiple photoinductive cycles the plants produce mature flowers in two weeks (Salisbury 1963). Partial induction is also found in the short-day plant *Impatiens balsamina*, which requires only three photoinductive cycles for floral bud initiation. For these buds to form flowers, however, more than eight photoinductive cycles are necessary. Formation of floral primordia by the long-day aquatic plant *Lemna gibba* (duckweed), requires a minimum of one long day. However, at least six long days are required to produce mature flowers; long days are apparently needed for the early stages of flower development in this plant (Cleland and Briggs 1967).

That different plants require different numbers of photoinductive cycles for complete flowering, suggests that floral induction involves the creation of some more or less permanent change in the induced plant, resulting in a continuously supplied stimulus to the flower. If the induction stimulus is very weak, some plants may revert to the vegetative form after a short period of flowering. One implication from this is that some factor involved in the flowering response is accumulated during the inductive cycle. In some plants, e.g. *Xanthium*, enough is accumulated after one cycle to promote flowering; in other plants, e.g. *Plantago*, a number of inductive cycles are needed. In long-day plants a non-inductive cycle does not appear to modify the effects of previous exposure to inductive cycles; in short-day plants, however, the non-inductive cycle appears to be inhibitory (Schwabe 1959).

Ripeness to Flower

Almost all plants must reach a certain age before they can be induced to flower; only a few plants respond to photoperiod as seedlings. As already mentioned, *Pharbitis nil* responds to short days when still in the cotyledonary stage and some species of *Chenopodium* have been shown to respond to photoperiod as small seedlings. Most species, such as *Xanthium*, must have a number of expanded leaves before they can respond to short days; *Hyoscyamus* must be 10–30 days old before it will respond to long days. After germination, most annual and perennial seedlings enter a rapidly growing phase in which flowering usually cannot be induced. In this early vegetative phase of growth, plants are said to be *juvenile*. Plants reach a *mature* or *adult phase* when they achieve the ability to flower. Annual plants need to be only several weeks old before they become competent to respond to a floral stimulus. The juvenile phase, with respect to flowering, can vary in perennials from one year in certain shrubs up to 40 years in beech (*Fagus sylvatica*). Klebs called the condition a plant must reach before it can flower in response to a stimulus *Bluhreife*, which is translated as 'ripeness to flower'. In many species, the number of required photoperiodic cycles decreases with increase in age of the plant. In other words, ripeness to respond to photoperiod increases with age; often the plant flowers independently of the photoperiod and becomes day-neutral. In contrast, leaves of *Anagallis arvensis*, the scarlet pimpernel, are most sensitive to long days when the plant is at the seedling stage; sensitivity declines in the most recently produced leaves and the plant becomes more difficult to induce as it gets older. Individual leaves must also reach a ripeness to respond; in *Pharbitis nil*, the cotyledons represent this stage. In some species, the leaf is maximally sensitive when it is fully expanded, but in *Xanthium* it is the half-expanded leaf, the one growing most rapidly, that is most sensitive.

The only proof of floral induction is the appearance of floral primordia at the meristem; this meristematic change is called *evocation* (Evans 1969). If evocation can occur, the meristem is said to be competent. A test for competence is to graft a meristem onto a flowering plant of the same species: if it flowers, it is competent. Virtually nothing is known about the metabolic difference between the juvenile stage when plants are incapable of being induced and the adult stage when the plant is competent and is sensitive to floral stimuli.

Perception of the Photoperiodic Signal

While the change from vegetative to reproductive growth occurs in the meristem, the main site for the perception of the photoperiodic stimulus is the leaf. This was first shown by Knott (1934) for spinach, a long-day plant in which exposing only the leaves to long photoperiods resulted in the formation of floral primordia at the shoot apex; when the apical bud alone was treated, the plants remained vegetative. Knott postulated that something is produced in the leaf in response to the photoinductive cycle and translocated to the apical tip, causing the induction of floral primordia. That a floral stimulus is translocated is further supported by experiments carried out by Chailakhyan (1968) in which as many as five *Xanthium* plants grafted together translocated the floral stimulus from one photoinduced leaf to the other four plants maintained on a non-inductive cycle (Fig. 7.13). Usually only a single leaf needs to be exposed to favourable photoperiods in order to achieve a response. Treatment of a single leaf of *Xanthium strumarium* with short days was shown to result in macroscopically visible buds (Naylor 1952), even though the remainder of the plant was growing in long-day conditions. In *Xanthium* it is not necessary to defoliate the remainder of the plant, but in many cases treating a single leaf will only result in flowering when the other leaves are removed. It is especially important to remove the leaves between the treated leaf and the apex, as these often interfere with the flowering response to the photoinduced leaf. Surprisingly, mature leaves also seem capable of neutralizing the flower-promoting effect of a photoperiodic stimulus. When a photoinduced leaf is grafted to a plant receiving a non-inductive cycle, mature leaves present on the receptor plant antagonize progress towards the flowering response. Defoliation of the receptor plant eliminates the antagonism. It has been shown that grafting individual photoinduced leaves from one plant to another plant on a non-inductive cycle promotes flowering in the receptor plant. Zeewaart (1958) showed that grafted donor leaves from the short-day plants *Perilla* and *Xanthium* cause a flowering response in receptor plants on a long-day treatment. In *Perilla* it was possible to remove leaves in which more than 3 months had elapsed since the last short-day cycle and regraft them and cause flowering in the receptor plant. That the leaf is independently capable of perceiving the photoinductive stimulus is shown in experiments in which leaves of *Perilla* were detached from a vegetative parent plant exposed in isolation to the appropriate photoinductive treatment and then grafted to receptor plants maintained on unfavourable daylengths. Such leaves in *Perilla* readily caused flowering (Zeewaart 1958). Induction of excised leaves does not appear to have been reported for long-day plants.

Phytochrome and Photoperiodism

The finding that a flash of light in the middle of a long dark period (night break) prevents flowering in short-day plants suggests that light has to be absorbed to be effective; there must be a photoreceptor involved. In the early 1950s, S.B. Hendricks and H.A. Borthwick, at USDA, compared the absorption spectrum of known constituents of the plant with the action spectrum of the photobiological process under investigation. If the absorption spectrum of the extracted plant constituent closely resembles the action spectrum of the process, it is a strong indication of the

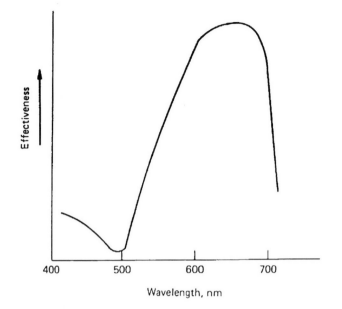

Figure 7.9 The action spectrum of photoperiodism

involvement of that constituent in the process; it is most likely the photoreceptor that initiates the process. For example, the most effective wavelengths of light in photosynthesis are in the blue and red parts of the spectrum and it is in these regions that chlorophyll absorbs maximally. Also, the action spectrum for *Avena* coleoptile curvature closely resembles the absorption spectrum for riboflavin; thus, riboflavin is thought to be the most likely photoreceptor in photoperiodism.

Hendricks and Borthwick (1954) examined the action spectrum for the inhibitory effects of the night breaks during the dark phase of photocycles. They exposed the leaves of the short-day soybean to appropriate short days but interrupted the dark period with a flash of light, of various wavelengths. The wavelengths most effective in preventing flowering were in the red part of the spectrum (Fig. 7.9). Unlike photosynthesis, shorter wavelengths (blue light) were not effective and there was also a rapid falling-off in the far-red region. Hendricks and Borthwick returned to some earlier work on the effect of light on light-sensitive Grand Rapids lettuce seed germination (Flint and McAlister 1935). These workers had found that germination was promoted by red light but not by blue or far-red (Fig. 7.10).

Hendricks and Borthwick tried experimentally to determine whether the inhibiting far-red light would reverse the promoting effect of red light and *vice versa*. They established that red light promoted germination and that far-red light nullified this promotion (Table 6.1). They then tried a similar experiment on the flowering of short-day plants and found that interrupting the long night with flashes of red or far-red light would inhibit or promote flowering respectively (Fig. 7.11). Further, a flash of far-red light following a flash of red light would nullify the effect of the red flash and the plant would flower as if the dark period was not interrupted.

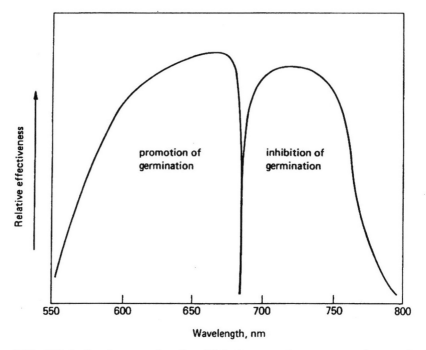

Figure 7.10   Effect of various wavelengths of light in promoting or suppressing germination in Grand Rapids lettuce seeds. (From Borthwick *et al.* 1952)

These findings were supported by work carried out by Downs (1956) with *Xanthium* and Biloxi soybean which showed that the radiation applied last in the sequence determines the response of the plant (Table 6.2). Red/far-red photoreversibility has also been demonstrated in some long-day plants in which a red nightbreak promoted flowering and a subsequent far-red exposure prevented flowering (Fig. 7.11).

The similarities of the absorption spectra of the two forms of phytochrome and the action spectra of plant responses confirm the involvement of phytochrome as the receptor for these responses (Fig. 7.12). That the response caused by red light is reversed by an immediate subsequent exposure to far-red light is further evidence. Other supporting evidence, as already mentioned, is that only low irradiance levels of either red or far-red light are capable of converting phytochrome from one form to the other to cause these responses. It is also interesting to note that induction occurs very rapidly on irradiation, despite the fact that floral primordia may not be detectable for days or even weeks afterwards. The rapidity of induction is such that sensitivity is lost within 30 minutes and application of far-red light after this time interval fails to restore flowering in short-day plants (Downs 1956).

Flowering Hormones

The fact that photoperiodic perception occurs in the leaf and the resulting signal takes effect some distance away in the meristem satisfies one important criterion for

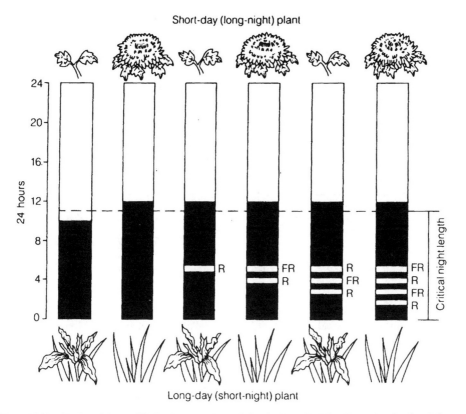

Figure 7.11 Red and far-red light interruption of the dark period for flowering in short-day and long-day plants. (From Taiz and Zeiger 1991)

a hormonal effect. Michael Chailakhyan (1968), working in Russia during the 1930s, coined the term 'florigen' for the unidentified, photoinduced, hypothetical, flowering hormone thought to be present in plants. Support for the concept of florigen and proof of its ease of translocation comes from early work (Naylor 1961) with two-branched *Xanthium* plants grafted in series. If the end branch of a series of plants is given a photoinductive cycle, while the second branch of the first plant and of the other five plants are kept on a non-inductive cycle, it will cause flowering in a chain-like manner in all plants of the series (Fig. 7.13).

These grafting experiments demonstrate that the flowering stimulus moves only through a living-tissue union between graft partners and probably only through phloem. Treatments such as girdling (removal of tissue external to the xylem), localized heating or cooling, which restrict phloem transport prevent translocation of the floral stimulus. Studies (Evans and Wardlow 1966; King *et al.* 1968) have shown that the rate of transport of the flowering signal is comparable to or slightly slower than the rate of translocation of sucrose in phloem tissue. These findings suggest that the stimulus is chemical (as against electrical) in nature and is probably carried along with assimilates in the translocation stream. The fact that in some

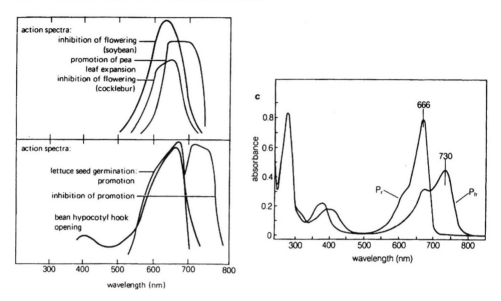

Figure 7.12 A comparison of the absorption spectra of both forms of phytochrome with action spectra of various physiological processes in plants. (a. Parker *et al.* 1949; b. Withrow *et al.* 1957 and Borthwick *et al.* 1954; c. Vierstra and Quail 1983)

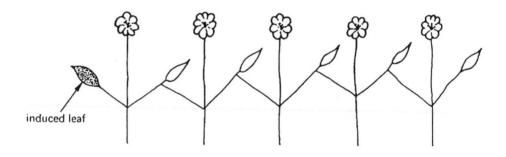

Figure 7.13 Translocation of a floral stimulus from a photoinduced plant to a series of plants maintained on a non-inductive cycle

species the flowering stimulus is exported from very young leaves that would be expected to be importing assimilates, indicates that it may move by other pathways in plants growing under different conditions. Other grafting experiments (Zeewaart 1958) also demonstrated that the flowering stimulus is not species-specific and that it has very similar properties in both short-day and long-day plants. Zeewaart grafted long-day plants to short-day plants and *vice versa*. When the long-day *Sedum spectabile* was grafted to the short-day *Kalanchoe blossfeldiana*, it flowered under short-day conditions. When the latter plant was grafted to the long-day plant, it flowered under long-day conditions. Also, Hodson and Hamner (1970) showed that extracts from flowering *Xanthium* could induce flowering in both vegetative duckweed (*Lemna purpusilla*) and vegetative *Xanthium*, but extracts from vegetative *Xanthium* could not. To obtain flowering in *Xanthium*, it is necessary to supplement the flower-inducing extract with gibberellic acid (GA_3); the *Xanthium* extract on its own is sufficient to induce flowering in *Lemna*. That the flowering stimulus might be common to many plants of differing response types is indicated from the results obtained from interspecific and intergeneric grafts. Since the stimulus is transmitted only through a successful graft union, transmissibility can only be demonstrated in dicots and then only in species that are closely related. Successful transmission of a flowering stimulus were recorded by Lang (1965); a selection of examples are given in Table 7.4.

The majority of experiments have been carried out with herbaceous plants but graft transmission of flowering has also been reported for woody species. Successful intergeneric and interspecific grafts suggest that the stimulus is interchangeable between different genera and species of plants. While a short-day plant may flower on exposure to a single long dark period, flowering is enhanced if the inductive dark period is preceded by a period of light. For example, for maximum flowering with a single inductive dark period *Xanthium* requires 8 hours of light. *Pharbitis* requires at least 6 hours, while in *Kalanchoe* a few seconds of light per day is sufficient. Where longer periods of light are required, it may be that photosynthesis is necessary to source a translocation stream; where only short periods of light are required, clearly

Table 7.4 Successful transfers of flowering by grafting in various members of the Solanaceae. (From Lang 1965)

Response type	Flowering donor	Receptor
Day-neutral	*Nicotiana tabacum*	*Nicotiana tabacum*
	Trapezond and others	Trapezond and others
Short-day	*Nicotiana tabacum* Maryland Mammoth	*Nicotiana tabacum* Maryland Mammoth
Long-day	*Nicotiana sylvestris*	*Nicotiana sylvestris*
	Hyoscyamus niger	*Hyoscyamus niger*

photosynthesis is not involved. In many cases, high-fluence light following the inductive dark periods is also important; this may be necessary to provide assimilates for the translocation stream and enhance transport from the leaf to the apical meristem.

There is a substantial amount of evidence showing that leaves on non-inductive cycles may exert an inhibitory effect on flowering. If a plant is induced to flower by exposing only one leaf to a favourable photoperiod, flowering is usually inhibited by non-induced leaves on the plant. Also, in order to successfully transfer the flowering stimulus by means of a graft union, the receptor plant must be defoliated. Inhibition by leaves on non-inductive cycles is usually not observed unless they occur between the induced leaf and the receptive meristem (Lang 1980). For example, a long-day leaf growing between the induced short-day leaf and the bud may competitively block the translocation of the flowering stimulus from the induced leaf to the meristem. Not all inhibitory effects can be explained on the basis of translocation. If the whole explanation were photosynthesis and translocation of assimilates, then light intensity and duration would be important. However, a brief night-break of low irradiance can cause inhibition in a single leaf on a short-day plant (King and Zeewaart 1973). If the inhibitory process originates in leaves on non-inductive cycles, it may, in some way, antagonize induction or may lead to the production of a transmissible substance (an 'antiflorigen') that subsequently inhibits flower initiation at the apex. Thus, it seems that photoperiodic control of flowering is regulated by more than one transmissible substance: a promoter from induced leaves and an inhibitor from non-induced leaves. In the pea, where both promoters and inhibitors appear to be present, results from grafting experiments suggest that an absolute amount of floral stimulus is required for floral primordia initiation rather than a simple balance between two opposing substances (Taylor and Murfet 1994).

Given the general success with the five major groups of plant hormones, it is surprising that the identity of the hypothetical floral hormone, florigen, or of the inhibitory antiflorigen, has not been elucidated. Numerous extraction procedures have been used and fractionated extracts from flowering plants applied in an attempt to evoke flowering in non-induced plants. Unfortunately, although several attempts have been made, only a few have shown even limited success and these have not been consistently repeated. Failure to isolate the hypothetical florigen has cast doubt on its existence. Sachs (1978) doubted the florigen concept and suggested that photoperiodic induction causes a diversion of assimilates within the plant, leading to floral initiation. He suggested that nutrients diverted towards the apex might cause evocation; inhibitory effects might be due to diversion to sinks other than those meristems competent to produce flowers. However, the defoliation and grafting studies mentioned above are difficult to explain on the basis of an assimilate-diversion hypothesis. In summary, there is a lot of circumstantial evidence that flower initiation is controlled by plant hormones – a flower-inducing florigen and a flower-inhibiting antiflorigen. Neither of these hypothetical substances has yet been isolated or characterized.

Growth Regulators and Floral Initiation

The transition from a vegetative to a flowering condition at the shoot apex results from biochemical and cellular changes that lead to the production of floral primordia. It is likely that flowering involves a complex system of interacting factors, especially growth regulators. Some of these factors may act in sequence since some of the cellular changes occur before the floral stimulus takes effect at the apical meristem. In an attempt to identify florigen much attention has been focused on the known plant hormones, many of which are known to modify flowering behaviour in a wide variety of plants. All plant growth substances, with the exception of ethylene, are found in phloem, that is, they are translocated in the same conduit as the floral stimulus.

Auxins have been shown to inhibit flowering in short-day species. The inhibition is thought to occur in the leaf where it may antagonize the synthesis of the floral stimulus and/or interfere with its translocation. In long-day plants, auxins have been shown to be ineffective or to promote flowering. Auxins cause flowering in some bromeliads, notably pineapple. In pineapple and cocklebur, applied auxin causes the production of ethylene, which, like auxins, inhibits flowering in short-day plants and promotes flowering in bromeliads. Since ethylene is a gas it is unlikely that it is part of any floral stimulus; this does not, however, rule out precursors of ethylene which could be transported from the leaf.

Sometimes changes usually associated with flowering may be due to treatments that predispose the plant to flowering but do not themselves bring about flowering. For example, one of the earliest events to occur following photoperiodic induction is a transitory increase in the number of cells undergoing mitosis. When cytokinins are applied to the stem apex of the long-day, white mustard plant (*Sinapsis alba*), they cause an increase in cell division similar to that caused by exposure to a single long day. However, the long day induces flowering but the cytokinin treatment does not. Thus, while cytokinin may be a component of the floral stimulus, it is insufficient by itself to cause flowering. Gibberellins appear to be the only group of naturally occurring plant regulators to have an important role in flowering in a wide variety of higher plants (Table 7.5).

It is well established that gibberellins (notably GA_3) can substitute for photoperiodic induction in long-day plants (Lang 1965); gibberellin application can also evoke flowering in non-inductive conditions in a number of qualitative short-day plants (Sawhney and Sawhney 1985). The induction of flowering in response to gibberellin application is almost entirely restricted to those long-day plants that normally grow as rosettes in short days. In these plants, the flowering response, either to gibberellin or to long days, is accompanied by the elongation ('bolting') of the flowering stem. However, it is important to note that flower production and stem elongation are distinct, independent processes (Zeewaart 1976, 1984) and have been shown (Wellensiek 1973) to be under the control of two separate but closely linked genes (Chapter 6). Gibberellins can also substitute, either partially or totally, for the over-wintering, cold-treatment requirement ('vernalization') of certain biennial species before they flower as long-day plants.

Table 7.5 Floral induction by gibberellins in plants with different environmental requirements for flowering

Flowering requirement	Plant	Effect of gibberellin
Long-day plants	*Lolium*	Promotes in SD
	Fuchsia	Inhibits in LD
Short-day plants	*Zinnia*	Promotes in LD
	Fragaria	Inhibits in SD
	Xanthium	No effect
Day-neutral plants	Many conifers	Promotes
	Many woody angiosperms	Inhibits
Plants requiring vernalization	*Daucus*	Promotes
	Oenothera	No effect

The meristems then undergo a cold treatment during winter. In the following spring the stem undergoes rapid internode elongation and the plants subsequently flower in response to long days. Applied gibberellin can substitute for the cold treatment and with long days allows stem elongation and flowering to occur in a single, growing season. Gibberellins, however, will not evoke flowering in most short-day plants (e.g. *Xanthium* and Biloxi soybean) maintained under long-day conditions. The main effect of gibberellins on rosette plants under short days is to stimulate cell division immediately below the apical meristem, just as GA_3 does in stems of dwarf cultivars and other elongating stems.

Gibberellic acid is essential in the stimulation of stem elongation growth (bolting), which is a prerequisite for flowering in many long-day plants; in most short-day plants it is ineffective. Thus, while gibberellic acid may be part of a complex floral stimulus in a wide range of plants, it is not in itself florigenic. Chailakhyan used the term 'anthesin' to denote a hypothetical flower-initiating hormone that works with gibberellin to produce florigenic activity. He suggested that gibberellin is nearly always present in sufficient amounts for flowering in short-day plants (regardless of daylength) since its addition seldom causes short-day plants to flower. Long-day plants, however, lack sufficient gibberellin so that it must be added either artificially or naturally (by inductive long days) for flowering to occur. Short-day plants, although they contain gibberellin, do not flower except in short days. Chailakhyan therefore suggested that another hypothetical hormone, anthesin, is necessary in addition to gibberellin in order to cause flowering. According to his theory, anthesin is formed in short-day plants as a result of inductive short days but is probably present in sufficient amounts in long-day plants regardless of daylength (like gibberellin in short-day plants). Day-neutral plants have sufficient quantities of both anthesin and gibberellin present in order to flower without induction (Table 7.6). However, as with florigen, this requires another hypothetical substance not yet isolated or characterized. The involvement of light and plant growth regulators in mediating growth and development and in evoking

Table 7.6   The anthesin hypothesis as elaborated by M. Chailakhyan

	Short days	Long days
Short-day plants	gibberellin + anthesin = flowering	gibberellin only = no flowering
Day-neutral plants	gibberellin + anthesin = flowering	gibberellin + anthesin = flowering
Long-day plants	anthesin only = no flowering	gibberellin + anthesin = flowering

cross-talk between these overlapping components of morphogenic responses in higher plants is discussed in Chapter 6.

Biological Clocks

Exposure to alternating periods of light and darkness is a feature of the environment of most organisms; plants and animals usually exhibit rhythmic behaviour in association with these changes (Brady 1982; Sweeney 1987). Many of these responses, such as leaf movements (day and night positions), stomatal movements (opening and closing) and metabolic processes (photosynthesis and respiration) are examples of adaptations to the fluctuating environment. Clearly the time-keeping of these fluctuations must involve some kind of clock in order to measure the duration of light and darkness; there must also be a photoreceptor to distinguish between light and its absence.

Timing devices fall into two categories – hourglass and oscillator. An hourglass device measures the time interval required to complete some operation and then it must be reset. This sort of clock is affected by temperature and is therefore not very accurate except under constant conditions. An oscillator, on the other hand, depends on the regular oscillation of a system of fixed period, such as the pendulum or the balance wheel of a clock. When an organism is removed from all rhythmic fluctuations such as the daily cycles of light and darkness and transferred to continuous darkness (or continuous dim light) the rhythmic processes continue, at least for some time. Since they continue in constant light or darkness, these rhythms cannot be direct responses to either light or darkness and so must be based on an internal pacemaker. This pacemaker is often referred to as an endogenous oscillator – a *biological clock*. One of the most striking manifestations of the biological clock is the rise and fall of the leaves (nyctinastic or sleep movements) of certain species (Fig. 7.14).

In the 1920s, Rose Stoppel, in Hamburg, studied the leaf movements of the bean *Phaseolus vulgaris*. Stoppel observed that when leaf movements were measured in a darkroom, at constant temperature, the maximum vertical position (night position) occurred at the same time every day, between 3 and 4 a.m. Because the period was almost 24 hours, Stoppel reasoned that a biological clock could probably not be so accurate unless some external factor in the environment was resetting the clock daily. Since the plants were kept in continuous darkness at constant temperature, this factor could not be light or darkness; Stoppel called it 'factor *x*'. Bünning and

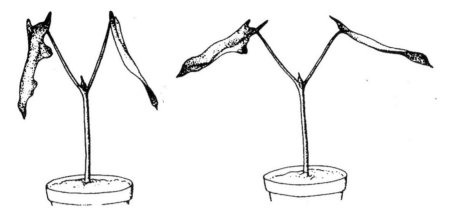

Figure 7.14   Sleep movements in leaves of *Phaseolus multiflorus*. Left: night position; right: day position. (From Bünning, 1953)

Stern (1930) repeated Stoppel's work. In contrast to Stoppel's practice of watering the plants each morning under a red 'safe' light (at that time it was believed that red light had no influence on plant movements or on photomorphogenesis), Bünning and Stern grew their plants in a cellar and watered them in the afternoon in virtual darkness. They found that most of the maximal night positions no longer occurred between 3 and 4 a.m. but rather between 10 a.m. and 12 midday. They concluded that Stoppel's factor x was red light and was not 'safe' but must synchronize the movements so that the night position always occurs about 16 h after the light's action. When the light was eliminated, the period was no longer exactly 24 h but rather 25.4 h, so that the leaf movement was soon out of phase with day and night outside the darkroom.

Terminology

Biological rhythms are based on physical wave phenomena and the terminology that describes physical wave oscillations is adopted for use with biological cycles. A rhythm is the repeated occurrence of some function and the period of the rhythm is the time taken for an oscillation to make a complete cycle and return to the original starting position. The term 'phase point' is used for any point in the cycle recognizable by its relationship to the rest of the cycle; the most obvious phase points are the maximum (peak) and minimum (trough) positions. The range is the difference between the maximum and the minimum positions while the magnitude of the oscillation (height of peak or depth of trough) is known as the amplitude and can vary while the period remains unchanged (Fig. 7.15a). Frequency is the number of cycles per unit time and thus describes period length. When rhythms are allowed to continue under constant environmental conditions (continuous darkness or dim light), they drift in relation to solar time, either gaining or losing time depending on whether the period is shorter or longer than 24 h, in other words the periodic clock

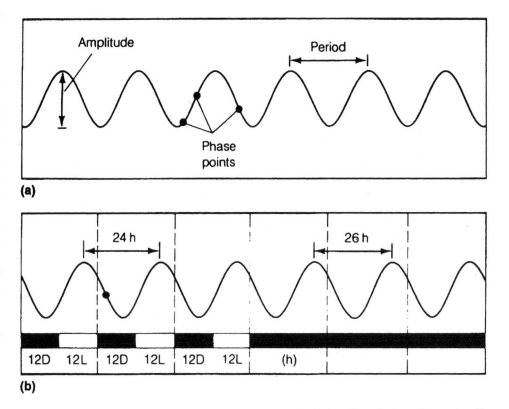

Figure 7.15 Features of circadian rhythms: (a) a typical circadian rhythm; (b) a circadian rhythm entrained to a 24 h light/dark cycle and its reversion to the free-running period (26 h in this example) following transfer to continuous darkness. (From Taiz and Zeiger 1991)

can 'give the wrong time'. Under these constant conditions, the rhythm is said to be 'free-running' (Fig. 7.15a) and its periodicity is then close to but not exactly 24 h (for example, bean leaf movements under constant conditions have a period of 25.4 h). In the 1950s, Franz Halberg (Halberg *et al.* 1959) coined the term *circadian* (Latin, *circa* = about + *dies* = day) to describe such rhythms. In addition to circadian rhythms (periods of 20 to 28 h) there are *ultradian* (periods less than 20 h) and *infradian* (periods longer than 28 h) rhythms.

Because the period of a free-running rhythm is not exactly 24 h and the rhythm drifts in relation to solar time, free-running periods longer than 24 h have a peak time later each day and would therefore gain time; rhythms with a period shorter than 24 h peak earlier and would similarly lose time. Under natural conditions, therefore, the endogenous oscillator has to be *entrained* (synchronized) to set the rhythm to a true 24 h cycle. This is achieved by means of periodic environmental signals, the most important being the light-to-dark transition at dusk and the dark-to-light transition at dawn. Such signals are termed *zeitgebers* (German for 'time-givers'). When the entraining signals are removed, for example by transfer of the organism to continuous darkness, the rhythm reverts to the free-running pattern with the endogenous period characteristic of the particular organism (Fig. 7.15b). Within limits, rhythms can be entrained to light–dark cycles either shorter or longer than 24 hours. Entrainment to very short or long cycles, however, is rare, since plants cannot easily adjust to such extreme programmes.

The amplitude of a free-running circadian rhythm usually diminishes with time until it eventually disappears altogether. This damping out of the rhythm may be due to declining energy reserves in prolonged darkness, since the amplitude can be maintained, for some time, by feeding sucrose. Although the rhythms are innate, they normally require an environmental zeitgeber, such as a light–dark transfer or change of temperature, to restart them. Temperature has little or no effect on the periodicity of free-running rhythms, the Q_{10} of which is close to 1 with observed values ranging from 0.8 to 1.3 (Sweeney 1987). While a change in temperature may affect the amplitude of the oscillation, it causes only a transient alteration in the periodicity and the normal, free-running period is usually re-established after a few cycles. Bünning (1930) showed that in bean seedlings raised from seed in the dark, leaf movements were small and unsynchronized. A zeitgeber of a single flash of light initiates large synchronized movements with a periodicity of 28.3 h at a constant 15 °C and 28 h at 25 °C. However, when seedlings are changed from 20 °C to 15 °C, the initial period is 29.7 h; after one or two cycles, periods of approximately 28 h become re-established. The clock mechanism is therefore temperature-compensated in some way. This insensitivity to temperature is an essential characteristic to allow the clock to keep accurate time under the fluctuating temperatures experienced under natural conditions.

Phase Responses

Discussions of endogenous rhythms are complicated by the fact that two time frames are involved: solar time and circadian time. Solar time is based on a 24 h day. Circadian time is based on the free-running time and one cycle is considered to be 24 hours long, regardless of the length in solar time. Thus, if the free-running period is 27 hours, an event that occurs at 13.5 h of darkness will have occurred at 12 h in circadian time. When an organism is entrained to a certain light/dark regime and then allowed to free-run in darkness, the phase of the free-running rhythm that corresponds to day in the previous entraining cycle, is called the *subjective day* and that which coincides with the dark period is called the *subjective night*.

Bünning (1936) suggested that the measurement of time in photoperiodism is dependent on an endogenous oscillation. He proposed that photoperiodic time-keeping involves a regular oscillation of phases of the rhythm with different sensitivities to light. He suggested that the transfer of a plant to light set in motion a 12 h *photophile* (light-loving) phase which is followed by a 12 h *skotophile* (dark-loving) phase. The involvement of circadian rhythms in flowering was discovered during the night-break experiments in which short-day plants were prevented from flowering by interrupting the long night with brief flashes of red light. Short-day (long-night) plants have an endogenous rhythm of sensitivity to night breaks. If the night break occurs at 6 hours into the dark period, at 30 hours (24 + 6) or at 54 hours (24 + 24 + 6) and so on, flowering is inhibited. If the night break is given when the endogenous rhythm is in the photophile phase corresponding to daytime, such as at 16 hours, 40 hours (24 + 16) or 64 hours (24 + 24 + 16) after the beginning of the dark period, flowering is not inhibited. Under experimental conditions, when a plant is placed under continuous darkness, the photophile phase is broadly equivalent to subjective day and the skotophile phase broadly equivalent to subjective night.

The operation of the endogenous oscillator sets a response to occur at a particular time of day, that is, the period of the endogenous rhythm is fixed but it may be 'fast' or 'slow' relative to the 24-hour solar day. Under natural conditions, the daily durations of night and darkness are continually changing with the seasons so that the entraining light zeitgebers do not always occur at the same solar time. Organisms must, therefore, adapt to entrainment under changing photoperiods and change the phase of the rhythm by moving the whole cycle forwards or backwards in time, without altering the period. In order to examine the response of the endogenous oscillator to light signals (dawn = light on, dusk = light off) the organism is placed in continuous darkness and given a brief exposure to light at different times during the free-running period (Fig. 7.16). The light pulses will reset the clock and the direction and magnitude of the resulting phase shift can be plotted to give a *phase-response curve* (Fig. 7.16) which shows the delay or advance of the rhythm of the flowering response to a night break.

A typical phase-shifting response to a light pulse given shortly after transfer to darkness is shown in Figure 7.17. If we consider a phase point that occurs 1 hour after the onset of darkness in a 12 h light/12 h dark cycle, i.e. if daylength is extended by 1 hour, this phase point will appear to have drifted forwards relative to the dusk (light off) zeitgeber and the response is a correcting backwards shift (phase delay) such that the phase point still occurs 1 hour after the end of the light period and the rhythm stays on local time. In the same way, the response to light perceived at a phase point that would normally occur 1 hour before dawn, would be a phase advance. In other words, if a light pulse is given during the first few hours of the subjective night, the rhythm is delayed, that is, the peak comes later than expected. It is as though the light pulse acted as dusk but by coming later it caused a delay. In contrast, a light pulse given towards the end of the subjective night advances the phase of the rhythm and the next peak comes earlier than expected; the flash of light then acts as dawn. Phase shifting in this way constantly synchronizes or entrains the rhythm to local time. Experimentally derived phase-response curves provide a fairly accurate picture of the relationships between the plant's circadian timekeeping and local, solar time.

Biological clocks and various rhythms have been found in virtually all eukaryotes that have been studied (Table 7.3). Several rhythms have been studied in single-celled organisms. The biological clock in the marine dinoflagellate *Gonyaulax polyedra*, which has rhythms for three important variables – bioluminescence, photosynthesis and cell division, has received much attention (Sweeney 1976). Most spectacular is the rhythm of bioluminescence, which can be seen when

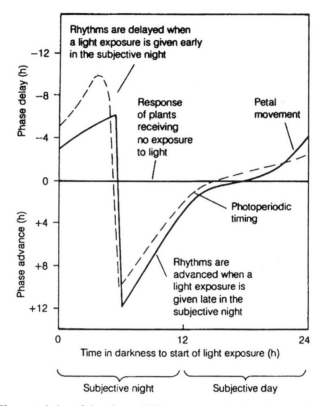

Figure 7.16   Characteristics of the phase-shifting response in circadian rhythms. (From Zimmer 1962)

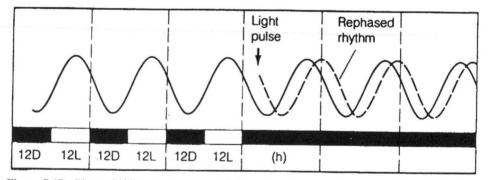

Figure 7.17   Phase-shifting response to a light pulse given shortly after transfer to darkness. (From Taiz and Zeiger 1991)

a suspension of *Gonyaulax* cells is mechanically disturbed, causing the cells to emit light. The amount of light emitted follows a circadian rhythm. Different rhythms have different phase relationships with the entraining light cycles resulting in maximum photosynthetic capacity shortly after midday, maximum 'flashing' luminescence near midnight and maximum 'glowing' luminescence towards the

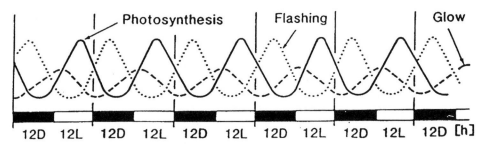

Figure 7.18 The phase relationships of different rhythms in *Gonyaulax polyedra* entrained to 12 h light/12 h dark cycles. (From Vince-Prue 1986)

end of the night (Fig. 7.18). The bioluminescence rhythm has been shown to match a daily rhythm in the activity of luciferase (Johnson *et al.* 1984; Morse *et al.* 1990), the enzyme that catalyses the reaction for light emission, and also in the cell content of luciferin, which is oxidized in the reaction (Fig. 7.19).

Endogenous circadian rhythms have been shown in amorphous tissue cultures. This is evidence that the rhythm is independent of the organization of tissues and organs. With callus tissue it has been shown that the release of CO_2 after culture in a 12 h light/12 h dark cycle, proceeds rhythmically after transfer to continuous darkness. The tissue culture thus shows an endogenous rhythm of CO_2-release very similar to that of an intact leaf under the same conditions. Although light plays an obvious role in entraining circadian rhythms, the photoreceptor is not the same in all organisms. For some rhythms such as CO_2-evolution in *Bryophyllum* and *Lemna*

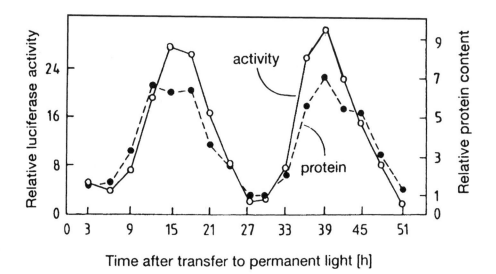

Time after transfer to permanent light [h]

Figure 7.19 Oscillation in the luciferase content of cells of *Gonyaulax polyedra* cultured in a 12 h light/12 h dark cycle and transferred to continuous weak light. (From Morse *et al.* 1990)

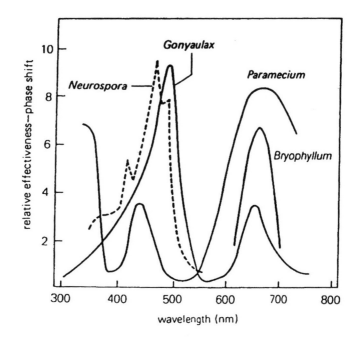

Figure 7.20 Action spectra for phase shifting in the rhythms of *Neurospora*, *Gonyaulax*, *Paramecium* and *Bryophyllum*. (From Ehret 1960; Munoz and Butler 1975; Sweeney 1987; Wilkins and Harris 1975)

and leaf movements in *Phaseolus*, phytochrome has been identified as the photoreceptor. Action spectra for the rhythms in *Gonyaulax* and certain fungi share a large peak in the blue region of the spectrum suggesting that a flavoprotein may be the photoreceptor (Fig. 7.20).

In higher plants, light-dependent development is a complex process involving the combined actions of several photoreceptor systems. During seedling development light stimulates leaf and chloroplast differentiation, inhibits the rate of hypocotyl growth and induces the expression of nuclear- and chloroplast-encoded genes. Later, light sensing allows plants to time the transition from vegetative to flowering and reproductive growth. The photoreceptors involved in these processes include red/far-red light-absorbing phytochrome (Quail *et al.* 1995), the blue/UV-A light-absorbing cryptochromes (Ahmad and Cashmore 1996) and distinct UV-A (Young *et al.* 1992) and UV-B (Beggs and Wellman 1985) light photoreceptors. The photoreceptors react with circadian systems at several levels in higher plants, although the details of these interactions are, as yet, not known. For example, red and/or blue light pulses control the phase of the clock, thereby mediating entrainment to the day/night cycle (Simon *et al.* 1976). The timing control of the biological clock on metabolism is evidenced by the observed circadian control of the expression level of many plant genes. Circadian rhythms have been reported in gene transcription in *Arabidopsis* (reviewed by McClung and Kay 1994), mRNA

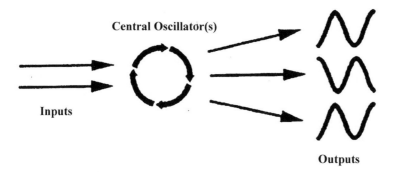

Figure 7.21   A model for circadian clock systems in bean. (From Kreps and Kay 1997). The output from the central oscillator produces rhythms that can differ in phase from one another (Zhong and McClung 1996; Kreps and Simon 1997)

translation (Mittag *et al.* 1994) and post-translational modification of proteins (Carter *et al.* 1991). In some pathways, such as the steps involved in photosynthesis, carbon fixation and nitrogen metabolism, there is a clear need for the proper regulation of timing. However, there are other examples in which the rationale for circadian cycling of expression is not so clear or where the gene products encode pioneer proteins which do not have a known function (Reimmann and Dudler 1993).

A single organism may have more than one biological clock. For example, Roenneberg and Morse (1993) found different oscillators within the same cell in *Gonyaulax polyedra*. In addition, Hennessey and Field (1992) showed that the bean *Phaseolus vulgaris* has circadian rhythms in stomatal opening and leaf movement, which have different free-running periods; this is possible only if there are different clocks (Pittenberg 1993). Kreps and Kay (1997) suggested that another possibility is that these two rhythms in bean have different free-running periods resulting from cell-specific modification of the same clock component. These authors propose that a model for circadian rhythm has input pathways, such as light (or temperature) which interact with at least one central oscillator, which, in turn, generates output rhythms via a range of cellular signalling pathways (Fig. 7.21).

Summary

Photoperiodism is a response to the duration and timing of light and dark periods. Plants can be divided into three general photoperiodic types – long-day plants (LDPs), short-day plants (SDPs) and day-neutral plants (DNPs). An important distinction between LDPs and SDPs is not the absolute duration of light received but when a plant will flower – before or after a critical length of time. Flowering in SDPs occurs only when the daylength is less than a specific critical length of time while flowering in LDPs occurs only after a critical daylength has been exceeded.

The absolute critical daylength varies from one species to another and the critical daylength for an LDP may be shorter than the critical daylength for an SDP.

By subjecting plants to daylengths other than 24 hours, flowering was shown to be more a response to darkness than to light. Interruption of the dark period with a flash of light (night break) nullifies the darkness and inhibits flowering; the most effective wavelengths in preventing flowering are in the red part of the spectrum. The response to the night break is mediated by phytochrome: P_{fr} inhibits flowering in SDPs but promotes flowering in LDPs.

Photoperiodic stimuli are perceived by the leaf but the response occurs elsewhere in the meristems. This transmission of a signal suggests the involvement of a floral hormone, which has been called florigen. While phytohormones such as gibberellin can modify flowering, attempts to isolate and characterize florigen have been unsuccessful.

Photoperiodism in plants is controlled by an endogenous oscillator. Time measurement in photoperiodism involves the interaction of phytochrome and a biological clock in order to regulate circadian rhythms. Maintaining a rhythm on local time depends on the phase response of the rhythm to environmental signals. The most important signals for shifting the phase ('setting the clock') are the light on/light off switches of dawn and dusk.

Photoperiodism enables plants to sense seasonal change and to synchronize their life cycles to the time of year. By this means, different plant species flower in their own optimum time for pollination and fertilization, in order to avoid competition and adverse winter weather conditions.

References

Ahmad, M. and Cashmore, A.R. (1996). Seeing blue: the discovery of cryptochrome. *Plant Mol. Biol.* **30**: 162–166.

Battey, N.H. and Lyndon, R.F. (1990). Reversion of flowering. *Bot. Rev.* **56**: 162–180.

Beggs, C.J. and Wellman, E. (1985). Analysis of light-controlled anthocyanin formation in coleoptiles of *Zea mays* L. The role of UV-B, blue, red and far-red light. *Photochem. Photobiol.* **41**: 481–486.

Bidwell, R.G.S. (1974). Organization in time, *Plant Physiology*, pp. 409–447. Macmillan, New York.

Borthwick, H.A., Hendricks, E.D., Parker, M.W., Toole, E.H. and Toole, V.K. (1952). A reversible photoreaction controlling seed germination. *Proc. Natl. Acad. Sci. USA* **38**: 662–666.

Borthwick, H.A., Hendricks, E.D., Toole, E.H. and Toole, V.K. (1954). Action of light on lettuce seed germination. *Bot. Gaz.* **115**: 205–225.

Brady, J. (ed.) (1982). *Biological Timekeeping*. Cambridge University Press, Cambridge.

Bünning, E. (1936). Die endogene Tagesrhythmik als Grundlage der photoperiodiaschen Reaktion. *Ber. Deut. Bot. Ges.* **54**: 590–607.

Bünning, E. (1953). *Entwicklungs- und Bewegungsphysiologie der Pflanze*. Springer, Berlin.

Bünning, E. and Stern, K. (1930). Uber die tagesperiodischen Bewegungen der Primarblatter von Phaseolus multiflorus II. Die Bewegungen bei Termokonstanz. (The diurnal movements of the primary leaves of Phaseolus multiflorus II. Movement at constant temperature.) *Berichte der Deutschen Botanischen Gesellschaft.* **48**: 227–252.

Carter, P.J., Nimmo, H.G., Fewson, C.A. and Williams, M.B. (1991). Circadian rhythms in the activity of a plant protein kinase. *EMBO J.* **10**: 2063–2068.

Chailakhyan, M.K. (1968). Internal factors of plant flowering. *Ann. Rev. Plant Physiol.* **19**: 1–36.

Cleland, C.F. and Briggs, W.R. (1967). Flowering responses of the long-day plant *Lemna gibba* G₃. *Plant Physiol.* **42**: 1553–1561.

Downs, R.J. (1956). Photoreversibility of flower initiation. *Plant Physiol.* **31**: 279–284.

Ehret, C.F. (1960). Action spectra and nucleic acid metabolism in circadian rhythms at the cellular level. *Cold Spring Harbor Symposia on Quantitative Biology*, **XXV**, pp. 149–158. Biological Clocks. The Long Island Biological Association, Cold Spring Harbor, NY.

Evans, L.T. (1969). The nature of flower induction. In: *The Induction of Flowering, Some Case Histories*, L.T. Evans (ed.), pp. 457–480, Macmillan of Australia, South Yarra, Vic.

Evans, L.T. and Wardlow, I.F. (1966). Independent translocation of ¹⁴C-labelled assimilates and of the floral stimulus in *Lolium temulentum*. *Planta* **68**: 310–326.

Flint, L.H. and McAlister, E.D. (1935). Wavelengths of radiation in the visible spectrum inhibiting the germination of light-sensitive lettuce seeds. *Smithson. Misc. Collns.* **94**: No. 5, 1–11.

Garner, W.W. and Allard, H.A. (1920). Effect of the relative length of day and night and other factors of the environment on growth and reproduction in plants. *J. Agric. Res.* **18**: 553–606.

Halberg, F., Halberg, E., Barnum, C.P. and Bittner, J.J. (1959). Physiologic 24-hour periodicity in human beings and mice, the lighting regime and daily routine. In: *Photoperiodism and Related Phenomena in Plants and Animals*, R.B. Withrow (ed.), pp. 803–878. Amer. Assoc. for the Advancement of Science, Washington, DC.

Hamner, K.C. (1940). Interaction of light and darkness in photoperiodic induction. *Bot. Gaz.* **101**: 658–687.

Hamner, K.C. and Bonner, J. (1938). Photoperiodism in relation to hormones as factors in floral initiation and development. *Bot. Gaz.* **100**: 388–431.

Hendricks, S.B. and Borthwick, H.A. (1954). Photoperiodism in plants. *Proc. 1st Int. Photobiol. Congr.*, The Netherlands, pp. 23–35.

Hennessey, T.L. and Field, C.B. (1992). Evidence of multiple circadian oscillators in bean plants. *J. Biol. Rhythms* **7**: 105–113.

Hodson, H.K. and Hamner, K.C. (1970). Floral-inducing extracts from *Xanthium*. *Science* **167**: 384–385.

Johnson, C.H., Roeber, J.F. and Hastings, J.W. (1984). Circadian changes in enzyme concentration account for rhythm of enzyme activity in *Gonyaulax*. *Science* **223**: 1428–1430.

King, R.W. and Zeewaart, J.A.D. (1973). Floral stimulus movement in *Perilla* and flower inhibition caused by non-induced leaves. *Plant Physiol.* **51**: 727–738.

King, R.W., Evans, L.T. and Wardlaw, I.F. (1968). Translocation of the floral stimulus in *Pharbitis nil* in relation to that of assimilates. *Z. Pflanzenphysiol.* **59**: 377–385.

Klebs, G. (1918). Uber die Blutenbildung bei *Sempervivum*. (On flower formation in *Sempervivum*.) *Flora* (Jena) **128**: 111–112.

Knott, J.E. (1934). Effect of a localized photoperiod on spinach. *Proc. Am. Soc. Hort. Sci.* **31**: 152–154.

Kreps, J.A. and Kay, S.A. (1997). Coordination of plant metabolism and development by the circadian clock. *Plant Cell* **9**: 1235–1244.

Kreps, J.A. and Simon, A.E. (1997). Environmental and genetic effects on circadian clock-regulated gene expression in *Arabidopsis*. *Plant Cell* **9**: 297–304.

Lang, A. (1965). Physiology of flower initiation. In: *Encyclopaedia of Plant Physiology*, **XV/1**, W. Ruhland (ed.), pp. 1380–1536. Springer-Verlag, Berlin.

Lang, A. (1980). Inhibition of flowering in long-day plants. In: *Plant Growth Substances 1980*, F. Skoog (ed.), pp. 310–322. Springer-Verlag, Berlin.

McClung, C.R. and Kay, S.A. (1994). Circadian rhythms in *Arabidopsis thaliana*. In: *Arabidopsis*, E.M. Meyerowitz and C.R. Somerville (eds), pp. 615–637. Cold Spring Harbor Laboratory Press, Cold Spring Harbor, NY.

Mittag, M., Lee, D.-H. and Hastings, W. (1994). Circadian expression of the luciferin-binding protein correlates with the binding of a protein to the 3' untranslated region of its mRNA. *Proc. Natl. Acad. Sci. USA* **91**: 5257–5261.

Morse, D.S., Fritz, L. and Hastings, J.W. (1990). What is the clock? Translational regulation of circadian bioluminescence. *Trends Biochem. Sci.* **15**: 262–265.

Munoz, V. and Butler, W. (1975). Photoreceptor pigment for blue light in *Neurospora crassa*. *Plant Physiol.* **55**: 421–426.

Naylor, A.W. (1952). The control of flowering. *Sci. Amer.* **186**: 49–56.

Naylor, A.W. (1961). The photoperiodic control of plant behaviour. In: *Encyclopaedia of Plant Physiology. Vol. 16*, W. Ruhland (ed.), pp. 331–389. Springer-Verlag, Berlin.

Parker, M.W., Hendricks, S.B., Borthwick, H.A. and Went, F.W. (1949). Spectral sensitivity for leaf and stem growth of etiolated pea seedlings and their similarity to action spectra for photoperiodoism. *Amer. J. Bot.* **36**: 194–204.

Pittenberg, C.S. (1993). Temporal organization: reflections of a Darwinian clock-watcher. *Ann. Rev. Physiol.* **55**: 17–54.

Quail, P.H., Boylan, M.T., Short, T.W., Xu, Y. and Wagner, D. (1995). Phytochromes: photosensory perception and signal transduction. *Science* **268**: 675–680.

Reimmann, C. and Dudler, R. (1993). Circadian rhythmicity in the expression of a novel light-regulated rice gene. *Plant Mol. Biol.* **22**: 165–170.

Roenneberg, T. and Morse, T. (1993). Two circadian oscillators in one cell. *Nature* **362**: 362–364.

Sachs, R.M. (1978). Nutrient diversion: an hypothesis to explain the chemical control of flowering. *Hort. Sci.* **12**: 220–222.

Salisbury, F.B. (1963). Biological timing and hormone synthesis in flowering of *Xanthium*. *Planta*. **59**: 518–534.

Sawhney, N. and Sawhney, S. (1985). Role of gibberellin in flowering of short-day plants. *Indian J. Plant Physiol.* **28**: 24–34.

Schwabe, W.W. (1959). Studies of long-day inhibition in short-day plants. *J. Exp. Bot.* **10**: 317–329.

Simon, E., Satter, R.L. and Galston, A.W. (1976). Circadian rhythmicity in excised *Samanea pulvini*. II. Resetting the clock by phytochrome conversion. *Plant Physiol.* **58**: 421–425.

Sweeney, B.M. (1976). Pros and cons of the membrane model for circadian rhythms in the marine algae, *Gonyaulax* and *Acetabularai*. In: *Biological Rhythms in the Marine Environment*, P.J. De Coursey (ed.), pp. 63–76. University of South Carolina Press, Columbia.

Sweeney, B.M. (1987). *Rhythmic Phenomena in Plants*. Academic Press, San Diego, CA.

Taiz, L. and Zeiger, E. (1991). The control of flowering, *Plant Physiology*, pp. 513–531. Benjamin/ Cummings.

Takimoto, A. (1960). Effect of sucrose on flower initiation of *Pharbitis nil*. *Plant Cell Physiol.* (Tokyo) **1**: 241–246.

Taylor, S.A. and Murfet, I.C. (1994). A short-day mutant in pea is deficient in the floral stimulus. *Flowering Newsletter* **18**: 39–43.

Tournois, J. (1914). Etudes sur la sexualité du houlbon. (Studies on the sexuality of hops.) *Annals des Sciences Naturelles (Botanique)* **19**: 49–191.

Vierstra, R.D. and Quail, D.H. (1983). Photochemistry of 124 kilodalton *Avena* phytochrome *in vitro*. *Plant Physiol.* **72**: 264–267.

Vince-Prue, D. (1986). The duration of light and photoperiodic responses. In: *Photomorphogenesis in Plants*, R.E. Kendrick and G.H.M. Kronenberg (eds), pp. 269–337. Martinus Nijhoff, Dordrecht.

Wellensiek, S.J. (1973). Genetics and flower formation of annual *Lunaria*. *Netherland J. Agric. Sci.* **21**: 163–166.

Wilkins, M.B. and Harris, P.J.C. (1975). Phytochrome and phase setting of endogenous rhythms. In: *Light and Plant Development*, H. Smith (ed.), pp. 399–417. Butterworth, London and Boston.

Withrow, R.B., Klein, W.H. and Elstad, V. (1957). Action spectra of photo-morphogenic induction and its photoinactivation. *Plant Physiol.* **32**: 453–462.

Young, J., Liscum, E. and Hangarter, R. (1992). Spectral-dependence of light-inhibited hypocotyl elongation in photomorphogenic mutants of *Arabidopsis*: evidence for a UV-A photosensor. *Planta* **188**: 106–114.

Zeewaart, J.A.D. (1958). Flower formation as studied by grafting. *Meded. LandbHoogesch. Wageningen* **58**: 1–88.

Zeewaart, J.A.D. (1976). Physiology of flower formation. *Ann. Rev. Plant Physiol.* **27**: 321–348.

Zeewaart, J.A.D. (1984). Photoperiodic induction, the floral stimulus and flower-promoting substances. In: *Light and the Flowering Process*, D. Vince-Prue, B. Thomas, and K.E. Cockshull (eds), pp. 137–142. Academic Press, London.

Zhong, H.H. and McClung, C.R. (1996). The circadian clock gates expression of two *Arabidopsis* catalase genes to distinct and opposite circadian phases. *Mol. Gen. Genet.* **251**: 196–203.

Zimmer, R. (1962). Phase shift and other light-interruption effects on endogenous diurnal petal movements. *Planta* **48**: 283–300.

8
Selected Photobiological Responses

Phototropism

During the later part of the nineteenth century, Charles Darwin and his son Francis studied plant tropisms. One of their interests was phototropism – the bending of plants towards light. The Darwins used seedlings of canary grass (*Phalaris canariensis*) for some of their studies. Grasses have their youngest leaf sheathed in a protective coleoptile. Coleoptiles are light-sensitive and when they are illuminated on one side for some time, they will bend towards the light source. The Darwins established that the tip of the coleoptile perceived the light: if the coleoptile was decapitated or the tip covered, no bending occurred. They concluded that some sort of growth signal is produced in the tip, travels downwards and causes the shaded side of the coleoptile to grow faster than the illuminated side, thus resulting in bending. The results of their work were published in 1881 in a book entitled *The Power of Movement in Plants*. The Darwins were aware that oat coleoptiles 'offer a strong exception to the rule that illumination of the upper part determines curvature of the lower part'. They noted that oat (*Avena*) coleoptiles bent towards light when the tips were covered and that sensitivity resides below the tips. Thus two kinds of curvature can be differentiated in coleoptiles – tip reaction and basal curvature; the tip response is about a thousand times more sensitive than the basal reaction. If dim light is used, most of the response is localized in the tip and curvature takes place in the subapical zone. If higher levels of light are used the reaction takes place over the entire length of the coleoptile.

The phototropic bending of plant hypocotyls and shoots towards unilateral light is one of the most widely studied low-fluence blue-light-mediated phenomena (Kaufman 1993; Liscum and Hangarter 1994; Poff *et al.* 1994; Short and Briggs 1994; Khurana *et al.* 1996). Green light, although significantly less effective, can also induce such phototropic responses (Steinitz *et al.* 1985). Evidence for the involvement of two distinct photoreceptors mediating responses to unilateral blue

Photobiology of Higher Plants. By M. S. Mc Donald.
© 2003 John Wiley & Sons, Ltd: ISBN 0 470 85522 3; ISBN 0 470 85523 1 (PB)

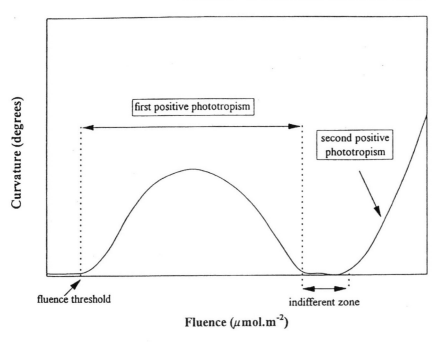

Figure 8.1   A diagrammatic representation of the fluence–response curvature for phototropism. (Adapted from Khurana and Poff 1989)

(450 nm) and green (500 nm) irradiations has been reported (Konjevic *et al.* 1989). The photoinductive curvature of a coleoptile is a function of the applied photon fluence (light dosage). The connection between the extent of phototropic curvature and unilaterally applied light fluence to the coleoptile tip is expressed in a fluence–response curve (Fig. 8.1). The fluence–response curve for phototropic curvature in hypocotyls consists of a first positive curvature induced by very low-fluence, unilateral blue light of short duration and a second positive curvature of relatively greater magnitude induced by prolonged exposure to unilateral blue light; the two curvatures are separated by a zone of indifference (Fig. 8.1).

The condition of induction implies that the Bunsen–Roscoe law of reciprocity is applicable. This law states that the extent of the reaction affected by light (photoresponse) depends only on photon fluence (mol photons m^{-2}) and not on the photon flux (mol photons m^{-2}s^{-1}) with which light is applied (i.e. the response depends only on the number of photons causing the response and is independent of the fluence rate used). The expression 'law of reciprocity' implies that to achieve a certain photoresponse, *a*, the necessary photon-fluence (*Fa*) can be applied either with a high photon flux (*J*) over a short irradiation time (*t*) or with a low photon flux over a long irradiation time:

$$Fa = Jt$$

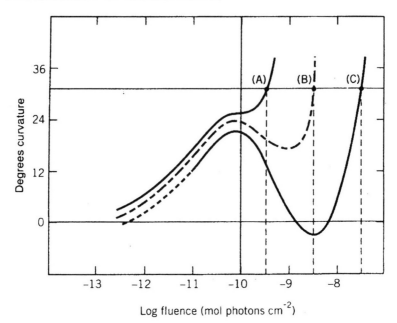

Figure 8.2 Phototropic photon fluence–response curves for *Avena* coleoptiles at three different fluence rates of blue light. (A) $1.4 \times 10^{-13} \, \text{mol cm}^{-2} \text{s}^{-1}$; (B) $1.4 \times 10^{-12} \, \text{mol cm}^{-2} \text{s}^{-1}$; (C) $1.4 \times 10^{-11} \, \text{mol cm}^{-2} \text{s}^{-1}$. (From Zimmerman and Briggs 1963)

Applicability of the law of reciprocity indicates that the photochemical primary reaction in phototropism is relatively simple. In a dose–response curve shown in Figure 8.2 the results obtained for curvature induced in *Avena* coleoptile by unilaterally applied blue light of three different fluence rates are plotted to show degrees of curvature as a function of light fluence.

First-positive curvature is generally described as bending of the apical tip of the coleoptile towards unilateral light delivered in brief pulses at very low fluences. At higher irradiances (unit radiation energy flux per unit area, $\text{J s}^{-1} \text{m}^{-2}$) curvature decreases with increasing fluence until the coleoptile bends away from the light source to give the first-negative curvature. With further increases in fluence, for longer durations, a second-positive curvature and even a third-positive curvature is achieved. Clearly reciprocity holds only for the first-positive curvature. The response is the product of the fluence rate and the exposure time, that is, the response depends only on the number of photons causing the response and is independent of the fluence rate used. Reciprocity does not apply to the first-negative or second-positive curvatures. Since reciprocity generally indicates involvement of a single photoreceptor, the absence of reciprocity in the second-positive response suggests that possibly more than one photoreceptor is involved. However, the determined action spectra for both first- and second-positive curvatures are identical, so it must be concluded that the same single photoreceptor acts for both. Phototropism is usually defined as a response to unilateral light.

Under natural conditions, however, the bending response occurs in plants that are receiving light from all sides. It is only necessary that there be an unequal distribution of the fluence rate (fluence per unit time, $\mu mol\, m^{-2}\, s^{-1}$); a 20% difference in fluence rate on the two sides of a coleoptile can induce bending. Thus, light can be applied unilaterally, bilaterally, from above or from all sides and a phototropic response is induced once a gradient in fluence rate is created across the plant organ. The optical properties of the tissue can modify the magnitude of the light gradient across plant cells. Plant pigments, including the photoreceptor pigment, can screen the applied light and attenuate the irradiance. Light can also be dissipated by scattering, reflection or diffraction within the cells. Gradients across individual cells, measured by micro-fibre-optic probes, may vary from 5:1 to 50:1 (Vogelmann and Haupt 1985). An additional complication is that plant organs, such as coleoptiles, appear themselves to function as fibre optics, transmitting light applied to the tip through cells further down the tissue.

The photoreceptor

Since the phototropic stimulus is light, a photoreceptor must be involved. In order to identify the pigment responsible for any photochemical process, it is essential to compare the action spectrum of the process with the absorption of the pigment(s) suspected to be involved. Most of the studies on phototropic response(s) have been carried out on *Avena* and *Zea* coleoptiles (Baskin and Iino 1987) and for the sporangiophores of the fungus *Phycomyces* (Presti *et al.* 1977). The action spectra for *Avena* (a monocot) and *Alfalfa* (a dicot) (Fig. 8.3) and for *Phycomyces*

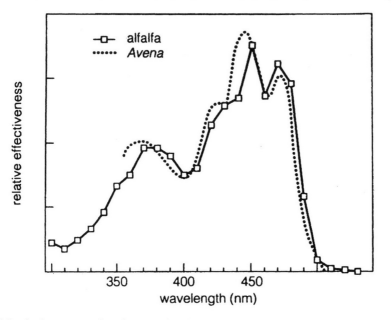

Figure 8.3   Action spectra for phototropism in coleoptiles of oat (*Avena sativa*), a monocot, and alfalfa (*Medicago sativa*), a dicot. (From Baskin and Iino 1987)

sporangiophores are virtually identical, suggesting that they share a common photoreceptor. All phototropic action spectra studied are similar and consistently show two peaks in the blue region (near 475 nm and 450 nm) and a shoulder at 420 nm (Fig. 8.3). In addition, there is a second peak in the UV-A region (near 370 nm).

On the basis of the action spectra, it is assumed that phototropism in higher plants is brought about by light absorbed by cryptochrome. Because the action spectrum shows a major peak in the blue region, the photoreceptor is likely to be a yellow pigment; two possible candidates are β-carotene and riboflavin (Fig. 8.4). However, neither pigment's absorption spectrum exactly matches the action spectrum for phototropism: β-carotene (in ethanol) is missing the secondary peak in the UV-A region, while riboflavin (in water) is missing the minor peaks (425, 445 and 475 nm) in the blue/violet range.

It was originally thought that the photoreceptor was a carotenoid. Carotenoids are ubiquitous in biological material where they are localized in tissues that show high phototropic sensitivity. Absorption of carotenoids exhibits the three-peak pattern that matches that of the phototropic action spectrum in the blue. Carotenoids, however, do not absorb significantly in the 300–400 nm range and so cannot match the action spectrum peak at 370 nm. Carotenoids are therefore thought not to be the blue-light receptor in phototropism. For example, albino mutants of *Phycomyces* (which do not contain carotenoids) exhibit a normal phototropic response (Presti *et al.* 1977).

In higher plants, albino mutants and seedlings treated with an inhibitor of carotenoid biosynthesis (the herbicide Flurozon) retain their phototropic ability

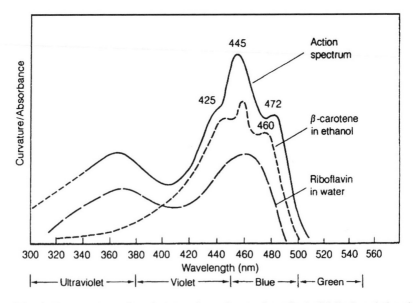

Figure 8.4 Action spectrum for phototropism of oat coleoptile (solid line) and the absorption spectra of β-carotene and riboflavin. (From Taiz and Zeiger 1991)

(Palmer *et al.* 1993). The fact that carotenoids in plant cells occur in plastids also seems inconsistent with a role in phototropism. The flavin molecule riboflavin was a later candidate. Flavins are also ubiquitous in biological material where they occur as prosthetic groups in the electron carriers flavin mononucleotide (FMN) and flavin adenine dinucleotide (FAD). Riboflavin absorbs blue light and has a peak in the 370 nm region as well (Fig. 8.4). Work by Konjevic *et al.* (1989) suggests that more than one blue-light photoreceptor may be involved; the coaction of a carotenoid and a flavin cannot be ruled out.

The first hint towards understanding phototropism at the biochemical level came when Gallagher *et al.* (1988) described the blue-light-dependent phosphorylation of a plasma membrane-associated protein in pea. The light-dependent phosphorylation of this protein, which ranges in size from 114 to 130 kDa depending on the species, seems to be ubiquitous in higher plants (Reymond *et al.* 1992). Other workers have revealed a strong correlation between the phosphorylation of this protein and phototropism with respect to tissue distribution and fluence–response characteristics (Short and Briggs 1994).

There is abundant physiological evidence suggesting that the light-induced phosphorylation of the plasma membrane protein is involved in phototropism: it occurs in the most photosensitive tissue, its action spectrum matches that of phototropism, it is sufficiently fast to precede curvature development, it shows the same dark-recovery kinetics as phototropism following a saturating light pulse, and both the phosphorylation and phototropism obey the rules of first-order photochemistry. In addition, Salomon *et al.* (1997b) have demonstrated a gradient in *in vivo* phosphorylation across coleoptiles unilaterally illuminated by certain fluences of blue light, with greater phosphorylation on the illuminated than on the shaded side. This is the first demonstration of a light-induced biochemical gradient that might be directly associated with phototropism.

It has been reported for maize that light-dependent protein phosphorylation is not restricted to the coleoptile tip but is also detected in basal parts of the coleoptile (Palmer *et al.* 1993). Salomon *et al.* (1997b) showed that while the protein phosphorylation levels in oat declined exponentially from tip to node, it was detectable along the entire length of the coleoptile. The results reported by Salomon *et al.* from oat coleoptile studies, suggest that the establishment of a pronounced gradient of protein phosphorylation might be a prerequisite for differential growth resulting in bending of the plant towards a light source. These workers hypothesize that the degree of asymmetric phosphorylation determines the degree of curvature. They also suggest that, at least for the basal part, the extent of asymmetric distribution of protein phosphorylation is regulated by time-dependent processes rather than by the magnitude of the light gradient, that is, the total number of photons available for the stimulation of phosphorylation on the irradiated and on the shaded sides of the coleoptile.

From these studies it would appear that the 114–130 kDa plasma membrane-bound protein is the possible photoreceptor for phototropism. The correlation between phosphorylation of this protein and phototropism is further supported by detailed biochemical and genetic analysis of an *Arabidopsis* mutant that lacks all phototropic responses (Liscum and Briggs 1995). The deficiency in phototropism in this mutant was due to a mutation in a single gene locus (*nph1*-non-phototropic hypocotyl). In young etiolated seedlings the *nph1* mutant displays severe phenotypic changes marked by loss of phototropic response to blue light and to UV-A and green light. It has been postulated that the NPH1 gene may encode the apoprotein for a multichromorphic holoprotein capable of absorbing blue, UV-A and green light and that this photoreceptor regulates all phototropic responses in *Arabidopsis* because *nph1* mutants have a null phenotype in all light qualities tested (Liscum and Briggs 1995). Christie *et al.* (1998) designated the *nph1* holoprotein, *phototropin*.

The *nph1* mutant is also deficient in the 120 kDa protein. This protein and the blue-light-induced phosphorylation response have been detected in wild-type (NPH1) *Arabidopsis* and in a number of other species, both monocot and dicot. It has been suggested that this protein is the apoprotein of the phototropic photoreceptor and that its activation by early phosphorylation may be the first step in the signal transduction chain (Briggs and Liscum 1997). While the nature of the NPH1 chromophore has not yet been established, it is possibly a flavin or a pterin. It is, however, biochemically and genetically distinct from the *Arabidopsis* HY4 gene-encoded product that mediates the inhibition of hypocotyl elongation (Ahmad and Cashmore 1993, 1996). The HY4 gene product, named CRY1 (cryptochrome 1) is a cytoplasmic protein with a mass of about 75 kDa, while the NPH1 protein is a plasma-membrane protein with a mass of about 120 kDa. *Arabidopis* seedlings that have a cryptochrome-deficient mutation (*hy4*) do not show suppression of hypocotyl elongation in blue and UV-A light but do show normal phototropism. Conversely, the non-phototropic mutant *nph1* shows suppression of hypocotyl elongation. Seedlings that carry both the *hy4* and the *nph1* mutations exhibit neither blue-light-suppression of hypocotyl elongation nor phototropism (Briggs and Liscum 1997). Clearly cryptochrome and the putative phototropic photoreceptor are two different genetically distinct blue-light photoreceptors. Light really has two distinct effects on phototropism. As discussed above, blue light triggers the bending response; it may also, however, decrease the sensitivity of the organ to subsequent applied light. This effect is non-directional and is referred to as a *tonic effect*. For example, if a coleoptile is given a short exposure to unilateral blue light ($0.03 \, \text{J s}^{-1}$), positive curvature occurs, but only if the coleoptile has previously been in the dark. If the coleoptile has previously been exposed to 10 seconds of the same unilateral irradiance, much less positive curvature occurs and the negative area of the curve (Fig. 8.1) is approached.

While the phototropic reaction is not triggered by long-wavelength light ($>520 \, \text{nm}$), it is modulated by it. The effective light acts via phytochrome. Pretreatment of a dicotyledonous seedling with red light causes a massive increase in the phototropic response (Fig. 8.5), even though this pretreatment lowers the phototropic sensitivity of the seedling towards the blue light, that is, the photon flux required for saturation of the phototropic reaction by blue light is increased. Pre-irradiation thus desensitizes the photoreceptor but at the same time stimulates curvature. In other words, red light changes the sensitivity of the tissue to the blue

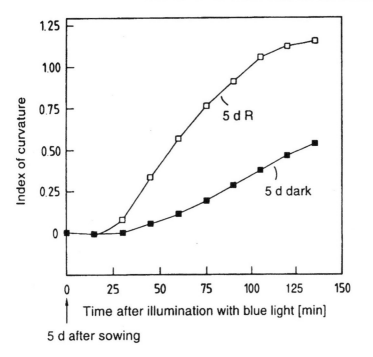

5 d after sowing

Figure 8.5   Phototropic curvature of the hypocotyl of sesame (*Sesamum indicum*) seedlings exposed to unilateral blue light after pretreatment with red light. (From Woitzik and Mohr 1988)

light that actually causes the bending; the pre-irradiated seedling is less sensitive to the stimulus but reacts more strongly after stimulation (Fig. 8.5).

Thus, the amplitude of phototropic curvature to blue light is enhanced by prior exposure of seedlings to red light. This enhancement is mediated by phytochrome (Janoudi *et al.* 1997). Fluence–response relationships constructed for red-light-induced enhancement in the PHYA null mutant, the PHYB-deficient mutant and in two transgenic lines of *Arabidopsis thaliana* that overexpress either PHYA or PHYB demonstrated the existence of two enhancement responses – one in the very-low-to-low fluence range and a second in the high-fluence range. Only the response in the high-fluence range is present in the PHYA null mutant; the PHYB-deficient mutant is indistinguishable from the wild-type parent in red-light responsiveness. Thus it appears that PHYA is necessary for the very-low-to-low but not for the high-fluence response and PHYB is not necessary for either response range (Janoudi *et al.* 1997). The involvement of multiple phytochromes in enhancement is not surprising; PHYA and PHYB have been shown to have overlapping functions in plant development (Reed *et al.* 1994). The mechanism by which PHYA can increase the phototropic response to a blue-light stimulus is not known, but clearly there is some form of interaction between these two distinct photosensory response systems. Early studies have shown that enhancement is maximized when the red-light stimulus precedes the blue-light phototropic stimulus by 2 h (Janoudi and Poff

1991) and it is suggested that phytochrome affects the phototropic response by modulating a component of the blue-light signal transduction sequence (Janoudi and Poff 1992).

Models of phototropism

Despite decades of research on phototropism there are still at least two alternative hypotheses, each of which may account for light-induced differential growth. The first of these, the Cholodny–Went hypothesis, postulates that unilateral light (or higher light flux on one side than on the other) induces a lateral redistribution of endogenous auxin near the apex of the organ, such that auxin apparently migrates from the irradiated to the shaded side of a unilaterally illuminated coleoptile. This lateral shift of auxin is maintained as the hormone moves basipetally from tip to base. Hence, the cells on the shaded side, which receive the greater amount of auxin, elongate more than those on the illuminated side. The differential growth results in curvature towards the light source. A second hypothesis, proposed by A.H. Blaauw as early as 1909, also assumes a light gradient across the organ. Blaauw proposed that phototropically effective light causes a local inhibition of cell growth. The light gradient causes a gradient of growth inhibition across the organ, the lowest inhibition occurring on the shaded side. Because the hormonal model was more fashionable at the time, Blaauw's light-inhibition model received little attention.

The Cholodny–Went hypothesis is based on results obtained from agar diffusion experiments carried out by Went. In Went's experiments, coleoptiles were first exposed to unilateral light. The coleoptile tips were excised and split longitudinally, and the two halves were placed on agar blocks so that the diffusible auxin was secreted at the cut surface and diffused into the agar blocks. The amount of auxin secreted was assayed, using an *Avena* curvature test. Went reported that the shaded half of the coleoptile secreted a significantly greater amount of auxin than did the illuminated side; he concluded that a greater proportion of the auxin was transported down the shaded side of the coleoptile. One early hypothesis put forward to explain this phenomenon suggested that the asymmetric distribution of auxin was due to the photodestruction of auxin on the illuminated side of the coleoptile. This was supported by evidence that light catalysed the oxidation of IAA *in vitro* in the presence of riboflavin. However, later work, described below, shows that phototropic stimulation does not significantly reduce the total yield of auxin *in vivo*. Numerous unsuccessful attempts were made to test the validity of the Cholodny–Went hypothesis; the application of ^{14}C-IAA to tropically stimulated coleoptiles failed to verify asymmetric distribution of auxin. Poor experimental technique may have been responsible for inconclusive results. More recent work shows that a large proportion of the radioactive auxin taken up by tissues does not enter the auxin transport stream. When non-diffusible auxin is taken into account, a clear differential in auxin distribution can be detected (Briggs 1963; Gillespie and Thimann 1961). For example, when ^{14}C-IAA was applied to the tips of maize coleoptiles, 65% of the radioactivity was recovered from the shaded side.

In the 1960s, W.R. Briggs and his colleagues systematically re-evaluated the Cholodny–Went hypothesis. Briggs (1963) repeated Went's original split-tip experiments, but unlike Went, he excised the tips and placed them on the agar blocks before applying the phototropic stimulus (Fig. 8.6). Briggs confirmed that auxin production in coleoptiles of *Zea mays* is confined to the apical 1–2 mm and that lateral photoinduced redistribution occurs within the uppermost 0.5 mm. He also confirmed that phototropic stimuli, tested over a wide range of fluence, do not reduce the yield of diffusible auxin. The results obtained by Briggs in his split-tip experiments (Fig. 8.6) clearly show that when the tip is partially split, leaving tissue continuity only at the very apex of the coleoptile, irradiation with unilateral light causes an increase in the amount of diffusible auxin in the shaded side and a corresponding decrease in the illuminated side. The total amount of auxin secreted into the agar blocks remains constant. When the tip is totally split, no asymmetric redistribution of auxin occurs. These results clearly support the Cholodny–Went concept that unilaterally applied light induces a redistribution of auxin, resulting in a greater concentration of auxin on the shaded side.

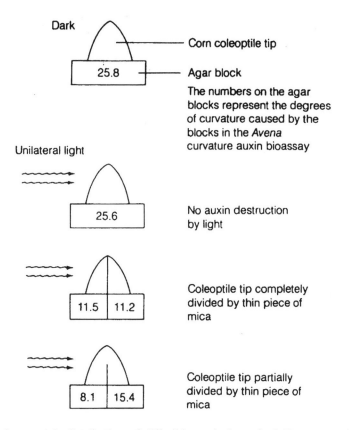

Figure 8.6 Asymmetric distribution of diffusible auxin in excised *Zea mays* coleoptile apices following phototropic stimulation by unilaterally applied light. (From Briggs 1963)

While the Cholodny–Went model thus appears to account for the basic transduction mechanism in phototropism, the model was again called into question in the 1970s. Firn and Digby (1980) proposed criteria by which the Cholodny–Went concept might be tested and they concluded that these criteria had not been met in the original hypothesis. Briggs and Baskin (1988), in a review, summarized Firn and Digby's proposed tests of the Cholodny–Went model, by listing four criteria: (i) during phototropically-induced growth, elongation growth on the shaded side of the organ should be accompanied by growth retardation on the illuminated side; (ii) the light-induced lateral auxin gradient must accompany or precede any differential growth; (iii) it must be shown that the endogenous auxin is indeed the limiting growth factor in the tissue; and (iv) it must be shown that the auxin concentration differential established is sufficient to account for the growth differential achieved.

The first criterion requires that growth promotion rates on the non-illuminated side be accompanied by growth retardation on the illuminated side. One approach to the study of this phenomenon is the use of time-lapse photography. Markers (small glass or styrene beads) are placed along both sides of coleoptiles or hypocotyls illuminated from one side and the photoinduced bending photographed at set time intervals. Distances between the marker beads displaced as cells elongate are measured on projected photographic negatives and growth rates are calculated. By this means it is possible to measure elongation rates simultaneously on the illuminated and shaded side of the photostimulated coleoptiles. A rapid cessation of growth on the illuminated side is a consistent finding of all kinetic studies. For the shaded side, results are less consistent – growth may be promoted, remain unchanged or increase after a brief period of decrease. Franssen et al. (1982) reported that when light levels are used to elicit a basal response, growth on the illuminated side stopped almost instantaneously, at the beginning of the illumination, whereas growth on the non-illuminated side continued. It is well established that red light (acting through phytochrome) and blue light can inhibit growth in stems and coleoptiles (as noted in the discussion on the tonic effect, above). Studies which show that the rate of elongation decreases on the illuminated side clearly support Blaauw's light-growth model. Blaauw's hypothesis equates blue-light inhibition of elongation with blue-light induction of phototropism and argues that a light gradient across the coleoptile would establish a gradient of photoinhibition. A number of workers (Iino and Briggs 1984; Baskin 1986; Rich et al. 1987) have shown that the two phenomena – light-dependent growth inhibition and phototropic response – can be separated. If coleoptiles are uniformly illuminated with red or blue light sufficient to saturate the growth inhibition response before exposure to the phototropic stimulus, then the increased rate of growth on the non-illuminated side is compensated for by the reduction of growth on the illuminated side. The result from these studies appears to be that that there is a general inhibition of growth due to light, as Blaauw suggested. However, support for the Cholodny–Went hypothesis seems more compelling at least where coleoptiles are stimulated by low-fluence rates for short durations. Under such conditions, there is compensatory growth, possibly caused by lateral auxin transport, as suggested by the Cholodny–Went hypothesis. The mechanism by which blue light causes a redistribution of auxin is not known, although phototropic mutants of *Arabidopsis* may provide clues to the signal transduction chain involved (Briggs and Liscum 1997).

The other criteria are more easily fitted to the Cholodny–Went model. Several studies (Pickard and Thimann 1964; Shen-Miller *et al.* 1969) have shown that the direction and rate of transport of ^{14}C-IAA in coleoptiles could be influenced by light. Iino and Briggs (1984) showed that when curvature began in the apex of the maize coleoptile and moved downwards, the rate of movement was consistent with known rates of auxin transport. Baskin *et al.* (1985) showed that coleoptiles to which auxin had been applied to one side near the tip, exhibited a rate of basipetal movement of growth stimulation similar to that found as a result of photostimulation. In other work, Baskin *et al.* (1986) applied auxin to coleoptile tips and found that auxin was indeed the limiting factor for growth, that there was a linear relationship between auxin concentration and growth rate over a range that spanned the rates occurring in phototropic bending, and that the auxin gradient established at the coleoptile tip was well sustained during the basipetal transport. In view of these facts, Briggs and Baskin (1988) concluded that there is strong evidence for the Cholodny–Went model for phototropism in a monocot coleoptile (maize). However, there is evidence that a different mechanism may exist in dicots. Franssen and Bruinsma (1981) found no auxin gradient across a phototropically stimulated sunflower hypocotyl but they did demonstrate a gradient in an inhibitor, xanthoxin, across the hypocotyl, with up to 70% of the inhibitor occurring on the illuminated side. Hasegawa *et al.* (1989) isolated and characterized three inhibitors that formed a similar gradient across a phototropically stimulated radish hypocotyl. These workers critically tested Went's experiment with the *Avena* coleoptile. They found the same result as Went for agar blocks placed on prepared coleoptiles: the block from the dark side caused the most bending. However, when they analysed the blocks with a physiochemical assay, they found equal amounts of IAA in the two blocks and that there was 2.5–7 times as much IAA overall as was indicated by the curvature test. This finding suggested that the blocks contained an inhibitor along with the auxin. They also found two unidentified inhibitors in the blocks, with more of each in the block from the illuminated side. These results strongly support the inhibitor model proposed by Blaauw.

One of the four criteria asks if the differential in auxin concentration induced by a phototropic stimulus is sufficient to cause the required changes in growth rates on both sides of the coleoptile. Straight growth of coleoptile segments incubated on IAA solutions is a function of the logarithm of IAA concentration. This would seem to indicate that large differences in IAA concentration would be needed to cause a significant difference in growth rate (Firn and Digby 1980; Trewavas 1981). Agar-diffusion studies with tropically stimulated coleoptiles show a differential of only 2:1 between the non-illuminated and the illuminated sides. Baskin *et al.* (1986) argue that, since agar-diffusion experiments average the auxin gradients on both the illuminated and non-illuminated sides of the coleoptile, they underestimate the true magnitude of the gradient. In addition, it has long been established that both the *Avena* curvature test, in which IAA is applied directly to the cut surface of the decapitated coleoptile, and IAA-induced curvature in intact maize coleoptiles, exhibit a near linear relationship with IAA concentration.

The Cholodny–Went model was based on the concept that the apex played a special role in controlling extension growth of the whole organ and that the sites of phototropic perception and phototropic response are separate. However, this concept has been questioned by many workers. It has been shown (Franssen *et al.* 1982) that the apex is not the site of phototropic perception in dicots and that the apex is not even required for the phototropic response of oat coleoptiles subjected to continuous unilateral light. Thus, the basic principle on which the Cholodny–Went model was based (the role of the apex), is now thought to apply only to coleoptiles exposed to very low fluence rates of short duration. Phototropism in coleoptiles exposed to high fluence rates for long durations do not require the presence of the apex; here the Cholodny–Went model must therefore be modified to allow for IAA redistribution throughout the entire elongation zone.

Stomatal Movement

Stomata

Stomata are small, ellipsoidal pores on the leaf epidermis. The pores are surrounded by two guard cells; the cells adjoining the guard cells are subsidiary cells – they are not part of the stomatal system but differ slightly (in the thickness of their walls) from the other epidermal cells. The walls of the guard cells are differentially thickened – walls next to and above and below the pore are thickened, whereas walls remote from the pore are thin. The stomata of dicots typically consist of two kidney-shaped guard cells (Fig. 8.7).

The cellulose microfibrils are oriented in a radial manner, thus allowing the guard cells to buckle outwards when they are turgid, causing the pore to open; when the cells lose water and become flaccid, the pore closes. The guard cells of grasses and most monocots are elongated or 'dumb-bell-shaped' (Fig. 8.7). The bulbous ends are thin-walled while the 'handle' has thick walls on the pore side; the pore is an elongated slit which opens when the bulbous ends push against each other as they swell. The guard cells are the only cells of the epidermis to contain chloroplasts, yet each guard cell contains only a few. Stomata occur mainly on the lower epidermis of the leaf or may be confined to it. Stomata are small (3–4 μm in width and 10–40 μm in length) and numerous (1000–2000 cm^{-2} to 100 000 cm^{-2}), yet the total pore area accounts for only 1–2% of the total leaf surface. Stomata function as hydraulic valves – guard cells absorb water, become turgid and cause the pore to open, thus allowing CO$_2$ into the leaf; they lose water, become flaccid and close the pore, thus preventing water loss from the leaf. Stomata generally open in daytime and close in

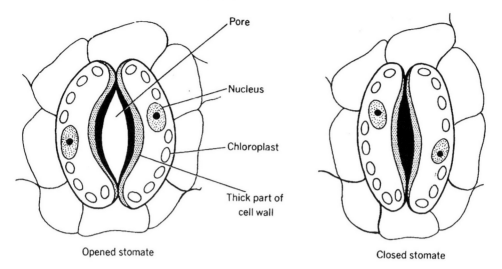

Opened stomate

Closed stomate

Figure 8.7   Structure of open and closed stomata. Note that the guard cells contain chloroplasts

darkness. Certain succulent plants, e.g. cacti, growing in hot, dry conditions, open their stomata at night and fix CO_2 into organic acids in darkness; they close their stomata during the day to prevent excessive water loss. All of the gas exchange in the leaf – intake of CO_2 for photosynthesis and extrusion of water vapour by transpiration – occurs via the stomata.

Comparison of the absorption of CO_2 by a free-absorbing surface (NaOH) and the surface of a leaf which has stomata only on the lower epidermis and of which the total stomatal pore area is 1% of the total leaf surface, shows that the leaf is 50 times more efficient. The explanation for this is that when gas diffuses through a pore, it does not do so in straight lines – it converges towards the pore and fans out (Fig. 8.8). The concentration of the diffusing molecules at the rim of the aperture is lower than at the central part, so that the rate of diffusion is greater at the rim. As the aperture becomes smaller, the marginal area forms a greater proportion of the total pore and thus the rate of diffusion tends to become proportional to the perimeter (or to the diameter, if the pore is circular) rather than to the cross-sectional area. Sayre (1926) compared the amount of water lost with pore diameter, perimeter and area in *Rumex patientia* (Table 8.1).

From these results it can be seen that the area of the smallest pore (0.01) equals 1% of the area of the largest pore (1.00); the perimeter of the smallest pore (0.13) equals 13% of the perimeter of the largest pore (1.00); the amount of water lost by the smallest pore (0.364) equals 14% of the amount of water lost by the largest pore (2.655). That is, diffusion is proportional to the perimeter of the pore. The efficiency of stomatal diffusion is added to by the spatial arrangement of pores on the epidermis which is such that diffusion shells do not overlap and optimum density of the diffusing molecules is achieved (Fig. 8.8).

Table 8.1   Diffusion of water vapour through stomatal pores of *Rumex patientia*. (From Sayre 1926)

Diameter of pores (mm)	Relative loss of water vapour (g)	Relative amounts of water lost	Relative areas of pores	Relative perimeters of pores
2.64	2.655	1.00	1.00	1.00
1.60	1.583	0.59	0.37	0.61
0.95	0.928	0.35	0.13	0.36
0.81	0.762	0.29	0.09	0.31
0.48	0.455	0.17	0.03	0.18
0.35	0.364	0.14	0.01	0.13

Stomatal opening and closing

Guard cells control the size of the stomatal pore. When guard cells take in water by osmosis, they become turgid and swell, causing the pore to open; when the cells lose water they become flaccid and sag, so that the pore closes. To understand what controls stomatal opening and closure it is necessary to know what regulates osmosis in the guard cells. Over the years, a number of hypotheses have been put forward to explain turgor changes in guard cells. Most of the older hypotheses were based on the observation that guard cells contain chloroplasts and are therefore competent to carry out photosynthesis. Light shining on guard cells initiates photosynthesis. This results in an accumulation of sugars which causes a decrease in the osmotic potential (becomes more negative). This leads to an intake of water by the guard cells which thereby become turgid, causing the pore to open. Another hypothesis stated that light had an indirect effect: light, by initiating photosynthesis, brought about a decrease in the intercellular CO_2, a decrease in starch and an increase in sugars, thus eliciting water intake which caused the pore to open. These concepts were initially attractive and formed the basis of the starch–sugar hypothesis of stomatal movement since they explained the accumulation of starch in guard cells at night and the loss of starch during the day. Conversion of sugars to starch (which is insoluble and therefore has no effect on osmosis) results in an increase (less negative) in water potential in the guard cells, thus causing water to leave them and the pores to close at night. However, it later became established that the guard cells contain only few chloroplasts and do not produce enough sugar to

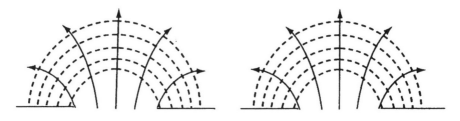

Figure 8.8   Diffusion shells

account for the change in water potential necessary to bring about stomatal movement. In addition, the guard cells of certain species, e.g. onion (*Allium cepa*), do not store starch and the guard cells of the achlorophyllous parts of variegated leaves such as those of *Pelargonium*, do not have chloroplasts, yet their stomata function normally. In these species, at least, the opening and closing of stomata is not caused by the production of sugars in light and their conversion to starch in darkness. In wheat leaves, a reduction of the CO_2 concentration in the substomatal cavity from 0.03% to 0.01% opens stomata. How such a small change in CO_2 concentration can bring about the massive change of stomatal opening is difficult to explain. In addition, RubisCO, the principal carbon-fixing enzyme in plants, has never been detected in significant amounts in stomatal guard cells, suggesting that carbon fixation does not occur in guard cells (Outlaw 1987). Photosynthetic carbon fixation is therefore unlikely as a general mechanism facilitating stomatal movement.

These older hypotheses became seriously questioned after the discovery of potassium ion (K^+) fluxes in guard cells. The accumulation of K^+ in the guard cells of open stomata was first observed in 1943 by S. Imamura in Japan. This work was not brought to the attention of western scientists until the publication of a report by M. Fujino in 1967. The findings of the Japanese workers were confirmed later in Europe and the USA (Willmer 1983). The changes in osmotic potential that open and close stomata result mainly from the reversible uptake and loss of K^+ by the guard cells. Stomata open when the guard cells actively accumulate K^+ from the neighbouring epidermal cells and close when K^+ is extruded. This uptake of K^+ lowers (makes more negative) the water potential of the guard cells, with the result that water enters by osmosis making the cells turgid and thus opening the pore. The level of K^+ that accumulates in the guard cells during stomatal opening is sufficient to account for the opening: K^+ concentration increases several-fold when stomata open – from 0.1 M in the closed stomata to 0.5 M in the open state, which is enough to lower the water potential by about 2.0 MPa.

It has long been accepted that the passive diffusion of water across the lipid bilayer accounted for water fluxes across biological membranes. This hypothesis was questioned in the early 1990s, when measurements were made which indicated that water transport across membranes is too specific and too rapid to be explained by simple diffusion. This gave rise to the concept of selective channels for water transport. Such channels have been discovered in the membranes of both plants and animals. The channels are water-specific, channel-forming proteins, called *aquaporins*. While these porins cannot actively transport water, they facilitate osmosis, acting as gated channels which allow an activated diffusion of water in response to signals such as turgor pressure changes in cells.

The cell's increase in positive charge due to the influx of K^+ is electrically balanced by the parallel uptake of chloride anions (Cl^-) and the build-up of organic malate anions. Like potassium, chloride is taken up during stomatal opening and extruded during closure. Malate accumulates in the guard cells as a result of the breakdown of starch. Hence, the decrease in starch concentration observed during stomatal opening is linked to the accumulation of K^+, by being the source of malate in the guard cells of open stomata. The guard cells' increase in positive charge due to the influx of K^+ is also offset by the pumping out of the cells of the H^+ released from the malate. The fate of malate during the closure of stomata is not known.

Stomatal closure results from K^+ extrusion from the guard cells with the accompanying increase in water potential (less negative) and consequent loss of water. Malate concentration in the guard cells must decrease in parallel with the loss of K^+; a drop in malate levels could occur through respiration which would convert malate to CO_2. Stomatal closure has not received as much attention as opening has but it cannot be assumed that the process is a simple reversal of the steps leading to opening. The rate of closure is often too rapid to be accounted for by the simple passive loss of ions; other metabolic pumps may be actively involved in the extrusion of ions when stomata close (MacRobbie 1987; Schroeder and Keller 1992). One possible mechanism is that signals which regulate stomatal closure may stimulate the opening of Ca^{2+} channels. One such signal, the plant growth regulator abscisic acid (ABA), inhibits secretion of H^+ and induces an efflux of K^+ and anions from guard cells. The ABA effect is probably caused via Ca^{2+} channels in the plasma membrane. These channels are activated to allow an increase in Ca^{2+} concentration in the cytoplasm. The uptake of Ca^{2+} into the cytosol would depolarize the plasma membrane, which may, in turn, lead to the opening of anion channels and facilitate the extrusion of Cl^- and malate ions. Loss of these anions would further depolarize the membrane, open K^+ channels and trigger the passive extrusion of K^+ into the adjacent epidermal cells.

It is widely accepted that the accumulation of ions by plant cells is driven by an ATP-powered proton pump located on the plasma membrane (Fig. 8.9). The proton pump hydrolyses ATP and uses the released energy to pump H^+ out of the cell; this results in a proton gradient, with the H^+ concentration higher outside the cell. The loss of H^+ from the cell generates a membrane potential, that is, a voltage or charge separation across the membrane. This charge separation is a form of stored potential energy that can be used to perform work in the cell.

Plant cells can use the energy stored in the proton gradient and membrane potential to drive the transport of different solutes into and out of cells. For example, membrane potential helps drive K^+ into the cell since these ions are positively charged and the inside of the cell is negatively charged. While the accumulation of K^+ occurs by passive transport, it is the active transport of H^+ that maintains the membrane potential and allows the cell to accumulate K^+. The role of proton pumps in transport processes of plant cells is a specific application of chemiosmosis, which links energy-releasing processes to energy-consuming processes by means of transmembrane proton gradients. The observation that protons are extruded out of the guard cells during stomatal opening has led to the hypothesis that a proton gradient in guard cells generates a proton-motive force that accomplishes ion uptake (Zeiger 1983). The most likely source of the ATP that powers the guard-cell proton pumps would be either photosynthesis in the guard-cell chloroplasts or cellular respiration. As already mentioned, the chloroplasts of guard cells are generally smaller, fewer in number and have less-well-developed

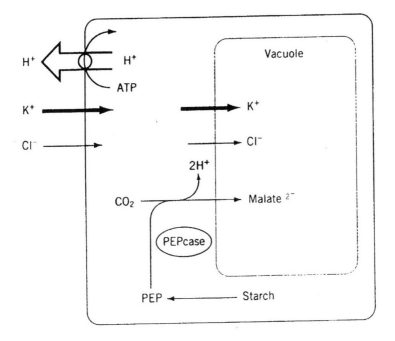

Figure 8.9   A simplified model of ion fluxes in guard cells during stomatal opening

thylakoids than the chloroplasts of mesophyll cells. While they also apparently lack the enzymatic machinery for photosynthetic carbon fixation, there is indirect evidence that they can produce ATP by the process of photophosphorylation. Since stomata may be made to open in darkness, photophosphorylation may not be the only source of energy. Guard cells contain large numbers of mitochondria and have high levels of respiratory enzyme activity. Thus, respiration may be an alternative source of energy for stomatal movement.

While the mechanism of stomatal movement and ion transport in guard cells is not thoroughly understood, it is probable that the rapid import and export of K^+ is powered by proton pumps and that it occurs through K^+-selective, voltage-dependent ion channels. It has been possible to prove directly the existence of such channels, using the patch-clamp method (Hedrich and Schroeder 1989). This technique, which isolates a small 'patch' of membrane from a guard-cell protoplast, allows the study of ion movement through a single type of channel, for instance a K^+ or H^+ channel. From measurements of the K^+ flux it has been possible to calculate that there are at least 300 such K^+ channels in the plasma membrane of *Vicia faba* guard cells; channels specific for Cl^- and malate ions have also recently been found in *Vicia*. Certain transport proteins, by virtue of alterations in their conformation, function as selective channels to allow ions such as K^+ and Cl^- to pass in and out of cells. Some such channels are gated, that is, they open and close in response to environmental stimuli, as in the regulation of K^+ gates in the membranes of guard cells during stomatal opening. Opening signals induce a hyperpolarization and closing signals a depolarization of the plasma membrane. The hyperpolarization preceding opening is coupled to the simultaneous extrusion of H^+ into the external medium. This has led to the belief that the activation of the proton pump, by thus lowering the electrical potential inside the cell relative to the outside, forms an electrochemical potential for the passive influx of K^+ through these selective channels. This concept has been substantiated experimentally. For example, the fungal phytotoxin, fusicoccin, which is a known activator of plasma membrane ATP-ases and which causes H^+ extrusion by plant cells leading to the enlargement of these cells, hyperpolarization and K^+ influx into isolated guard cells, induces stomatal opening in intact leaves (Marré 1979).

As already mentioned, the excess K^+ accumulation in the guard cells of open stomata is electrically balanced due to the extrusion of protons and partly due to the influx of Cl^- and the production of organic acid anions such as malate, within the cell. In most species, malate probably accounts for the bulk of the counterion required, whereas in others, K^+ intake is accompanied by Cl^- uptake. In those species whose guard cells lack chloroplasts or starch, Cl^- is the main counterion involved (Nobel 1991). A number of findings support malate as a counterion. Malate levels in the guard cells of open stomata are five to six times higher than that found in the guard cells of closed stomata. Guard cells contain high levels of phosphoenolpyruvate carboxylase (PEPCO), a key enzyme in the synthesis of malate from carbon skeletons resulting from the hydrolysis of starch. The amount of malate formed correlates with the decrease in starch content of the guard cells of open stomata. In addition, factors that influence stomatal movement also influence PEPCO activity. As mentioned above, fusicoccin, which stimulates stomatal opening, also causes an increase both in PEPCO activity and in malate concentration (Zhirong *et al.* 1997). Abscisic acid, on the other hand, which promotes stomatal closure, antagonizes the effect of fusicoccin.

Light and stomatal movement

In general, stomata are open during the day and closed at night. This protects the plant from wastefully transpiring water when there is not enough light for photosynthesis. The contributions of light and CO_2 are closely coupled in their effects on stomatal movement. It has long been considered that stomatal movement was a response to the intracellular concentration of CO_2 in guard cells: low concentrations of CO_2 and light stimulate opening while high CO_2 levels cause rapid closure, even in light. The response to light was seen as an indirect effect which, by initiating photosynthesis, lowers the level of CO_2. This indirect effect of light requires high fluence rates, and the reduction in CO_2 levels was attributed to photosynthesis in the mesophyll cells. This conclusion is supported by the observation that the action spectrum for moderate to high fluence rates resembles that for photosynthesis, with peaks in both the red and blue parts of the spectrum. Thus, it appears that CO_2 is a primary trigger, at least in intact leaves, where the indirect effect of light is to stimulate photosynthesis in mesophyll cells. It is difficult to interpret how the light signal is transduced into a stomatal response in intact leaves because of the many other internal and external factors that can affect the response. Light may act on the mesophyll cells and the signal transduced to the guard cells or the photoreceptor for the light may reside in the guard cells themselves. The influence of mesophyll cells may be obviated by separating portions of the leaf epidermis containing intact guard cells, from the rest of the leaf. These preparations are called *epidermal peels* (Mc Donald 1977). When epidermal peels, floated on a solution of potassium chloride, are illuminated, stomata are observed to open. Such experiments show that guard cells do not require the presence of mesophyll to respond to a light stimulus.

Stomata, both in the intact leaf and in isolation, have been shown to respond directly to light. Sharkey and Raschke (1981a) demonstrated that stomata responded to light even in leaves in which photosynthesis had been inhibited by the application of the herbicide cyanazine; they showed that stomata opened in light in the absence of CO_2 fixation by the mesophyll. They suggested that light absorbed by the guard cells (rather than by the mesophyll cells) is primarily responsible for the effect. Sharkey and Raschke (1981b) also measured the wavelengths of light that were most effective in causing stomata to open. Blue light (wavelength 430–460 nm) was almost 10 times as effective as red light (630–680 nm) in causing a given stomatal response. The red wavelengths are the same as those that are active in photosynthesis and the response to red light is inhibited by inhibitors of photosynthesis. Thus, the red-light response is apparently caused by light absorbed by chlorophyll but the blue-light effect is independent of photosynthesis. Direct evidence for a specific light response of isolated stomata arises from work with isolated protoplasts. Zeiger and Hepler (1977) showed that guard-cell protoplasts behave as osmometers, swelling under conditions of stomatal opening and shrinking when exposed to closing signals such as the water-stress growth regulator, abscisic acid. Guard-cell protoplasts respond to blue light by swelling. The swelling is peculiar to guard cells and requires the presence of K^+ in the bathing solution. The capacity of guard-cell protoplasts to respond to blue light provides evidence for a specific light response of stomata, independent of any other component of the leaf.

The blue-light effect has been demonstrated in a variety of ways. As already mentioned, stomatal opening is promoted by both red and blue light but is generally more sensitive to blue light. At low fluence rates ($10–15\,\mu\mathrm{mol\,m^{-2}\,s^{-1}}$) blue light causes stomatal opening but red does not. Hsiao and his co-workers (1973) studied the wavelength dependence of stomatal opening and K^+ uptake in *Vicia faba* and showed that, at low fluence rates, only a single peak was observed at 455 nm. At higher fluence rates, the peak due to blue light was unchanged but a second, smaller peak at 650 nm also appeared. These spectral features are strikingly similar to those reported for stomatal conductance in intact leaves of *Xanthium strumarium*, which show a major peak at around 450 nm and a small shoulder at 660 nm (Sharkey and Raschke 1981b). At higher fluence rates the blue-light effect is considerably greater than that of red light at the same fluence rate. Schwartz and Zeiger (1984), using epidermal peels of *Commelina communis*, applied a protocol of dual-beam laser irradiance to distinguish the two (red and

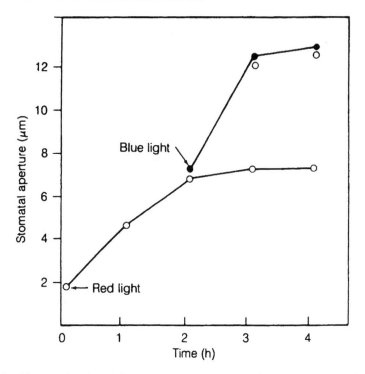

Figure 8.10 Changes in stomatal aperture as a function of time in stomata from detached epidermis of *Commelina communis* treated with red light at saturating fluence rates and followed by blue light added at the time indicated by the arrow. (From Schwartz and Zeiger 1984)

blue light) responses. They applied monochromatic red light of sufficient fluence rate to saturate the photosynthetic response. After the rates of both photosynthesis and conductance (a measure of stomatal opening) had reached a steady state, a short (30 s) pulse of blue light was applied over the background of continuous red light. A clear increase in stomatal conductance was observed without any further increase in photosynthesis, indicating that stomata were responding directly to blue light (Fig. 8.10).

The sharp increase in stomatal opening due to the added blue light provides evidence for the operation of two distinct photoreceptors in guard cells. Further evidence for the dual photoreceptor hypothesis comes from work with stomata from the orchid *Paphiopedilum*; this genus of the Orchidaceae is unique in that its guard-cell chloroplasts lack chlorophyll. Nelson and Mayo (1975) measured transpiration in intact leaves of *Paphiopedilum*. They found responses to both blue and red light and concluded that guard-cell chloroplasts played no role in the light responses of the stomata. However, it is difficult to interpret transpiration measurements in intact leaves due mainly to the presence of chlorophyll-containing mesophyll. In order to circumvent this problem, Zeiger (1983) used epidermal peels of *Paphiopedilum* to measure the stomatal response to blue and to red light and found that stomata opened in response to blue but not to red light. These results are consistent with the dual photoreceptor hypothesis, that is, in the absence of guard-cell chlorophyll, stomata would be expected to lack a red-light response but be capable of reacting to the blue light because of the presence of the blue-light photoreceptor.

The demonstrated involvement of both the red- and blue-light photoreceptors of guard cells in the stomatal response to light points to the need to characterize the role of each photosystem in the signal transduction pathway leading to stomatal opening. Studies of guard-cell chloroplasts have been hampered by the fact that guard cells constitute only a small fraction of total leaf tissue and that they contain only a fraction of the chlorophyll found in mesophyll cells. Zeiger *et al.* (1981) obviated this problem by using the variegated leaves of *Chlorophytum comosum*, the spider plant. The mesophyll of the albino portion of these leaves is completely devoid of chlorophyll, but because of the different embryonic origin of the epidermis, the guard cells have the normal chloroplast complement. Zeiger *et al.* (1981) found that the guard-cell chloroplasts of *C. comosum*, like their mesophyll counterparts, possess photosystem I and photosystem II and have the capacity for electron transport. Several lines of evidence suggest that the major role of guard-cell chloroplasts in stomatal movement is to supply ATP for active ion uptake during stomatal opening. Scwartz and Zeiger (1984) have shown that, under red light, stomatal opening in epidermal peels is inhibited by the photosynthesis inhibitor dichloromethyl-dimethylurea (DCMU). The simplest explanation of this finding is that under red light, with only the guard-cell chloroplasts responding to irradiation, photophosphorylation by these organelles provides the energy for stomatal opening. This notion is consistent with the inability of the achlorophyllous stomata from *Paphiopedilum* to open under red light.

The main feature of the response of stomata to blue light is the greater degree of opening compared to the red-light response. There are several other responses which are specific to blue light, including protoplast swelling in onion (Zeiger and Hepler 1977), enhanced malate synthesis in *Vicia* (Ogawa *et al.* 1978) and increased transpiration in grasses (Johnsson *et al.* 1976). The action spectrum of the blue-light responses is typical of other blue-light responses, such as phototropism. The identity and location of the chromophore is not known but the response is probably mediated by cryptochrome. The mode of action of blue light is not certain. Blue light does cause swelling in isolated guard-cell protoplasts, which indicates that blue light acts directly on the guard cells. A chemiosmotic model for stomatal movement may help to explain the function of the two photoreceptor systems of guard cells in terms of the regulation of proton extrusion and ATP supply. Several investigators have reported that blue light activates proton extrusion by the guard cells, thus allowing potassium ions to enter passively. In addition, blue light stimulates malate biosynthesis in guard cells, another prerequisite for stomatal opening. It has been suggested that the blue-light response may have a role in the early morning opening of stomata. Stomata can open before sunrise (and remain open after dusk) when fluence rates are much lower than those required for photosynthesis. Because of the high sensitivity of the blue-light response to low fluence rates together with the relatively high proportion of blue light at dawn (and at dusk), blue light may function as a 'light-on' signal. Thus the blue-light response may serve to anticipate the need for atmospheric CO_2 and predispose the leaf for photosynthesis.

The photoreceptor

Blue-light activation of stomatal opening is a well-established response of higher plants to light. Stomata respond to blue light at very low fluence rates with sufficient energy to drive stomatal opening. The blue-light pulse added to a strong red-light background saturating for photosynthesis caused transient additional opening in *Commelina communis* (Schwartz and Zeiger 1984) (Fig. 8.10). Similar results were obtained with *C. communis* seedlings placed in darkness and CO_2-free air (Lascève *et al.* 1993). Thus, under saturating red light or CO_2-free air and darkness, blue light acts as a signal independent of phytochrome or chlorophyll to promote stomatal opening.

A fundamental component of the blue-light response in guard cells is H^+ extrusion by H^+-ATPase into the apoplast. There is evidence that a plasma-membrane redox chain exists that may modulate the activity of the H^+-ATPase following blue-light stimulation (Gautier *et al.* 1992). The resulting hyperpolarization of the plasma membrane creates an electrical gradient that provides the driving force for the uptake of K^+ in exchange for the H^+ ions. Accumulation of K^+ in the guard cells allows the intake of water by osmosis, causing the guard cells to swell with consequent opening of the stomatal pore. Under certain conditions blue light also stimulates sucrose accumulation in guard cells (Talbott and Zeiger 1998). Sucrose, though osmotically inactive, may accumulate from starch breakdown that occurs under blue light (Lu *et al.* 1997).

None of the established blue-light photoreceptors, CRY1, CRY2 or NPH1, have been implicated in the stomatal response to blue light. The carotenoid zeaxanthin has received much attention as a possible guard-cell blue-light photoreceptor mediating stomatal opening (Zeiger and Zhu 1998). Zeaxanthin is a component of the photoprotective xanthophyll cycle in chloroplasts (Chapter 1) in which excess light energy is dissipated in the conversion of violaxanthin through antheraxanthin to zeaxanthin. The absorption spectrum of zeaxanthin correlates reasonably well with the action spectrum for guard-cell blue-light response (Quinones *et al.* 1996). In addition, there is good correlation between the levels of guard-cell zeaxanthin and the extent of blue-light-stimulated stomatal opening. The *Arabidopsis* mutant *npq1* (non-photochemical quenching) has been shown to be defective in violaxanthin de-epoxidase and thus cannot convert violaxanthin to zeaxanthin (Niyogi *et al.* 1998). Zeiger and Zhu (1998) tested the responses of wild-type and *npq1* mutant seedlings for blue-light-induced stomatal opening under increasing background levels of red light. Wild-type stomata showed a strong increase in the blue-light-activated opening response with an increasing fluence rate of background red light, while the *npq1* failed to show any response.

Blue-light-stimulated stomatal opening in detached epidermis of *Vicia faba* has been shown to be reversed by a 30 s pulse of green light (Frechilla *et al.* 2000). The magnitude of the green-light reversal dependent on the fluence rate, full reversal is being achieved at a green-light fluence rate twice that of the blue light. Action spectrum for the green light showed a maximum at 540 nm and minor peaks at 490 and 580 nm. This spectrum is similar to the action spectrum for blue-light-stimulated stomatal opening red-shifted by about 90 nm. It is suggested that blue/green reversibility may be explained by a pair of interconvertible zeaxanthin isomers, one absorbing in the blue and the other in the green, with the green-absorbing form being the physiologically active one (Frechilla *et al.* 2000). Thus, while zeaxanthin may, at present, be a candidate for the chromophore of a blue-light photoreceptor that mediates stomatal opening, no definite conclusion can be drawn regarding the nature of the photoreceptor.

Summary

Phototropism is a growth response to a gradient in blue and UV-A light; organs grow towards (positive phototropism) or away from (negative phototropism) the higher irradiance. Comparison of action spectra with the absorption of known pigments present in coleoptiles has indicated that either carotenoids or flavins might act as photoreceptors in the process. However, from work carried out with carotenoid-deficient plants and phototropic mutants of *Arabidopsis*, it appears that a flavoprotein located on the plasma membrane is the most likely candidate. Studies at the biochemical level suggest that light-dependent phosphorylation of a plasma membrane-associated protein may be responsible for phototropism with the degree of asymmetric phosphorylation across a coleoptile determining the degree of curvature. The photoreceptor was named phototropin.

There are at least two alternative hypothesis which may account for phototropism. The more popular Cholodny–Went hypothesis of the 1920s proposes the lateral redistribution of auxin as it flows basipetally from the apex, where it is synthesized. The higher concentration of auxin flows down the non-illuminated side of the organ, causing cells on that side to elongate, resulting in curvature towards the light source. The second hypothesis, proposed by Blaauw in 1909, also assumed a light gradient across the organ – phototropically effective light causes a local inhibition of cell growth. The light gradient causes a gradient in growth inhibition across the tissue, the lowest inhibition occurring on the non-illuminated side.

Stomata are small ellipsoidal pores in the leaf epidermis and account for 1–2% of the total leaf surface. Stomata function as hydraulic valves controlled by two guard cells which can absorb water and swell and thus cause the pore to open, allowing gas exchange. When the guard cells lose water they become flaccid, allowing the pore to close and thus prevent water loss from the leaf. Generally stomata are open in light and closed in darkness. Stomatal opening and closing is regulated by osmosis which is controlled by the exchange of potassium ions between the guard cells and the surrounding epidermal cells. Potassium ion uptake by the guard cells is followed by the intake of water by osmosis which is facilitated by water-specific channel-forming proteins (aquaporins). Loss of potassium ions from the guard cells is followed by water loss and closure of the pore. Stomatal movement is controlled by a number of environmental factors, including light, internal leaf CO_2 concentration, leaf water status and temperature. While zeaxanthin may, at present, be a candidate for the chromophore of a blue-light photoreceptor that mediates stomatal opening, no definite conclusion can be drawn regarding the nature of the photoreceptor.

References

Ahmad, M. and Cashmore, A.R. (1993). HY4 gene of *A. thaliana* encodes a protein with characteristics of a blue-light photoreceptor. *Nature* **366**: 162–166.

Ahmad, M. and Cashmore, A.R. (1996). Seeing blue: the discovery of cryptochrome. *Plant Molec. Biol.* **30**: 851–861.

Baskin, T.I. (1986). Redistribution of growth during phototropism and mutation in the pea epicotyl. *Planta* **169**: 406–414.

Baskin, T.I. and Iino, M. (1987). An action spectrum in the blue and ultraviolet for phototropism in alfalfa. *Phytochem. Phytobiol.* **46**: 127–136.

Baskin, T.I., Iino, M., Green, P.B. and Briggs, W.R. (1985). High resolution measurement of growth during first positive phototropism in maize. *Plant Cell and Environ.* **8**: 595–603.

Baskin, T.I., Briggs, W.R. and Iino, M. (1986). Can lateral distribution of auxin account for phototropism in maize coleoptiles? *Plant Physiol.* **81**: 306–309.

Blaauw, A.H. (1909). Die perzeption des lichtes. (The perception of light.) *Rec. Trav. Botica Neerlandica* **5**: 209–272.

Briggs, W.R. (1963). Mediation of phototropic responses of corn coleoptiles by lateral transport of auxin. *Plant Physiol.* **38**: 237–247.

Briggs, W.R. and Baskin, T.I. (1988). Phototropism in higher plants – controversies and caveats. *Botanica Acta* **101**: 133–139.

Briggs, W.R. and Liscum, E. (1997). The role of mutants in the search for the photoreceptor for phototropism in higher plants. *Plant Cell Environ.* **20**: 768–772.

Christie, J.M., Reymond, P., Powell, G.P., Bernasconi, P. and Raibekas, A.A. (1998). *Arabidopsis* NPH1: a flavoprotein with the properties of the photoreceptor for phototropism. *Science* **282**: 1698–1701.

Firn, R.D. and Digby, J. (1980). The establishment of tropic curvatures in plants. *Ann. Rev. Plant Physiol.* **31**: 131–148.

Franssen, J.M. and Bruinsma, J. (1981). Relationships between xanthoxin, phototropism and elongation growth in *Helianthus annuus* L. *Planta* **151**: 365–370.

Franssen, J.M., Firn, R.D. and Digby, J. (1982). The role of the apex in the phototropic curvature of *Avena* coleoptiles: positive curvature under conditions of continuous illumination. *Planta* **155**: 281–286.

Frechilla, S., Talbott, L.D., Bogomolni, R.A. and Zeiger, E. (2000). Reversal of blue-light-stimulated stomatal opening by green light. *Plant Cell Physiol.* **41**: 171–176.

Gallagher, S., Short, T.S., Ray, P.M.. Pratt, L.H. and Briggs, W.R. (1988). Light-mediated changes in two proteins found associated with plasma membrane fractions from pea stem sections. *Proc Natl. Acad. Sci. USA* **85**: 8003–8007.

Gautier, H., Vavasseur, A.G.L. and Boudet, A.M. (1992). Redox processes in the blue light response of guard cell protoplasts of *Commelina communis* L. *Plant Physiol.* **98**: 34–38.

Gillespie, B. and Thimann, K.V. (1961). The lateral transport of indoleacetic acid-C^{14} in geotropism. *Experimentia* **17**: 126–129.

Hasegawa, K., Sakado, M. and Bruinsma, J. (1989). Revision of the theory of phototropism in plants: a new interpretation of a classical experiment. *Planta* **178**: 540–544.

Hedrich, R. and Schroeder, J.I. (1989). The physiology of ion channels and electrogenic pumps in higher plants. *Ann. Rev. Plant Physiol. Plant Mol. Biol.* **40**: 539–569.

Hsiao, T.C., Allaway, W.G. and Evans, L.T. (1973). Action spectra of guard cell Rb^+ uptake and stomatal opening in *Vicia faba*. *Plant Physiol.* **51**: 82–88.

Iino, M. and Briggs, W.R. (1984). Growth distribution during first positive phototropic curvature of maize coleoptiles. *Plant Cell Environ.* **7**: 97–104.

Janoudi, A.K. and Poff, K.L. (1991). Characterization of adaptation in phototropism of *Arabidopsis thaliana*. *Plant Physiol.* **95**: 517–521.

Janoudi, A.K. and Poff, K.L. (1992). Action spectrum for enhancement of phototropism by *Arabidopsis thaliana* seedlings. *Photochem. Photobiol.* **56**: 655–659.

Janoudi, A.K., Gordon, W.R., Wagner, D., Quail, P. and Poff, K.L. (1997). Multiple phytochromes are involved in red-light-induced enhancement of first-positive phototropism in *Arabidopsis thaliana*. *Plant Physiol.* **113**: 975–979.

Johnsson, M., Issaias, S., Brogard, T. and Johnsson, A. (1976). Rapid, blue-light induced transpiration response restricted to plants with grasslike stomata. *Physiol. Plant* **36**: 229–232.

Kaufman, L.S. (1993). Transduction of blue-light signals. *Plant Physiol.* **102**: 333–337.

Khurana, J.P. and Poff, K.L. (1989). Mutants of *Arabidopsis thaliana* with altered phototropism. *Planta* **178**: 400–406.

Khurana, J.P., Kochar, A. and Jain, P.K. (1996). Genetic and molecular analysis of light-regulated plant development. *Genetica* **97**: 349–361.

Konjevic, R., Steinitz, B. and Poff, K. (1989). Dependence of the phototropic responses of *Arabidopsis thaliana* on fluence rate and wavelength. *Proc. Natl. Acad. Sci. USA* **86**: 9876–9880.

Lascève, G., Gautier, H., Jappé, J. and Vavasseur, A. (1993). Modulation of the blue light response of stomata of *Commelina communis* by CO_2. *Physiol. Plant* **88**: 453–459.

Liscum, E. and Hangarter, R.P. (1994). Mutational analysis of blue-light sensing in *Arabidopsis*. *Plant Cell Environ.* **17**: 639–648.

Liscum, E. and Briggs, W.R. (1995). Mutations in the NPH1 of *Arabidopsis* disrupt the perception of phototropic stimuli. *Plant Cell* **7**: 473–485.

Lu, P., Outlaw, W.H., Smith, B.G. and Freed, G.A. (1997). A new mechanism for the regulation of stomatal aperture size in intact leaves. *Plant Physiol.* **114**: 109–118.

MacRobbie, E.A.C. (1987). Ionic relations of guard cells. In: *Stomatal Function*, E. Zeiger, G.D. Farquhar and I.R. Cohen (eds), pp. 125–162. Stanford University Press, Stanford, CA.

Marré, E. (1979). Fusicoccin: a tool in plant physiology. *Ann. Rev. Plant Physiol.* **30**: 273–288.

Mc Donald, M.S. (1977). Preparation of stomatal impressions from leaf epidermis using a cellulose acetate 'peel' technique. *Lab. Practice* **26**: 691.

Nelson, S.D. and Mayo, J.M. (1975). The occurrence of functional, non-chlorophyllous guard cells in *Paphiopedilum* spp. *Can. J. Bot.* **53**: 1–7.

Niyogi, K.K., Grossman, A.R. and Björkman, O. (1998). *Arabidopsis* mutants define a central role for the xanthophyll cycle in the regulation of photosynthetic energy conversion. *Plant Cell* **10**: 1121–1134.

Nobel, P.S. (1991). *Physiochemical and Environmental Plant Physiology*. Academic Press, New York.

Ogawa, T., Ishikawa, H., Shimada, K. and Shibata, K. (1978). Synergistic energy of red and blue light action spectra for malate formation in guard cells of *Vicia faba* L. *Planta* **42**: 61–65.

Outlaw, W.H. (1987). An introduction to carbon metabolism in guard cells. In: *Stomatal Function*, E. Zeiger, G.D. Farquhar and I.R. Cohen (eds), pp. 115–123. Stanford University Press, Stanford, CA.

Palmer, J.M., Short, T.W. and Briggs, W.R. (1993). Correlation of blue light-induced phosphorylation to phototropism in *Zea mays*. *Plant Physiol.* **102**: 1219–1225.

Pickard, B.G. and Thimann, K.V. (1964). Transport and distribution of auxin during tropistic response. II. The lateral migration of auxin in phototropism of coleoptiles. *Plant Physiol.* **39**: 341–350.

Poff, K.L., Janoudi, A.-K., Orbovic, V., Konjevic, R., Fortin, M.-C. and Scott, T.K. (1994). The physiology of tropisms. In: *Arabidopsis*, E.M. Meyerowitz and C.R. Sommerville (eds), pp. 639–664. Cold Spring Harbor Laboratory Press, Cold Spring Harbor, NY.

Presti, D., Hsu, W.-J. and Delbruck, M. (1977). Phototropism in Phycomyces mutants lacking β-carotene. *Phytochem. Phytobiol.* **26**: 403–405.

Quinones, M.A., Lu, Z. and Zeiger, E. (1996). Close correspondence between the action spectra for the blue light responses of the guard cell and coleoptile chloroplasts and the spectra for blue light-dependent stomatal opening and coleoptile phototropism. *Proc. Natl. Acad. Sci. USA* **93**: 2224–2228.

Reed, J.W., Nagatani, A., Elich, T.D., Fagan, M. and Chory, J. (1994). Phytochrome A and phytochrome B have overlapping but distinct functions in *Arabidopsis* development. *Plant Physiol.* **104**: 1139–1149.

Reymond, P., Short, P.W. and Briggs, W.R. (1992). Blue light activates a specific protein kinase in higher plants. *Plant Physiol.* **100**: 655–661.

Rich, T.C., Whitelam, G.C. and Smith, H. (1987). Analysis of growth rates during phototropism: modifications by separate light–growth responses. *Plant Cell Environ.* **10**: 303–311.

Salomon, M., Zacherl, M. and Rudiger, W. (1997a). Asymmetric, blue light-dependent phosphorylation of a 116-kilodalton plasma membrane protein can be correlated with the first- and second-positive phototropic curvature in oat coleoptiles. *Plant Physiol.* **115**: 485–491.

Salomon, M., Zacherl, M. and Rudiger, W. (1997b). Phototropism and protein phosphorylation in higher plants: unilateral blue light irradiation generates a directional gradient of protein phosphorylation across the oat coleoptile. *Botanica Acta* **110**: 214–216.

Sayre, J. D. (1926). Physiology of stomata of *Rumex patientia*. *Ohio J. Sci.* **26**: 233–266.

Schroeder, J.L. and Keller, B.U. (1992). Two types of anion channel currents in guard cells with distinct voltage regulation. *Proc. Natl. Acad. Sci. USA,* **89**: 5025–5029.

Schwartz, A. and Zeiger, E. (1984). Metabolic energy for stomatal opening. Roles of photophosphorylation and oxidative phosphorylation. *Planta* **161**: 129–136.

Sharkey, T.D. and Raschke, K. (1981a). Separation and measurement of direct and indirect effects of light on stomata. *Plant Physiol.* **68**: 33–40.

Sharkey, T.D. and Raschke, K. (1981b). Effect of light quality on stomatal opening in leaves of *Xanthium strumarium*. *Plant Physiol.* **68**: 1170–1174.

Shen-Miller, J., Cooper, P. and Gordon, S.A. (1969). Phototropism and photoinhibition of basipolar transport of auxin in oat coleoptiles. *Plant Physiol.* **44**: 491–496.

Short, T.W. and Briggs, W.R. (1994). The transduction of blue light signals in higher plants. *Ann. Rev. Plant Physiol. Plant Mol. Biol.* **45**: 143–171.

Steinitz, B., Ren, Z. and Poff, K.L. (1985). Blue and green light-induced phototropism in *Arabidopsis thaliana* and *Lactuca sativa* L. seedlings. *Plant Physiol.* **77**: 248–251.

Taiz, L. and Zeiger, E. (1991). Auxins: growth and tropisms, *Plant Physiology*, pp. 398–425. Benjamin/Cummings.

Talbott, L.D. and Zeiger, E. (1998). The role of sucrose in guard cell osmoregulation. *J. Exp. Bot.* **49**: 329–337.

Trewavas, A. (1981). How do plant growth substances work? *Plant Cell Environ.* **4**: 203–228.

Vogelmann, T.C. and Haupt, W. (1985). The bluelight gradient in unilaterally irradiated maize coleoptile: measure with a fiber-optic probe. *Photochem. Photobiol.* **41**: 569–576.

Willmer, C.M. (1983). *Stomata*. Longman, Harlow, Essex.

Woitzik, F. and Mohr, H. (1988). Control of hypocotyl phototropism by phytochrome in a dicotyledonous seedling (*Sesamum indicum* L.). *Plant Cell Environ.* **11**: 653–661.

Zeiger, E. (1983). The biology of stomatal guard cells. *Ann. Rev. Plant Physiol.* **34**: 441–475.

Zeiger, E. and Hepler, P.K. (1977). Light and stomatal function: blue light stimulates swelling of guard cell protoplasts. *Science* **196**: 887–889.

Zeiger, E. and Zhu, J. (1998). Role of zeaxanthin in blue-light photo-reception and the modulation of light–CO_2 interactions in guard cells. *J. Exp. Bot.* **49**: 433–442.

Zeiger, E., Armond, P. and Melis, A. (1981). Fluorescence properties of guard cell chloroplasts. *Plant Physiol.* **67**: 17–20.

Zhirong, D., Aghoram, K. and Outlaw, W.H. (1997). In vivo phosphorylation of phospho-enolpyruvate carboxylase in guard cells of *Vicia faba* L. is enhanced by fusicoccin and suppressed by abscisic acid. *Archives Biochem. Biophys.* **337**: 345–350.

Zimmermann, B.K. and Briggs, W.R. (1963). Phototropic dosage–response curves of oat coleoptiles. *Plant Physiol.* **38**: 248–253.

9
Signal Transduction

Introduction

Plants are constantly exposed to a wide range of changing environmental conditions such as light, stress, nutrition and temperature fluctuations. Plants possess sophisticated sensory systems which perceive these stimuli and trigger signal transduction pathways that initiate a complex series of events within plant cells. These pathways amplify and coordinate the environmental stimuli to produce a range of physiological and developmental responses.

Higher plants possess photosensory systems which monitor many aspects of light in their environment – quantity, direction and wavelengths, which may regulate gene expression at every stage in the life cycle of the plant, from seed germination through seedling development to flowering. These light-regulated developmental processes are collectively called photomorphogenesis. Seedling development illustrates the effects of light on morphogenesis. Seedlings may follow either of two pathways of development – skotomorphogenesis in darkness and photomorphogenesis in light (Fig. 6.1).

Seedlings that are grown in continuous darkness have a characteristic etiolated appearance – elongated hypocotyls, unopened apical hooks and unexpanded cotyledons that contain etioplasts. When these seedlings are exposed to light, hypocotyl elongation ceases immediately, the apical hook opens and the cotyledons turn green as chloroplasts develop and chlorophyll is synthesized. These changes result from the stimulation of a large number of genes, including those which encode the small subunit of RubisCO (RBCS), chlorophyll a/b-binding proteins (CAB) and ferredoxin-NADP$^+$ reductase (FNR), all of which are required for photosynthesis. In addition, light enhances the expression of the genes that encode enzymes such as chalcone synthase (CHS) which synthesizes the light-protecting anthocyanins and thereby saves chlorophyll from destruction by photo-oxidation.

Many photomorphogenetic changes in plants are regulated at fluence rates which are too low to initiate photosynthesis. These changes include phototropism, stomatal movement, seed dormancy and germination, floral induction, growth,

Photobiology of Higher Plants. By M. S. Mc Donald.
© 2003 John Wiley & Sons, Ltd: ISBN 0 470 85522 3; ISBN 0 470 85523 1 (PB)

anthocyanin synthesis and movement of chloroplasts within cells. The photosensors responsible for mediating these changes include the red/far-red light-absorbing phytochromes (Quail *et al.* 1995), the blue/UV-A light-absorbing cryptochrome (Ahmad and Cashmore 1996), UV-A absorbing (Young *et al.* 1992) and UV-B absorbing (Beggs and Wellman 1985; Christie and Jenkins 1996) light photo-receptors. Plants can detect and respond to light intensities that vary over several orders of magnitude (Kendrick and Kronenberg 1994). Some light responses, such as the induction of nuclear-encoded genes encoding the light-harvesting chlorophyll proteins of photosystem II, can be initiated by fluence rates as low as $0.1\,\mathrm{nmol\,m^{-2}s^{-1}}$. These are classified as very-low-fluence responses (VLFR). Other responses, such as light-sensitive lettuce seed germination, require fluence rates of at least $1\,\mathrm{mmol\;m^{-2}}$ and are referred to as low-fluence responses (LFR). Light responses such as stem growth inhibition and floral induction require prolonged irradiation by fluence rates greater than $10\,\mathrm{mmol\,m^{-2}}$ and are called high-irradiation responses (HIR). Phytochromes mediate the VLF, LF and HI responses to red and far-red light (Mancinelli and Rabino 1978; Kaufman *et al.* 1984). In addition, certain LF and HI responses are mediated by blue-light-absorbing cryptochromes (Warpeha and Kaufman 1990).

The identity and biochemical characteristics of the UV-B receptor are still unknown, whereas cryptochrome, the blue/UV-A photoreceptor, has recently been cloned and its biochemical characteristics are being studied. Phytochrome, the red/far-red light photoreceptor, is, by far, the best-characterized photoreceptor, its physiological role having been reviewed (Furuya and Schäfer 1996). Phytochrome itself was identified about fifty years ago and has been purified and cloned within the past twenty years. The regulation of numerous responses has been ascribed to phytochrome, including the regulation of transcription of its own PHYA genes. Most plants contain several related phytochromes (isoforms) that absorb light at similar wavelengths. The downstream responses triggered by each isoform of phytochrome can vary, with the specific contribution of each individual phytochrome to the overall response depending on the light conditions. Work with mutants deficient in various species has yielded the identification of some of the roles played by each isoform. Identification of the downstream signalling molecules utilized by phytochrome to detect changes in growth and gene expression, has proven more difficult. Recent biochemical and genetic approaches used to characterize such signalling components will be dealt with later in this chapter.

Stages in Signal Transduction

Present-day understanding of signal transduction is based on the pioneering work of Earl W. Sutherland, whose study of how the animal hormone epiniphrine (adrenaline) stimulates the breakdown of glycogen (animal starch) in liver cells and in the muscle cells of blood vessels, earned him the Nobel Prize in 1971. He showed that when epiniphrine is added to a mixture of phosphorylase and its substrate

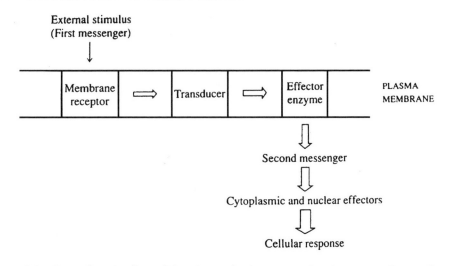

Figure 9.1   General mechanism of signal transduction across the plasma membrane of a cell. (From Horton *et al.* 1996)

glycogen *in vitro*, no breakdown of glycogen occurred. However, when epiniphrine was added to a solution containing intact cells, depolymerization of glycogen readily occurred. Sutherland and his co-workers concluded that something in the plasma membrane of the intact cells mediated the transmission of the epiniphrine stimulus. The sequence of events initiated by an external stimulus can generally be resolved into three stages – reception, transduction and response. A general mechanism for signal transduction is shown in Fig. 9.1. The plasma membrane of all cells contains specific receptors that allow the cell to respond to external stimuli that cannot cross the membrane. Signal perception involves the reaction of the stimulus with a receptor site on the surface of the target cell; receptor molecules are glycoproteins. As a result of signal perception, the receptor molecule is induced to change in conformation and become activated.

The second stage in the pathway is the transduction stage. The activated receptor passes the signal on to a membrane protein called a G-protein, which acts as a transducer. The transducer relays the signal to an effector enzyme, located on the inner surface of the plasma membrane. The action of the effector enzyme is to generate an intracellular second messenger, which is usually a small molecular ion. The second messenger relays the information from the stimulus (first messenger) to the nucleus, an intracellular compartment such as the endoplasmic reticulum, or to the cytosol. This results in a response, the third stage of the signal transduction pathway. An important feature of signalling pathways is amplification. A single stimulus–receptor complex can interact with many transducer molecules, each of which can activate several molecules of an effector enzyme. Similarly, second messengers can amplify the original signal by activating enzymes that can phosphorylate many target proteins. The series of amplification steps is called a cascade (Fig. 9.4).

Plant Hormones and Signal Transduction

The transfer of the stimulus induced by plant hormones provides a clear model for the study of the signal transduction pathway (Fig. 9.2). In this model, the hormone (ligand) binds to its specific receptor on the surface of the target cell and forms a hormone–receptor complex. Most hormone molecules are water-soluble and too large to pass through the plasma membrane. The presence of a receptor determines which cells are targeted by the hormone. Different cell types have different receptors that can elicit different responses to the same hormone, or a single cell may contain more than one type of receptor for each hormone. This may help to explain the multiple responses elicited by some hormones.

The hormone–receptor complex activates a membrane-bound G-protein (*transducer*) which in turn binds to a third membrane-bound protein (*effector*). In this model the enzyme adenylate cyclase, which is loosely attached to the cytoplasmic side of the membrane, is the third protein. Binding of the activated G-protein to adenylate cyclase activates the enzyme and thus stimulates the formation of cyclic adenosine monophosphate (cAMP), a second messenger, in the cytoplasm. Many hormone receptors in the plasma membrane rely on guanine nucleotide proteins (G-proteins) as transducers. These G-proteins, which consist of α, β and γ subunits, exist in two interconvertible forms – an inactive GDP-bound

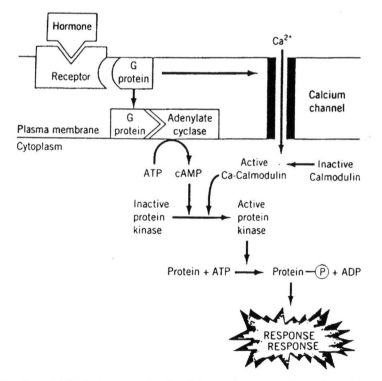

Figure 9.2 A model for hormone action involving a plasma membrane-bound receptor. (From Hopkins 1995, reproduced with permission)

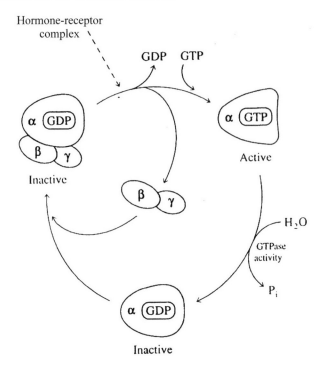

Figure 9.3   G-protein cycle. G-proteins undergo activation after binding to a receptor–ligand complex and are slowly inactivated by their own GTPase activity. (From Horton 1996)

(guanosine-diphosphate) form and an active GTP-bound (guanosine triphosphate) form (Fig. 9.3).

When a hormone–receptor complex, diffusing laterally in the membrane meets the inactive G-protein, it binds to it and induces the G-protein to change to the active conformation. Bound GDP is exchanged for GTP and the subunits β and γ dissociate from the G-GTP. The activated G-protein then binds to another protein, usually an enzyme and alters the activity of the enzyme. These changes are only transient as the GTPase activity of the G-protein acts as a built-in timer and slowly hydrolyses GTP to GDP. This hydrolysis reaction deactivates the G-protein, halting its activity when the extracellular signal is no longer present. Alternatively, the activated G-protein may interact with a ligand-gated ion channel. These channels, which are protein pores in the plasma membrane, open or close in response to a chemical signal. One such ion channel controls the flow of Ca^{2+} into the cytoplasm. Once in the cytoplasm, Ca^{2+}, a second messenger, binds to a number of calcium-binding proteins, such as calmodulin (CaM). The effect of either of the second messengers, cAMP or Ca^{2+}-CaM, is to activate proteins called protein kinases.

Protein kinases phosphorylate other proteins by transferring a phosphate group from ATP. Phosphorylation activates the protein, thereby altering the metabolism of the cell and triggering the cell's ultimate response to the initial stimulus. Protein

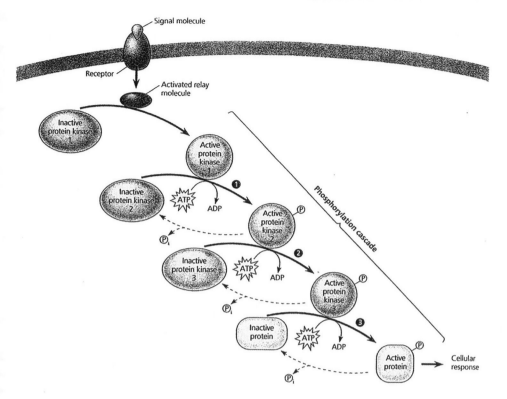

Figure 9.4   A phosphorylation cascade. (From Campbell *et al.* 1999)

kinases act on each other so that the external signal is transmitted by a cascade of protein phosphorylation, each reaction producing a conformational change and with the addition of a phosphate group, changes the protein from an inactive to an active form (Fig. 9.4).

For a cell to respond normally to an external signal it must have a mechanism for switching off the signal-transduction pathway when the initial signal is no longer present. The phosphorylation activity of protein kinases is rapidly reversed in the cell by the action of protein phosphatases – enzymes that remove phosphate groups from proteins. The activity of a protein regulated by phosphorylation depends on the balance in the cell between kinase and active phosphatase. Protein kinases were first characterized by Edwin Krebs and Edmund Fischer in the 1950s, and in 1992 they were awarded the Nobel Prize for their work on the fundamental role of protein kinases in cell metabolism. More than seventy protein kinases, which regulate many different aspects of metabolism, growth and second-messenger changes have been isolated and characterized (Stone and Walker 1995). There is abundant evidence that plants use reversible enzyme phosphorylation to regulate the rate of various metabolic pathways. The light-harvesting chlorophyll–protein complex has been shown to undergo phosphorylation–dephosphorylation, and this probably plays a role in regulating photosynthetic electron flow (Ranjeva and

Boudet 1987). There is evidence that many other plant enzymes may be controlled *in vivo* by reversible incorporation of covalently bound phosphate (Pooviah and Reddy 1987; Budde and Cholet 1998). These findings, along with the existence in plants of several types of protein kinase and at least one type of protein phosphatase (Hepler and Wayne, 1985; Pooviah and Reddy 1987), suggest that phosphorylation–dephosphorylation is a common control mechanism in plants for signal transduction of external stimuli.

Second Messengers in Plants

Second messengers are essential components of signal transduction pathways. Second messengers such as calcium ions (Ca^{2+}), cyclic adenosine and guanosine monophosphate (cAMP and cGMP), inositol triphosphate (IP_3) and 1,2-diacylglycerol (DAG) were first described in animal systems (Berridge and Ervine 1984). The last decade has been marked by substantial progress in elucidating the mechanisms of intracellular signalling in plants (Gilroy and Trewavas 1994). Comparisons with mammalian systems have shown that Ca^{2+}-calmodulin, cAMP (Neuhaus *et al.* 1993) and phosphoinositide signal mechanisms (Alexandre *et al.* 1990; Allen *et al.* 1995) also exist in plants. The involvement of cAMP as a second messenger in plant signal transduction pathways has been controversial for a long time (Assman 1995; Bolwell 1995). According to Trewavas (1997), however, 'plant cyclic AMP has come in from the cold'. While cAMP is an important signalling molecule in both prokaryotes and eukaryotes (Francis and Corbin 1996), its significance in higher plants has been generally doubted because plant cells have low adenylyl cyclase activity and barely detectable amounts of cAMP (Assman 1995; Ichikawa *et al.* 1997). Activation T-DNA tagging has been used to create tobacco cell lines that can proliferate in the absence of phytochrome in the culture medium (Ichikawa *et al.* 1997). The sequence tagged in one line was used to isolate complementary DNA encoding adenylyl cyclase – the first from a higher plant. Expression of the cDNA in *Escherichia coli* resulted in an increase in endogenous cAMP levels. Tobacco protoplasts treated with cAMP or the adenylyl cyclase activator forskolin, were shown to no longer require auxin to divide. This finding, together with the observation that the adenylyl cyclase inhibitor didoxy-adenosine inhibits cell proliferation in the presence of auxin, suggests that cAMP is involved in auxin-triggered cell division in higher plants.

Amplification

Enzyme cascades amplify the cell's response to an external signal. At each step in the cascade the number of activated products is much greater than in the previous step (Fig. 9.4). For example, each adenylyl cyclase molecule catalyses the formation of many cAMP molecules, each of which, in turn, activates a protein kinase which phosphorylates the next kinase in the pathway and so on. The amplification process ensures that proteins remain in the active form long enough to process numerous molecules before they become inactive again. As a result of signal amplification a small amount of external stimulus can lead to the production of a large number of metabolites that can elicit a response in an organism. Plant cells can initiate new developmental programmes in response to extremely weak stimuli. In these instances the physiological response must result from substantial amplification of the original signal and almost always does not have sufficient energy on its own to elicit a response.

Calcium and Signal Transduction

Calcium reportedly controls numerous physiological processes in plants and calcium ion concentration can have a great effect on cell behaviour. Free Ca^{2+} is recognized as having an important role in plants as a second messenger which couples various extracellular stimuli such as hormones, light or stress with intracellular metabolic activity. In general, the concentration of Ca^{2+} is normally millimolar. The concentration gradient of Ca^{2+} between the cytoplasm and the cell compartments is maintained by membrane-bound calcium-dependent ATPases believed to be localized in the endoplasmic reticulum and plasma membrane (Kaus 1987; Pooviah and Reddy 1987). This system has a very high affinity for Ca^{2+} and may be the main pump which maintains the low, free cytoplasmic Ca^{2+} concentration found in cells. A second mechanism for Ca^{2+} transport uses proton gradients as the driving force and has been characterized as a Ca^{2+}/H^+ antiport (Kaus 1987). This Ca^{2+}/H^+ antiport has been found in the tonoplast and possibly in the plasma membrane. Plastid and mitochondrial membranes are also believed to contain Ca^{2+} transport systems. External stimuli, such as light and hormones, relayed by a signal transduction pathway, affect the activity of ATPase and increase the cytoplasmic Ca^{2+} levels. The elevated calcium level is transient as the cytosolic calcium is pumped back out of the cell into one or more of the calcium stores (Pooviah and Reddy 1987; Gilroy and Trewavas 1994; Bush 1995). For example, gibberellin stimulates the synthesis and secretion of α-amylase by barley seed aleurone cells in the presence of millimolar amounts of external calcium (Fincher 1989; Jones and Jacobsen 1991). This response is triggered by a three-fold, gibberellin-induced increase in Ca^{2+} concentration. Ca^{2+} influx into the cytoplasm can occur because of the opening of Ca^{2+} channels in the membranes with Ca^{2+} entering by means of the concentration gradient. The cytoplasmic concentration of Ca^{2+} in barley and wheat aleurone cells is elevated from 100 nM to 500 nM by gibberellic acid, whereas abscisic acid reduces the Ca^{2+} levels (Gilroy and Jones 1992; Bush 1996; Gilroy 1996). Calcium-calmodulin (CaM) is also increased in response to gibberellic acid, implicating Ca^{2+} and CaM in the GA-response of these cells. Cytoplasmic Ca^{2+} levels can be measured by injecting Ca^{2+}-fluorescent dyes into the cytoplasm. These dyes fluoresce when complexed with Ca^{2+} and the amount of fluorescence is proportional to the concentration of Ca^{2+}. A simple and effective method of measuring changes in cytosolic Ca^{2+} concentration in whole plants uses the calcium-sensitive luminescent protein aequorin. This method involves the expression of the Ca^{2+}-activated photoprotein in transgenic plants (Knight *et al.* 1993; Knight and Knight 1995). This approach has been adopted for use in protoplasts, which provide an excellent system for the calibration of the Ca^{2+} signal (Haley *et al.* 1995; Knight *et al.* 1996). Volotovski *et al.* (1998) showed that increases in the second messengers cAMP and cGMP provoke increases in the Ca^{2+} concentration of plant cells. They suggest that a potential communication point for 'cross-talk' between signal transduction pathways, using these second messengers, is the release of Ca^{2+} from intracellular and extracellular stores.

Calcium-modulated Proteins

Plant cells contain a number of Ca^{2+}-binding proteins, referred to as 'Ca^{2+}-modulated proteins', which are targets or receptors for Ca^{2+}. These proteins can bind Ca^{2+} in a reversible manner. Stimulation of the cell increases cytoplasmic free Ca^{2+} concentration up to $10\,\mu M$, allowing the formation of a Ca^{2+}-modulated protein complex, which also acts as a second messenger. The most important and best studied of all known Ca^{2+}-modulated proteins is calmodulin (CaM). Calmodulin is a ubiquitous protein that has been isolated from a number of higher plants, yeast, fungi and green algae. Calmodulin from a number of plant sources has been characterized. It is a heat-stable, small-molecular-weight (17–19 kDa) protein, composed of a single unit and 148 amino acids. It contains four Ca^{2+}-binding sites per molecule and the binding of Ca^{2+} induces a conformational change which allows the protein to interact with and alter the activity of calmodulin-dependent enzymes. While the calcium–calmodulin complex (Ca^{2+}-CaM) has no enzyme activity itself, it may act as a regulatory subunit for other enzyme complexes such as Ca^{2+}-CaM-dependent NAD-kinase and protein kinases (Pooviah and Reddy 1987). NAD-kinase is found in the plastids, cytosol and mitochondria of plant cells and catalyses the ATP-dependent conversion of NAD to NADP. Because many redox enzymes are specific for one of these cofactors, regulating the balance between NAD and NADP effectively regulates metabolism, particularly reductive biosynthetic pathways which require NADPH. Several Ca^{2+}-dependent and Ca^{2+}-CaM-dependent NAD and protein kinases which result in the phosphorylation of key enzymes have been isolated from both soluble and membrane fractions of a large number of plants. This has added to the concept of signal 'amplification' and multienzyme 'cascades' (Ranjeva and Boudet 1987).

Phosphoinositides

Other second messengers that operate in plants are diacylglycerol (DAG) and inositol triphosphate (IP_3) (Boss 1989). These two second messengers are produced by the stimulus-induced cleavage of phosphatidylinositol-4,5-bisphosphate (PIP_2), a phospholipid on the cytoplasmic side of the plasma membrane (Fig. 9.5). The stimulus–receptor complex acting through a G-protein activates phospholipase C, an enzyme located on the cytoplasmic side of the plasma membrane. Phospholipase

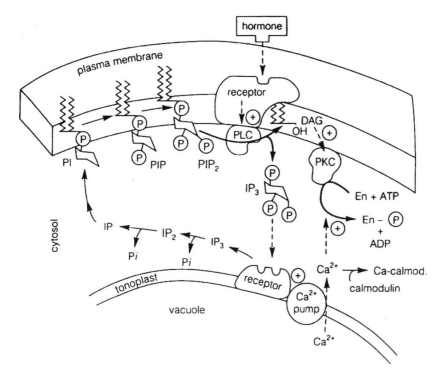

Figure 9.5 Hormone induced hydrolysis of a membrane lipid, phosphatidylinositol-4,5-bisphosphate (PIP_2) to release inositol-1,4,5-trisphosphate (IP_3) and a diacylglycerol (DAG). (From Salisbury and Ross 1992)

C, an effector, hydrolyses PIP_2 to inositol triphosphate (IP_3) and diacylglycerol (DAG). IP_3 is water-soluble and rapidly diffuses through the cytoplasm, whereas DAG is lipid-soluble and remains in the membrane. IP_3 triggers a transient increase in the level of cytoplasmic Ca^{2+} by opening calcium channels in the Ca^2 storage compartments such as the endoplasmic reticulum or plant cell vacuole; the Ca^{2+} then activates the next protein in one or more signalling pathways. Since IP_3, a second messenger, acts before Ca^{2+} in the pathway, Ca^{2+} could be considered as a 'third messenger'. However, by convention, the term 'second messenger' refers to all small, non-protein components of signal transduction pathways. DAG activates a specific calcium-dependent protein (protein kinase C) on the inside of the plasma membrane.

Early analysis of phospholipids in plant cells suggested that the quantities of phosphatidylinositol-4,5-bisphosphate (PIP_2), the precursor of IP_3 are too low to permit a signalling role. Subsequent measurements of IP_3 levels in guard cells indicated that quantities sufficient for signalling are present (Boss 1989). Moreover, IP_3-binding proteins have been detected in membrane fractions from several plant sources; phospholipase C, which hydrolyses PIP_2 to produce IP_3 and DAG, has been purified from plant tissues (Boss 1989). Thus it seems that the original problem regarding PIP_2 concentration may have been simply one of dilution: only a small subset of cells may be competent to transduce a specific signal at a particular time and only those cells will have appropriate levels of signalling intermediates.

Cyclic AMP

Many stimuli exert their effects on target cells by activating the adenylyl cyclase signalling pathway. Adenylyl cyclase is an effector enzyme located on the cytosol side of the plasma membrane (Fig. 9.6). It catalyses the formation of the second messenger cyclic AMP (cAMP). Cyclic AMP then diffuses from the membrane surface into the cytoplasm and activates an enzyme called protein kinase C, which in turn, initiates a phosphorylation cascade (Francis and Corbin 1996). Although genes with similarities to those encoding protein kinase A have been isolated from plants (Assman 1995) no plant protein kinase has yet been identified.

Protein kinase A, a serine–threonine protein kinase, catalyses the phosphorylation of the hydroxyl groups of specific serine and threonine residues in target enzymes. Phosphorylation of amino acid side-chains on enzymes is reversed by the action of protein phosphatase. Binding of a hormone to stimulatory membrane receptor (R_s) leads to the activation of a stimulatory G-protein (G_s); other hormones may bind to inhibitory receptors (R_i) which are coupled to adenylyl cyclase by an inhibitory G-protein (G_i) (Fig. 9.6). The opposing effects of the two G-proteins in regulating adenylyl cyclase enables the cell to fine-tune its metabolism in response to slight changes in the proportions of antagonizing hormones available. Published work in the 1980s contained only scattered reports of the presence of cAMP in plants and there were no plant responses known to be sensitive to cAMP. Furthermore, neither adenylyl cyclase nor cAMP-dependent

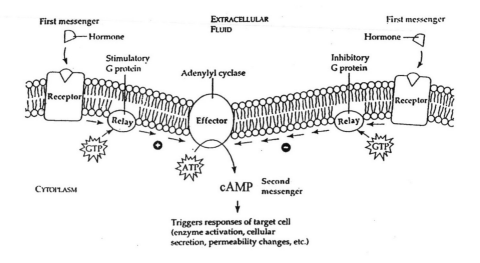

Figure 9.6 Signal transduction pathways involving cAMP as a second messenger. (From Campbell 1996)

protein kinases had been reported for plant tissues; there is no evidence that cAMP functions in signal transduction pathways in plants as it does in animal tissues.

In the 1990s, attempts to draw comparisons with mammalian systems had been unsuccessful in showing that Ca^{2+}-calmodulin (Neuhaus et al. 1993) and phosphoinositide signal mechanisms (Alexandre et al. 1990; Allen et al. 1995) also exist in plants. The long-standing controversial involvement of cAMP in signal transduction pathways in plants (Assman 1995; Bolwell 1995) are analogous to those found for IP_3 and DAG. While cAMP is an important signalling molecule in prokaryotes and eukaryotes (Francis and Corbin 1996), its significance in higher plants has been generally doubted because plant cells have low adenylyl cyclase activity and barely detectable amounts of cAMP (Assman 1995). The problem may be resolved if cAMP assays were to be carried out on the same cells and protoplasts that had been used to resolve the IP_3 question.

Phytochrome and Light Signalling

Throughout the life of the plant, phytochrome photoreceptors play a vital role in the plant's adaptation to its light environment. Phytochromes are involved in seed germination, seedling de-etiolation, gene expression, chloroplast differentiation, floral induction or suppression and senescence (Chory 1991). In addition, phytochrome plays a role in gravitropism (Parks et al. 1996) and is responsible for sensing the proximity of nearby plants (Smith 1995). Phytochrome also contributes to the diurnal and seasonal time-keeping mechanism of plants by detecting light quality and photoperiod (Kreps and Kay 1997). As already mentioned, phytochromes are the best-characterized plant photoreceptors. The phytochrome molecule contains a linear, covalently bound, chromophore that mediates light perception. The photoreceptor is synthesized in the red-absorbing P_r form and absorption of red light changes its conformation to the far-red-absorbing form P_{fr}, which is considered to be the physiological form. P_{fr} responses can be reversed by far-red light, which converts P_{fr} back to P_r. This ability to switch from one molecular form to another is unique to phytochrome. Experimental evidence has shown that there are at least two pools of phytochrome in plants, designated type I and type II (Furuya 1989; Quail 1991). These two phytochrome groups differ in their behaviour after they are exposed to red light. Type I P_{fr} is unstable and turns over very rapidly with a half-life of about 60 minutes. Type II phytochrome occurs in relatively high concentration in etiolated tissues (in dark-grown plants) and at substantially lower levels in green tissue (light-grown plants). On the other hand, type II phytochrome is light-stable and is present in approximately the same relatively low levels in light-grown green tissues as in etiolated tissues. Phytochrome apoproteins encoded by small, multigene families of five phytochrome genes (PHYA–PHYE) have been identified in Arabidopsis (Sharrock and Quail 1989; Clack et al. 1994), and each phytochrome is thought to have a different physiological role (Whitelam and Harberd 1994). Preliminary characterization of the protein products (Somers et al. 1991) has shown that PHYA encodes the light-labile type I phytochrome. This phytochrome is designated phytochrome A; PHYB and PHYC encode the light-stable, type II phytochrome.

How the phytochrome molecule transduces light stimuli is not clearly understood. Although the N-terminal region, which contains the chromophore-binding region, is involved in light perception, it appears that the C-terminal region is more likely to be involved in mediating biological activity, that is, signal transduction (Chory 1993; Boylan *et al.* 1994). Biochemical and genetic approaches have been used to characterize the components of the phytochrome-mediated signal transduction pathway. However, as yet, there is little or no overlap between the biochemical and genetic models of phytochrome signalling.

Genetic approach

The genetic approach is to screen for mutants that mimic the phenotype of the receptor-deficient mutant and yet have normal receptors. In *Arabidopsis*, two types of mutant have been studied (Chory 1993): those that have a dark-grown phenotype in the light, that is, exhibit a long hypocotyl, and those that have a light-grown phenotype in the dark and are characterized by a short hypocotyl. The former are light-insensitive mutant seedlings. When grown in the light their signalling pathway is impaired such that the seedlings fail to undergo photomorphogenesis and result in elongated hypocotyls (*hy* – long hypocotyl and *blu* – blue-light-inhibited hypocotyl elongation). Briefly, *hy1*, *hy2* and *hy6* mutants are defective in phytochrome chromophore biosynthesis and attachment to apoprotein and are thus deficient in all functional phytochromes (Parks and Quail 1991; Nagatani *et al.* 1993). The *hy3* and *hy8* mutants are deficient in phytochrome B and phytochrome A (Dehesh *et al.* 1993; Reed *et al.* 1993; Whitelam *et al.* 1993) and therefore do not exhibit red or far-red light inhibition, respectively, of hypocotyl elongation (Parks and Quail 1993). The *hy5* mutant is unique in that it is defective in inhibition of hypocotyl elongation mediated by red, far-red and blue light (Koorneef *et al.* 1980). This property suggests that the HY5 locus probably encodes a signal pathway component that functions downstream of the red, far-red and blue light photoreceptors (Whitelam *et al.* 1993). In other words, HY5 activates rather than represses photomorphogenesis. The second class of mutants exhibit a light-grown phenotype when grown in the dark. In this group the mutation causes the dark-grown seedlings to undergo photomorphogenesis, resulting in de-etiolated (*det*) or constitutive photomorphogenic (*cop*) mutants. Most of these mutants belong to a class of mutations known as *fusca*, which are characterized by short hypocotyls, open cotyledons and high concentrations of anthocyanins when grown in the dark. In addition, cells exhibit differentiation that is typical of light-grown plants and their light-regulated genes are active in the dark. Mutations in the three loci DET1 (Chory *et al.* 1989), DET2 (Chory *et al.* 1991) and COP1 (Deng *et al.* 1991; Deng and Quail 1992) lead to constitutive expression of normally light-regulated genes in dark-grown seedlings. These phenotypes suggest that these genetic loci are directly involved in coupling light perception to morphogenesis and gene expression throughout the plant's development. The recessive nature of all three mutations implies that their wild-type gene products act to repress photomorphogenesis in darkness (Wei and Deng 1992) and that light reverses this action. It has been suggested (Deng 1994) that since mutations in some of the COP and DET loci result in pleiotropic phenotypes, these loci may be involved in the early steps of the light-signalling pathway before it divides into branches that could control individual developmental processes. Although genetic screens for constitutive mutants select predominantly for mutations with defects in negative regulators, the large number of mutants now isolated indicate that negative regulation may be a common mechanism for controlling photomorphogenesis (Bowler and Chua 1994). Even though the *cop* and *det* mutants do not appear to have defects specific to phytochrome signalling, their phenotypes imply that the mutations lie within negatively acting global regulators that are able to receive signals from multiple stimuli and integrate them into a response. Photomorphogenic traits are influenced not only by light but also by phytohormones such as cytokinin and abscisic acid (Millar *et al.* 1994). For example, it has been shown that *det1* can be phenocopied in wild-type seedlings by the addition of cytokinin (Chory *et al.* 1991, 1994).

Mutants defective in the perception of light signals, namely *hy* and *blu* mutants, provide genetic evidence that there are distinct red- and blue-light photoreceptors and that the red-/far-red- and blue-light signal transduction pathways are genetically separable (Ang and Deng 1994). It is well established that photo-morphogenic seedling development in wild-type plants requires the concerted action of multiple photoreceptors (Kendrick and Kronenberg 1994). Thus, these individual pathways probably converge downstream along common regulatory steps of the light signal transduction pathway.

Approaches using transgenic plants have been developed to isolate mutants that are defective in light-regulated gene expression. *Arabidopsis* plants have been transformed with reporter genes under the control of light-responsive promoters. The activity of the reporter gene product can then be assayed in the mutagenized progeny to isolate mutants with deficiencies in signalling based on gene expression. The transgenic approach has been employed to create mutants that overproduce, rather than lack individual, functional phytochromes. Photoresponses initiated by overexpressed phytochrome may be 'exaggerated', thus allowing the responses that are regulated by a particular phytochrome species to be identified (Whitelam *et al.* 1994). This approach has been used to yield mutants that may identify signalling components. Screening transformed, mutagenized *Arabidopsis* seedlings for plants that overexpressed a light-inducible CAB3 (chlorophyll *a/b* binding protein) fusion gene in the dark, resulted in the isolation of mutants that have been named *doc* (dark overexpression of CAB) (Li *et al.* 1994). These *doc* mutants possess elevated CAB-mRNA levels but normal RBCS- (small subunit of RubisCO) mRNA levels, indicating a branch point in the signalling pathways that regulate the expression of the two genes. However, there are a number of potential drawbacks with the transgenic plant approach. For example, it is not certain that the responses mediated by the overexpressed phytochrome species in a transgenic plant are the true reflection of the responses mediated by the same phytochrome species in the wild-type plant. Ectopic expression of the transgenic phytochrome may allow a particular phytochrome species overexpressed to a high level to arrogate some or all of the functions that are normally performed by other phytochromes (Whitelam *et al.* 1994).

Phytochrome mutants and light intensity

Phytochrome mutants have also been useful as a means to determine which phytochromes are responsible for VLF, LF and HI responses. Studies using *phy*A, *phy*B or *phy*A*phy*B double mutants suggest that PHYA is the photoreceptor responsible for the VLF responses that control seed germination, hypocotyl growth inhibition and LHCP (light-harvesting chlorophyll proteins of PSII) expression (Botto *et al.* 1996; Shinomura *et al.* 1996; Mazzella *et al.* 1997). These studies also indicate that PHYA induces LHCP expression and seed germination over a broad spectrum (UV, blue, green, red and near far-red) of VLF light (Shinomura *et al.* 1996). These PHYA responses are not photoreversible and are therefore not typical phytochrome responses. PHYA also mediates the HIR that is responsible for hypocotyl growth inhibition in *Arabidopsis* (Parks and Quail 1993). Thus, PHYA appears to operate via two distinct action modes – HIR under continuous far-red light and VLF under multiple wavelengths of light.

In contrast, PHYB requires about 1000-fold more light to elicit seed germination and the induction of LHCP expression (Reed *et al.* 1994). It does so in a blue-light- or red-light-specific manner that is far-red-light reversible. These findings suggest that phytochrome acts through three different pathways in etiolated seedlings – one mediated by $P_{fr}A$ in VLF light, one by $P_{fr}B$ in LF red light and one by an unknown component of PHYA in HI responses. In addition, interactions between these three pathways can occur. For example, in the PHYA-mediated inhibition of hypocotyl growth in HIR, far-red light can be amplified by the $P_{fr}B$ that is formed after a red light pulse (Botto *et al.* 1996). PHYA and PHYB can also counteract each other's activity – PHYA can antagonize the activity of PHYB during shade avoidance (Smith 1995).

Biochemical approach

A number of biochemical approaches have been employed to identify and characterize the components of the phytochrome signal transduction pathway. These include a biochemical complementation technique (Neuhaus *et al.* 1993) in which the phytochrome responses are examined in single cells following a microinjection of putative signal transduction molecules into hypocotyl cells of mutant tomato. A second method uses pharmacological agonists and antagonists to interfere with phytochrome-mediated events; a third approach is to examine phosphorylation reactions (Millar *et al.* 1994). In the biochemical complementation study, seedlings of the phytochrome-deficient *aurea* mutant of tomato, which contains less than 5% of the amount of phytochrome found in wild-type tomato seedlings are used in order to avoid interference from responses due to endogenous phytochrome (Parks *et al.* 1987). The *aurea* has a defect, possibly in chromophore biosynthesis, that causes a deficiency in all phytochromes (Quail *et al.* 1995; Whitelam and Harberd 1994). Microinjection of purified oat phytochrome A (PhyA) into single cells of the *au*-mutant causes light-dependent anthocyanin accumulation and chloroplast development in the injected cells. In other words, microinjection of the oat PhyA into *au* cells rescues only PhyA-regulated processes and results in the biochemical complementation of the *au* phenotype – low anthocyanin and poor chloroplast development. In addition, co-injection of PhyA with a plasmid containing the phytochrome-responsive promoter from a gene encoding the wheat chlorophyll *a/b*-binding protein fused to the *Escherichia coli* β-glucuronidase gene, resulted in the light-dependent expression of the *cab*-GUS reporter construct in phytochrome-injected cells but not in non-injected cells (Neuhaus *et al.* 1993). Microinjection of relatively high concentrations of Ca^{2+} ($>5\,\mu M$) stimulated *cab*-GUS expression and partial chloroplast development responses in the absence of PhyA but did not affect anthocyanin levels. This result suggests that Ca^{2+} may play a role in phytochrome signal transduction and that the pathway is branched such that one branch uses Ca^{2+} as a component and the other does not (Whitelam and Harberd 1994). Further evidence that Ca^{2+} has a role in phytochrome-mediated signal transduction is the finding that transient increases in cytosolic free Ca^{2+} are involved in phytochrome-regulated swelling of oat protoplasts (Shacklock *et al.* 1992). It was shown that protoplast swelling in response to red-light pulses was preceded by Ca^{2+} transients. Both protoplast and Ca^{2+} transients are reversible by a far-red pulse given after the red pulse. Results from similar experiments suggest a role for GTP-binding proteins in phytochrome pathways leading to *cab*-GUS expression, chloroplast development and anthocyanin accumulation.

Co-injection of putative pathway intermediates which should complement the mutation, with a range of reporter constructs, thus provides a means of identifying active signalling molecules. While these results clearly suggest that Ca^{2+} and GTP-binding proteins are involved in signal transduction from microinjected phytochrome A, it should be noted that phytochrome A null mutants of tomato do not exhibit the phenotypic characteristics of *au* seedlings (Whitelam *et al.* 1993). Thus, absence of phytochrome A cannot be the cause of the *au* phenotype, unless it does so via routes not used in the normal plant. Therefore the signal transduction pathway analysed in this work may have only limited relevance to the pathways coupled to phytochrome photoreceptors in normal plants (Whitelam and Harberd 1994). Results from complementation studies in *aurea* have led to the proposal of a biochemical model for phytochrome signalling (Barnes *et al.* 1997). In this model (Fig. 9.7), light stimulation of PhyA (P_{fr}) activates at least one heterotrimeric G-protein, which in turn stimulates three biochemically distinct signalling cascades. First, as Ca^{2+} levels rise, CaM is activated, which results in the induction of PSII genes such as CAB and ATP-synthase and partial chloroplast development.

In the second branch, a cGMP-dependent pathway regulates expression of CHS and the production of anthocyanins. The third pathway arises from the convergence of the other two pathways (Fig. 9.7), using both Ca^{2+} and cGMP

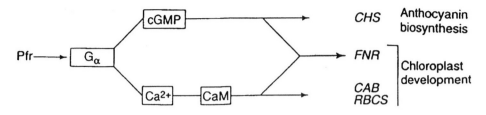

Figure 9.7 Phytochrome (P_{fr}) is mediated by heteromeric G-proteins (G_a) and subsequently by three pathways dependent upon Ca^{2+}, cGMP or on Ca^{2+} and cGMP together. (From Barnes *et al.* 1997)

to stimulate genes encoding PSI components such as FNR (Ferredoxin-NADP$^+$-reductase) and to direct full chloroplast development. The specificity of different phytochrome species for each branch of the Phytochrome signalling pathway has been studied by microinjecting recombinant PhyA and PhyB adducts (Kunkel *et al.* 1993) into *aurea* cells. Results have shown that although microinjection of a rice PhyA adduct and purified oat PhyA were able to activate CHS, CAB and FNR expression, the tobacco PhyB adduct was unable to activate CHS, whereas CAB and FNR induction was normal.

Potentially one of the simplest light-signalling pathways to visualize is the light-regulated expression of nuclear genes. The best studied of these target genes encode chloroplast-destined proteins, such as chlorophyll *a/b* binding protein of PSII (CAB) and the small subunit of RubisCO (RBCS). Both CAB and RBCS are positively regulated by light and show cell-type specificity, being expressed exclusively in chloroplast-containing cells (Chory and Susek 1994). Using pharmacological agonists and antagonists that disrupt phytochrome-mediated events, it has been shown that phytochrome may act through a signal transduction pathway involving heterotrimeric G-proteins and Ca^{2+}-activated calmodulin to derepress CAB and RBCS expression (Bowler and Chua 1994). Other nuclear genes whose expression is positively regulated by light are not involved in chloroplast functions and are expressed in cells that do not contain chloroplasts. These include anthocyanin genes such as CHS and the gene encoding chalcone synthase. The CHS gene is highly expressed in the epidermis very early during leaf development (Chory and Peto 1990). The pharmacological studies of Bowler and Chua (1994) implicate a heterotrimeric G-protein and cGMP in mediating phytochrome-regulated CHS gene expression. These studies suggest that phytochrome acts through different downstream effectors, which may be specific to the cell type in which light is eliciting a response.

The discovery that purified oat phytochrome contains an associated protein kinase activity raised the possibility that phytochrome initiated its signal transduction chain by phosphorylation (Wong *et al.* 1986). It has been shown, however, that this associated kinase activity could be separated from phytochrome (Grimm *et al.* 1989; Kim *et al.* 1989), clearly indicating that oat phytochrome is not a protein kinase. Nevertheless, the involvement of phosphorylation in phytochrome responses has been shown: red-light-regulated protein phosphorylation (Datta *et al.* 1985; Fallon *et al.* 1993), phosphorylation-regulated binding of factors to the promoters of phytochrome-regulated genes (Datta and Cashmore 1989; Klimczak and Hind 1990; Sarokin and Chua 1992) and light-regulated expression of genes encoding putative protein kinases (Li *et al.* 1993). In addition there are sequence similarities between the C-terminus of higher-plant phytochrome and bacterial histidine kinase catalytic domains (Schneider-Poetsch 1992). Furthermore, preparations of highly purified oat phytochrome have been shown to contain polycation-stimulated serine/threonine kinase activity (Wong *et al.* 1986) capable of phosphorylating the serine-rich N-terminus of the photoreceptor (McMichael and Lagarias 1990). These results, coupled with the report that phytochrome has an ATP-binding site (Wong *et al.* 1986), support the hypothesis that phytochrome is a protein kinase. There is evidence (Algarra *et al.* 1993; Thummler *et al.* 1992), however, that phytochrome from the moss *Cerotodon purpureus* possesses kinase activity. The phytochrome gene of this moss contains three exons. The first two exons encode a phytochrome chromophore domain similar to that of fern phytochrome (Thummler *et al.* 1992). The third exon encodes protein sequences that are not related to phytochrome but rather show striking similarities to the conserved regions of protein kinase catalytic domains (Thummler *et al.* 1992). One hypothesis suggests that the two functional domains were separated during evolution. Homologues of the *Cerotodon* phytochrome kinase domain in higher plants have yet to be found.

Summary

Higher plants possess photosensory systems which monitor environmental light that regulates gene expression at every stage of the life cycle of the plant. Plants grown in darkness are etiolated; in the presence of light they undergo photomorphogenesis. The main photoreceptors in plants are the, as yet unidentified, UV-B photoreceptors, the blue/UV-A cryptochromes and the red/ far-red phytochromes.

The main stages in the signal transduction pathway are reception, transduction and response. The plasma membrane of cells contains receptors through which the cell reacts with external stimuli; receptor molecules are glycoproteins. The signal (first messenger) is passed from the activated receptor through a G-protein (transducer) in the membrane and on to an effector enzyme located on the inner surface of the membrane. This activity results in the creation of a second messenger, usually a small ion, which relays the information to the interior of the cell where it elicits a response. In the transduction pathway the signal is amplified in a series of steps called a cascade. The transfer of the signal induced by plant hormones provides a good model of the process. In this model cyclic AMP (cAMP) is formed as a second messenger in the cytoplasm. Alternatively, the activated G-protein may interact with ion channels in the membrane, allowing the inflow of Ca^{2+} ions that may act as second messengers on their own or with specific Ca^{2+}-binding proteins such as calmodulin. The effect of second messengers is to activate protein kinases that drive a cascade of phosphorylation reactions. Second messengers in plants have been shown to include Ca^{2+}, Ca^{2+}-modulated proteins, cyclic adenosine- and guanosine-monophosphate (cAMP and cGMP respectively), inositol triphosphate (IP_3) and diacylglycerol (DAG).

Phytochromes are the best-characterized of all plant photoreceptors. Phytochrome is synthesized in the red-absorbing form (P_r) and is converted to the far-red-absorbing form (P_{fr}) – a switching ability unique to phytochrome. How the phytochrome molecule transduces light is not yet clearly understood. Genetic and biochemical approaches have been used to characterize the components of the phytochrome-mediated signal transduction pathway. The genetic approach is to screen for mutants that mimic the phenotype of the receptor-deficient receptor and yet have the normal receptor; most of these studies have been carried out on mutants of *Arabidopsis thaliana*. A number of biochemical approaches have been employed to identify and characterize the components of the pathway. These methods include mutant complementation studies in which phytochrome responses are examined in single cells following microinjection of putative signal transduction molecules into hypocotyl cells of mutant tomato.

At present, there is a lack of overlap between the biochemical and genetic models of phytochrome signalling. Biochemical pathways have been defined mainly by their positive elements, whereas most of the gene products inferred from mutant studies are likely to be repressors. This is simply a result of the different experimental approaches – the microinjection experiments were designed to identify positively acting signal intermediates, whereas the genetic screens were expected to yield predominantly negative regulators. Another important difference between the two approaches is that the biochemical pathways have defined very specific cellular reactions (chloroplast development and anthocyanin biosynthesis); most mutants, by contrast, have been isolated based on their developmental aberrations. Together these two approaches are beginning to unravel the details of a variety of signalling networks by which plants perceive and respond to environmental stimuli.

References

Ahmad, M. and Cashmore, A.R. (1996). Seeing blue: the discovery of cryptochrome. *Plant Mol. Biol.* **30**: 851–861.

Alexandre, J., Lassalles, J.P. and Kado, R.P. (1990). Opening of Ca^{2+} channels in isolated red beet root vacuole membrane by inositol 1,4,5-triphosphate. *Nature* **343**: 567–570.

Algarra, P., Linder, S. and Thummler, F. (1993). Biochemical evidence that phytochrome of the moss *Ceratodon purpureus* is a light-regulated protein kinase. *FEBS Lett.* **315**: 69–73.

Allen, G.J., Muir, S.R. and Sanders, D. (1995). Release of Ca^{2+} from individual plant vacuoles by both InsP3 and cyclic ADP-ribose. *Science* **268**: 735–737.

Ang, L.-H. and Deng, X.-W. (1994). Regulatory hierarchy of photo-morphogenic loci: allele-specific and light-dependent interaction between HY5 and COP1 loci. *Plant Cell* **6**: 613–628.

Assman, S.M. (1995). Cyclic AMP as a second messenger in higher plants. *Plant Physiol.* **108**: 885–889.

Barnes, S.A., McGrath, R.B. and Chua, N.-H. (1997). Light signal transduction in plants. *Trends Cell Biol.* **7**: 21–26.

Beggs, C.J. and Wellman, E. (1985). Analysis of light-controlled anthocyanin formation in coleoptiles of *Zea mays* L. The role of UV-B, blue, red and far-red light. *Photochem. Photobiol.* **41**: 481–486.

Berridge, M.J. and Ervine, R.F. (1984). Inositol triphosphate, a novel second messenger in cellular signal transduction. *Nature* **312**: 315–321.

Bolwell, G.P. (1995). Cyclic AMP, the reluctant messenger in plants. *Trends Biochem. Sci.* **12**: 492–495.

Boss, W.F. (1989). Phosphoinositide metabolism: its relation to signal transduction in plants. In: *Second Messengers in Plant Growth and Development*, W.F. Boss and D.J. Morré (eds), pp. 29–56. A.R. Liss, New York.

Botto, J.F., Sanchez, R.A., Whitelam, G.C. and Casal, J.J. (1996). Phytochrome A mediates the promotion of seed germination by very low fluences of light and canopy shade in *Arabidopsis*. *Plant Physiol.* **110**: 439–444.

Bowler, C. and Chua, N.-H. (1994). Emerging themes in plant signal transduction. *Plant Cell* **6**: 1529–1541.

Boylan, M., Douglas, N. and Quail, P.H. (1994). Dominant negative suppression of *Arabidopsis* photoresponses by mutant phytochrome A sequences identifies spatially discrete regulatory domains in the photoreceptor. *Plant Cell* **6**: 449–490.

Budde, R.J.A. and Cholet, R. (1998). Regulation of enzyme activity in plants by reversible phosphorylation. *Physiol. Plant* **72**: 435–439.

Bush, D.S. (1995). Calcium regulation in plant cells and its role in signalling. *Ann. Rev. Plant Physiol. Plant Mol. Biol.* **46**: 95–122.

Bush, D.S. (1996). Effects of gibberellic acid and environmental factors on cytosolic calcium in wheat aleurone cells. *Planta* **199**: 89–99.

Campbell, N.A. (1996). Chemical signals in animals, *Biology*, pp. 912–936. Benjamin/Cummings, Redwood City, CA.

Campbell, N.A., Reece, J.B. and Mitchell, L.G. (1999). Cell communication, *Biology*, pp. 188–205. Benjamin/Cummings, San Francisco.

Chory, J. (1991). Light signals in leaf and chloroplast development: photoreceptors and downstream responses in search of a transduction pathway. *New Biol.* **3**: 538–554.

Chory, J. (1993). Out of the darkness: mutants reveal pathways controlling light-regulated development in plants. *Trends Genet.* **9**: 167–172.

Chory, J. and Peto, C.A. (1990). Mutations in the DET1 affect cell-type-specific expression of light-regulated genes and chloroplast development in *Arabidopsis*. *Proc. Natl. Acad. Sci. USA* **87**: 8776–8780.

Chory, J. and Susek, R. (1994). Light signal transduction and the control of seedling development. In: *Arabidopsis*, E. Meyerowitz and C. Sommerville (eds), pp. 579–614. Cold Spring Harbor Press, Cold Spring Harbor, NY.

Chory, J., Peto, C.A., Feinbaum, R., Pratt, L. and Ausubel, F. (1989). *Arabidopsis thaliana* mutant that develops a light-grown plant in the absence of light. *Cell* **58**: 991–999.

Chory, J., Nagpal, P. and Peto, C.A. (1991). Phenotypic and genetic analysis of det2, a new mutant that affects light-regulated seedling development in *Arabidopsis*. *Plant Cell* **3**: 445–459.

Chory, J., Reinecke, D., Sim, S., Washburn, T. and Brenner, M. (1994). A role for cytokinins in de-etiolation in *Arabidopsis*. *Plant Physiol.* **104**: 339–347.

Christie, J.M. and Jenkins, G.I. (1996). Distinct UV-B and UV-A/blue light signal transduction pathways induce chalcone synthase gene expression in *Arabidopsis* cells. *Plant Cell* **8**: 1555–1567.

Clack, T., Mathews, S. and Sharrock, R.A. (1994). The phytochrome apoprotein family in *Arabidopsis* is encoded by five genes: the sequences and expression of PHYD and PHYE. *Plant Mol. Biol.* **25**: 413–427.

Datta, N. and Cashmore, A.R. (1989). Binding of a pea nuclear protein to promoters of certain photoregulated genes is modulated by phosphorylation. *Plant Cell* **1**: 1069–1077.

Datta, N., Chen, Y.-R. and Rou, S.J. (1985). Phytochrome and calcium stimulation of protein phosphorylation in isolated pea nuclei. *Biochem. Biophys. Res. Commun.* **128**: 1403–1408.

Dehesh, K., Franci, C., Parks, B.M., Seely, K.A., Short, T.W., Tepperman, J.M. and Quail, P.H. (1993). *Arabidopsis* HY8 locus encodes phytochrome A. *Plant Cell* **5**: 1081–1088.

Deng, X.-W. (1994). Fresh view of light signal transduction in plants. *Cell* **76**: 423–426.

Deng, X.-W. and Quail, P.H. (1992). Genetic and phenotypic characterization of cop1 mutants of *Arabidopsis thaliana*. *Plant J.* **2**: 83–95.

Deng, X.-W., Caspar, T. and Quail, P.H. (1991). COP1: a regulatory locus involved in the light-controlled development and gene expression in *Arabidopsis*. *Genes Dev.* **5**: 1172–1182.

Fallon, K.M., Shaclock, P.S. and Trewavas, A.J. (1993). Detection *in vivo* of very rapid red-light induced calcium-sensitive protein phosphorylation in etiolated wheat (*Triticum aestivum*) leaf protoplasts. *Plant Physiol.* **101**: 1039–1045.

Fincher, G.B. (1989). Molecular and cellular biology associated with endosperm mobilization in germinating cereal grains. *Ann. Rev. Plant Physiol. Plant Mol. Biol.* **40**: 305–346.

Francis, S.H. and. Corbin, J.D. (1996). In: *Signal Transduction*, C.-H. Heldin and M. Purton (eds), 223–240. Chapman and Hall, London.

Furuya, M. (1989). Molecular properties and biogenesis of Phytochrome I and II. *Adv. Biophys.* **25**: 133–167.

Furuya, M. and Schäfer, E. (1996). Photoreception and signalling of induction reactions by different phytochromes. *Trends Plant Sci.* **1**: 301–306.

Gilroy, S. (1996). Calcium-dependent and independent signal transduction in barley aleurone. *Plant Cell* **8**: 2193–2209.

Gilroy, S. and Jones, R.L. (1992). Gibberellic acid and abscisic acid coordinately regulate cytoplasmic calcium and secretory activity in barley aleurone protoplasts. *Proc. Natl. Acad. Sci. USA* **89**: 3591–3595.

Gilroy, S. and Trewavas, A.J. (1994). A decade of plant signals. *BioEssays* **16**: 677–682.

Grimm, R., Gast, D. and Rudiger, W. (1989). Characterization of a protein-kinase activity associated with phytochrome from etiolated oat (*Avena sativa* L.) seedlings. *Planta* **178**: 199–206.

Haley, A., Russel, A.J., Wood, N., Allan, A.C., Knight, M.R., Campbell, A.K. and Trewavas, A.J. (1995). Effects of mechanical signalling on plant cell cytosolic calcium. *Proc. Natl. Acad. Sci. USA* **92**: 4124–4128.

Hepler, P.K. and Wayne, R.O. (1985). Calcium and plant development. *Ann. Rev. Plant Physiol.* **36**: 397–439.

Hopkins, W.G. (1995) Biochemistry and mode of action of hormones, *Introduction to Plant Physiology*, pp. 311–399. Wiley, New York.

Horton, H.R., Moran, L.A., Ochs, R.S., Rawn, J.D. and. Scrimgeour, K.G. (1996). Lipids and membranes, *Principles of Biochemistry*, pp. 261–295. Prentice-Hall, Upper Saddle River, NJ.

Ichikawa,T., Suzuki, Y., Czaja, I., Schommer, C., Lennick, A., Schell, J. and Walden, R. (1997). Identification and role of adenylyl cyclase in auxin signalling in higher plants. *Nature* **390**: 698–701.

Jones, R.L. and Jacobsen, J.V. (1991). Regulation of the synthesis and transport of secretion proteins in cereal aleurone. *Int. Rev. Cytol.* **126**: 49–88.

Kaufman, L.S., Thompson, W.F. and Briggs, W.R. (1984). Different red light requirements for phytochrome-induced accumulation of cab and rbcs RNA. *Science* **226**: 1447–1449.

Kaus, H. (1987). Some aspects of calcium-dependent regulation in plant metabolism. *Ann. Rev. Plant Physiol.* **38**: 47–72.

Kendrick, R.E. and Kronenberg, G.H.M. (1994). *Photomorphogenesis in Plants*, 2nd edn. Kluwer Academic Publishers, Dordrecht.

Kim, I.-S., Bai, U. and Song, P.-S. (1989). A purified 124-kDa oat phytochrome does not possess a protein kinase activity. *Phytochem. Phytobiol.* **49**: 319–323.

Klimczak, L.J. and Hind, J. (1990). Biochemical similarities between soluble and membrane-bound calcium-dependent protein kinases of barley. *Plant Physiol.* **92**: 919–923.

Knight, H. and Knight, M.R. (1995). Recombinant aequorin methods for intracellular calcium measurements in plants. *Methods Cell Biol.* **49**: 201–216.

Knight, H., Trewavas, A.J. and Knight, M. (1996). Cold calcium signalling in *Arabidopsis* involves two cellular pools and a change in calcium signature after acclimation. *Plant Cell* **8**: 489–503.

Knight, M.R., Read, N.D., Campbell, A.K. and Trewavas, A.J. (1993). Imaging calcium dynamics in living plants using semi-synthetic recombinant aequorin. *J. Cell Biol.* **121**: 83–90.

Koorneef, M., Rolff, E. and Spruit, C.J.P. (1980). Genetic control of light-inhibited hypocotyl elongation in *Arabidopsis thaliana* (L.) Hyenh. *Z. Pflanzenphysiol.* **100**: 147–160.

Kreps, J.A. and Kay, S.A. (1997). Coordination of plant metabolism and development by the circadian clock. *Plant Cell* **9**: 1235–1244.

Kunkel, T., Tomizawa, K.-I., Kern, R., Furuya, M., Chua, N.-H. and Schäfer, E. (1993). *In vitro* formation of a photoreversible adduct of phycocyanobolin and tobacco apophytochrome B. *Eur. J. Biochem.* **215**: 587–594.

Li, E., Altschmied, L. and Chory, J. (1994). *Arabidopsis* mutants define downstream branches in the phototransduction pathway. *Genes Dev.* **18**: 339–349.

Li, H., Washburn, T. and Chory, J. (1993). Regulation of gene expression by light. *Opin. Cell Biol.* **5**: 455–460.

Mancinelli, A.L. and Rabino, I. (1978). The high irradiance responses of plant photomorphogenesis. *Bot. Rev.* **44**: 129–180.

Mazzella, M.A., Magliano, T.M. and Casal, J.J. (1997). Dual effect of phytochrome A on hypocotyl growth under continuous red light. *Plant Cell Environ.* **20**: 261–267.

McMichael, R.W. and Lagarias, J.C. (1990). Phosphopeptide mapping of *Avena* phytochrome phosphorylated by protein kinase *in vitro*. *Biochemistry* **29**: 3872–3878.

Millar, A.J., McGrath, R.B. and Chua, N.-H. (1994). Phytochrome phototransduction pathways. *Ann. Rev. Genet.* **28**: 325–349.

Nagatani, A., Reed, J.W. and Chory, J. (1993). Isolation and initial characterization of *Arabidopsis* mutants that are deficient in phytochrome A. *Plant Physiol.* **102**: 269–277.

Neuhaus, G., Bowler, C., Kern, R. and Chua, N.-H. (1993). Calcium/calmodulin-dependent and independent phytochrome signal transduction pathways. *Cell* **73**: 937–952.

Parks, B.M. and Quail, P.H. (1991). Phytochrome-deficient *hy1* and *hy2* long hypocotyl mutants *Arabidopsis* are defective in phytochrome chromophore biosynthesis. *Plant Cell* **3**: 1177–1186.

Parks, B.M. and Quail, P.H. (1993). *hy8*, a new class of *Arabidopsis* long hypocotyl mutants deficient in functional phytochrome A. *Plant Cell* **5**: 39–48.

Parks, B.M., Jones, A.M., Adamse, P., Koorneef, M., Kendrick, M.E. and Quail, P.H. (1987). The *aurea* mutant of tomato is deficient in spectrophotometrically and immunochemically detectable phytochrome. *Plant Mol. Biol.* **9**: 97–107.

Parks, B.M., Quail, P.H. and Hangarter, R.P. (1996). Phytochrome A regulates red-light induction of phototropic enhancement in *Arabidopsis*. *Plant Physiol.* **110**: 155–162.

Pooviah, B.W. and Reddy, A.S.N. (1987). Calcium messenger system in plants. *CRC Crit. Rev. Plant Sci.* **6**: 47–103.

Quail, P.H. (1991). Phytochrome: a light-activated molecular switch that regulates plant gene expression. *Ann. Rev. Genet.* **25**: 389–409.

Quail, P.H., Boylan, M.T., Parks, B.M., Short, T.W., Xu, Y. and Wagner, D. (1995). Phytochromes: photosensory perception and signal transduction. *Science* **268**: 675–680.

Ranjeva, R. and Boudet, A.M. (1987). Phosphorylation of proteins in plants: regulatory effects and potential involvement in stimulus/response coupling. *Ann. Rev. Plant Physiol.* **38**: 73–93.

Reed, J.W., Nagpal, P., Poole, D.S., Furuya, M. and Chory, J. (1993). Mutations in the gene for the red/far-red light receptor phytochrome B alter cell elongation and physiological responses throughout *Arabidopsis* development. *Plant Cell* **5**: 147–157.

Reed, J.W., Nagatani, A., Elich, T.D., Fagan, M. and Chory, J. (1994). Phytochrome A and phytochrome B have overlapping but distinct functions in *Arabidopsis* development. *Plant Physiol.* **104**: 1139–1149.

Salisbury, F.B. and Ross, C.W. (1992). Hormones and growth regulators, *Plant Physiology*, pp. 357–381. Wadsworth, Belmont, CA.

Sarokin, L.P. and Chua, N.-H. (1992). Binding sites for two novel phosphoproteins, 3AF5 and 3AF3 are required for rbcs-3A expression. *Plant Cell* **4**: 473–483.

Schneider-Poetsch, H.A.W. (1992). Signal transduction by phytochrome: phytochromes have a module related to the transmitter modules of bacterial sensor proteins. *Photochem. Photobiol.* **56**: 839–846.

Shacklock, P.S., Read, N.D. and Trewavas, A.J. (1992). Cytosolic free calcium mediates red light-induced photomorphogenesis. *Nature* **358**: 753–757.

Sharrock, R.A. and Quail, P.H. (1989). Novel phytochrome sequences in *Arabidopsis thaliana*: structure, evolution and differential expression of a plant regulatory photoreceptor family. *Genes Dev.* **3**: 1745–1757.

Shinomura, T., Nagatani, A., Hanzawa, H., Kubota, M., Watanabe, M. and Furuya, M. (1996). Action spectra for phytochrome A- and B-specific photoinduction of seed germination in *Arabidopsis thaliana*. *Proc. Natl. Acad. Sci. USA* **93**: 8129–8133.

Smith, H. (1995). Physiological and ecological function within the phytochrome family. *Ann. Rev. Plant Physiol. Plant Mol. Biol.* **46**: 289–315.

Somers, D.E., Sharrock, R.A., Tepperman, J.M. and Quail, P.H. (1991). The *hy3* long hypocotyl mutant of *Arabidopsis* is deficient in phytochrome B. *Plant Cell* **3**: 1263–1274.

Stone, J.M. and Walker, J.C. (1995). Plant protein kinase families and signal transduction. *Plant Physiol.* **108**: 451–457.

Thummler, W.F., Duffner, M., Kreisl, P. and Dittrich, P. (1992). Molecular cloning of a novel phytochrome gene of the moss *Ceratodon purpureus* which encodes a putative light-regulated protein kinase. *Plant Mol. Biol.* **20**: 1003–1017.

Trewavas, A.J. (1997). Plant cyclic AMP comes in from the cold. *Nature* **390**: 675–678.

Volotovski, I.D., Sokolovsky, S.G., Molchan, O.V. and Knight, M.R. (1998). Second messengers mediate increase in cytosolic calcium in tobacco protoplasts. *Plant Physiol.* **117**: 1023–1030.

Warpeha, K.M.F. and Kaufman, L.S. (1990). Two distinct blue-light responses regulate epicotyl elongation in pea. *Plant Physiol.* **92**: 195–199.

Wei, N. and Deng, X.-W. (1992). COP9: a new genetic locus involved in light-regulated development and gene expression in *Arabidopsis*. *Plant Cell* **4**: 1507–1518.

Whitelam, G.C. and Harberd, N.P. (1994). Action and function of phytochrome family members revealed through the study of mutant and transgenic plants. *Plant Cell Environ.* **17**: 615–625.

Whitelam, G.C., Johnson, E., Peng, J., Carol, P., Anderson, M.L., Cowl, J.H. and Harberd, N.P. (1993). Phytochrome A null mutants of *Arabidopsis* display a wild-type phenotype in white light. *Plant Cell* **5**: 757–768.

Wong, Y.-S., Cheng, H.-C., Walsh, D.A. and Lagarias, J.C. (1986). Phosphorylation of *Avena* phytochrome *in vitro* as a probe of light-induced conformational changes. *J. Biol. Chem.* **261**: 12089–12097.

Young, J., Liscum, E. and Hangarter, R. (1992). Spectral-dependence of light-inhibited hypocotyl elongation in photomorphogenic mutants of *Arabidopsis*: evidence for a UV-A photosensor. *Planta* **188**: 106–114.

Glossary

abscisic acid (ABA): a phytohormone (growth regulator) responsible for growth inhibition and seed dormancy; associated with stress conditions, stomatal closure.

absorption spectrum: the range of a pigment's capacity to absorb wavelengths of light.

actinic light: light that produces a biological response.

action spectrum: the range of wavelengths of light that elicits a specific response.

active transport: movement of an ion or small molecule across a membrane against its concentration gradient or electrochemical gradient by a process directly coupled to the hydrolysis of ATP.

adenylyl cyclase: an enzyme that catalyses the conversion of ATP to cyclic AMP (cAMP), a second messenger, in response to an external stimulus.

aequorin: a Ca^{2+}-sensitive luminescent protein.

aerobic: referring to a cell, organism or metabolic process that utilizes O_2.

aleurone: a layer of protein-rich cells located at the outer edge of the endosperm of monocot seeds.

allele: an alternative form of a gene.

all-or-none event: an action that occurs either completely or not at all.

amino acid: an organic molecule possessing an amino group and a carboxyl group; a building block of proteins.

amphipathic: a molecule with both hydrophilic and hydrophobic properties.

amphotoperiodic: a plant that flowers on either short-day or long-day cycles but not on intermediate daylengths.

amplification: multiplication of an original signal in a signal transduction pathway by the action of activating enzymes.

amylopectin: a highly branched polymer of glucose molecules connected by $\alpha[1-4]$ and $\alpha[1-6]$ linkages.

amylose: a straight-chain polymer of glucose molecules connected by $\alpha[1-4]$ linkages.

anaerobic: referring to a cell, organism or metabolic process that functions in the absence of O_2.

Photobiology of Higher Plants. By M. S. Mc Donald.
© 2003 John Wiley & Sons, Ltd: ISBN 0 470 85522 3; ISBN 0 470 85523 1 (PB)

angle of incidence: the angle at which light enters a medium.

annual plant: a plant whose life-cycle is completed within one growing season.

antenna chlorophyll: molecules of chlorophyll that absorb incoming photons of light and funnel their excitation energy on to the reaction centre of a photosystem.

anthesin: a hypothetical flowering hormone.

anthocyanidin: anthocyanin from which the sugar moiety has been removed.

anthocyanin: a group of compounds (flavonoids) responsible for various shades of pink, purple and blue colours in a variety of plant parts.

antibody: an antigen-binding immunoglobulin that functions as the effector in an immune response.

antigen: a macromolecule foreign to its host and which elicits an immune response.

antiport: the cotransport of two ions in opposite dirctions across cell membranes.

apical meristems: the undifferentiated, actively dividing cells at the growing tip of a plant shoot.

apoenzyme: the protein portion of an enzyme, exclusive of any organic or inorganic cofactors or prosthetic groups that might be required for catalytic activity.

apoprotein: the protein part of a holoprotein such as chromoprotein.

aquaporin: water-channel protein that facilitates the osmotic flow of water across the plasma membrane.

aspartate formers: plant species that use aspartic acid to transport CO_2 from mesophyll cells to bundle sheath cells.

ATP (adenosine triphosphate): a nucleotide containing adenine, ribose and three phosphate bonds, the hydrolysis of which is accompanied by the release of free energy. It is the most important molecule for capturing and transferring free energy in cells.

ATPase: one of a large group of enzymes that catalyse the hydrolysis of ATP to ADP and inorganic phosphate with the release of free energy.

ATP synthase: a membrane-bound enzyme that phosphorylates ADP to ATP by using energy from the diffusion of protons through the enzyme; also called CF_0–CF_1 complex.

aurea **mutant:** a phytochrome-deficient mutant of tomato that contains less than 5% of the amount of phytochrome found in wild-type tomato seedlings.

autocatalysis: a method by which a product can be recycled in order to generate more substrate. For example, in the PCR cycle, the product triose phosphate can be recycled to generate more substrate RuBP.

autotroph: an organism that synthesizes its own food from inorganic moecules; photosynthetic organisms are photoautotrophs.

auxin: a plant hormone that regulates cellular elongation, apical dominance and rooting; also referred to as indole-3-acetic acid (IAA).

Avogadro's number: the number of molecules in a gram molecular weight of a substance (6.022×10^{23} mol^{-1}); hence in photobiology, a *mole of light* contains 6.022×10^{23} photons.

basipetal movement: movement towards the base of a plant.

biennial: a plant that requires two growing seasons to complete its life-cycle; a cold winter temperature in the first year is necessary for reproduction.

bilayer: a two-layered arrangement of lipid molecules that make up membranes within and at the surface of cells.

biliprotein: a peripheral antenna pigment of the photosystems of blue-green and red algae.

biochemical complementation technique: the determination of responses following a microinjection of putative signal transduction molecules into tissue.

biological clock: an internal biological timing system that influences cyclic phenomena.

bioluminescence: light emission from an organism resulting from the oxidation of luciferin by O_2 in a reaction catalysed by the enzyme luciferase.

biomass: the combined dry or wet weight of all the organisms in a habitat.

biosphere: the entire portion of the earth that is inhabited by living things; the sum of the earth's communities and ecosystems.

Blaauw's light–growth model: phototropically effective light leads to a local inhibition of cell growth; the light gradient causes a growth-inhibition gradient through the organ whereby the lowest inhibition is at the non-illuminated side.

blue-light photoreceptors: the receptors that absorb radiation in the blue and near-ultraviolet part of the spectrum; they include carotenoids and flavins.

blue-light responses: light absorbed in the blue/violet (450 nm) part of the spectrum controls phototropism, stem elongation, stomatal opening and activation of flavoenzymes in higher plants.

Blühreife: the condition a plant must achieve before it will flower in response to the environment; translates as *ripeness to flower*.

bolting: during the second summer of biennial plants the apical meristem forms stem cells that elongate (bolt) into a flowering stalk.

brassinosteroids: a class of plant growth regulators that can suppress or rescue mutant dwarf phenotypes

bundle sheath cells: a layer of cells surrounding the vascular bundle in leaves; in C4 plants the bundle contains chloroplasts and is conspicuous.

Bunsen–Roscoe law of reciprocity: the extent of a photoresponse depends only on photon fluence (mol photon m^{-2}) and not on photon flux (mol photon $m^{-2}s^{-1}$) and is independent of the fluence rate used.

C2 glycolate pathway: also known as *the photosynthetic carbon oxidation (PCO) cycle* is a pathway to salvage the carbons from the phosphoglycolate produced by the oxygenase activity of RubisCo on RuBP.

C3 plant: a plant that uses the Calvin cycle to fix CO_2 and forms the three-carbon compound phosphoglycerate as the first stable intermediate.

C4 plant: a plant that prefaces the Calvin cycle with the formation of four-carbon compounds that provide the CO_2 for the Calvin cycle.

CAB: light-inducible chlorophyll *a/b* binding protein.

Ca^{2+}-channels: gated pores which facilitate the transport of Ca^{2+} ions across cell membranes.

caliche: a hard, impermeable soil containing high concentrations of Na^+ (sodic clay soil).

callus: an unorganized mass of dividing, undifferentiated cells, originally produced as wound-healing tissue, that can be maintained on a nutrient medium.

calmodulin: a small cytosolic protein that binds to four Ca^{2+} ions to form a Ca^{2+}–calmodulin complex that activates many enzymes.

Calvin cycle: the cyclic pathway used by plants to fix carbon dioxide and yield triose phosphates (synonym for C3 cycle).

CAM (crassulacean acid metabolism): a plant adaptation for photosynthesis in arid conditions. CO_2 enters via stomata that are open at night-time and is converted into organic acids that are the source of carbon for the Calvin cycle during the daytime.

canopy: the upper layer of forest vegetation.

carbon cycle: a natural cycle in which atmospheric CO_2 is fixed into carbohydrate via photosynthesis and in which energy and CO_2 are released by respiration.

carbon fixation: the incorporation of carbon from CO_2 into organic compounds by autotropic organisms.

carboxyl terminal domain: the amino acid residue at one end of a peptide and containing a free a-carboxyl ($-COOH$) group; the amino acid residue at the other end containing a free a-amino ($-NH_2$) group is termed the amino-terminal domain.

cardinal points: minimum, optimum and maximum.

carotene: mainly orange or orange-red pigment of the carotenoid family.

carotenoid: lipid-soluble photosynthetic pigment composed of isoprene units.

cascade: the series of amplification steps that occurs as part of a signal transduction pathway.

cDNA (complementary DNA): DNA molecule copied from an mRNA molecule by reverse transcriptase.

cell cycle: an ordered sequence of events in the life of a dividing eukaryotic cell.

CF_0–CF_1 complex: (coupling factor) part of ATP synthase. CF_0 is a proton channel; CF_1 is a set of enzymes that phosphorylates ADP to ATP.

cGMP (cyclic guanosine monophosphate): a second messenger compound.

chalcone synthase (CHS): a key enzyme in the synthesis of the flavonoid-like structure of anthocyanins.

chaperone: any protein that binds to an unfolded or partially folded target protein, thus preventing misfolding, aggregation and/or degradation of the target protein and facilitating its proper folding.

charge separation: the transfer of the light energy of photons to the special pair of chlorophyll molecules in the reaction centre results in the formation of a negatively charged electron-acceptor in the stroma and a positively charged chlorophyll molecule in the lumen of the chloroplast.

chemiosmosis: coupling of ATP synthesis and an electrochemical gradient (pH plus electric potential) across a cell membrane to pump metabolites across the membrane against their concentration gradient.

chlorophyll: a light-absorbing pigment essential for photosynthesis; it is composed of a porphyrin head that contains a Mg^{2+} ion at the centre of a tetrapyrrole ring that is connected to a long (C20) hydrophobic, phytol tail.

chloroplast: chlorophyll-rich plastid surrounded by a double membrane and which is the site of photosynthesis, the light-absorbing step of which occurs in a third, internal membrane called the thylakoid.

Cholodny–Went hypothesis: unilateral illumination induces a lateral redistribution of endogenous auxin near the apex of a coleoptile, causing a higher concentration of auxin in the cells on the non-illuminated side of the organ, resulting in their elongation and a differential growth that causes curvature towards the light source.

chromophere: when the sugars are removed from anthocyanins (which occur as glycosides) the part remaining, termed *anthocyanidin* is the chromophere.

chromophore: the tetrapyrrole portion of a chromoprotein; it is the part of the molecule that is responsible for absorbing light and, hence, colour.

chromoplast: a plastid that contains coloured pigments and is located in flower and fruit tissues.

chromoprotein: a pigment that contains protein as an integral part of the molecule: the pigment part is called the *chromophore* and the protein part is called the *apoprotein*.

circadian rhythm: an endogenous rhythm whose period is approximately 24 hours. (Latin, *circa* = about + *dies* = day.)

cis–trans isomers: isomers related by rotation about a double bond. In a carbon–carbon double bond the two groups of interest lie on the same side of the bond to form a *cis*-isomer and on opposite sides of the bond to form a *trans*-isomer.

clone: a single individual organism that is genetically identical to another individual.

CO_2-compensation point: the atmospheric concentration of CO_2 at which photosynthesis just compensates for respiration.

coleoptile: a specialized sheath that protects the growing shoot tissue of a monocot embryo.

conductance: a measure of stomatal opening; it is the inverse of resistance to diffusion (usually in units of mmol $m^{-2} s^{-1}$) and is used in preference to resistance since it is directly proportional to transpiration rather than inversely proportional.

consensus sequence: the nucleotides or amino acids most commonly found at each position in the sequences of related DNAs, RNAs or proteins.

constitutive: cellular production of a molecule at a constant rate which is not regulated by internal or external stimuli.

constructive interference: the superimposition of two waves on each other so that they coincide (are in phase) and reinforce each other to produce a new wave of exactly twice the amplitude of the original.

***cop* mutant:** constitutive photomorphogenic mutant resulting from a mutation that causes dark-grown seedlings to undergo photomorphogenesis.

corpuscular theory: Newton's explanation of the nature of light in which light can be considered to be made up of tiny particles (*corpuscles*) which travel in straight lines from the source.

cotyledon: seed leaf; monocots have one cotyledon and dicots have two.

coupling factor: see CF_0–CF_1.

cryptochrome: the name given to the as-yet-unidentified blue/UV-A photoreceptor, candidates for which include carotenoids and flavins or both (literally, *hidden pigment*).

cyanobacteria: photosynthetic eubacteria, formerly known as *blue-green algae*.

cyclic electron transport: the path of electron transport during the light reactions of photosynthesis that involves only PSI and yields ATP but not NADPH or oxygen.

cyclic photophosphorylation: the production of ATP by means of cyclic electron transport.

cytochrome complex: a group of haem proteins that act as electron carriers during cellular respiration and photosynthesis.

cytokinin: a plant hormone that regulates cell division.

cytoplasm: the entire contents of a cell, exclusive of the nucleus and bounded by the plasma membrane.

cytosol: the semi-fluid portion of the cytoplasm.

DAG (diacylglycerol): a second messenger in signal transduction pathways produced by the catalytic breakdown of phosphotidylinositol bisphosphate (PIP_2).

dalton: the atomic mass unit; a measure of mass of atoms and subatomic particles.

dark reactions: all the reactions that fix CO_2 into sugar during photosynthesis in a process that is indirectly dependent on the *light reactions*; also called *Calvin cycle*.

dark reversion: the P_{fr} form of phytochrome can spontaneously revert to the P_r form in darkness both *in vivo* and *in vitro*.

day-neutral plants: plants that will flower irrespective of daylength.

DCMU (dichlorophenyldimethylurea): a herbicide that, applied to the soil, moves through the xylem into the leaves where it blocks electron transport and thus inhibits photosynthesis.

dephosphorylation: removal of a phosphate group from a molecule by hydrolysis. The activity of many phosphorylated proteins is modulated by removal of the phosphate group by various phosphoprotein phosphatases.

depolarization: change in membrane potential resulting in a less negative membrane potential.

destructive interference: two waves at $180°$ out of phase dampen or cancel each other.

***det* mutant:** de-etiolated mutant which arises from a mutation that causes dark-grown seedlings to exhibit a light-grown phenotype.

devernalization: plants that have been induced to flower by vernalization (cold treatment) can have the effects of the cold treatment reversed if followed immediately by a high ($30\,°C$) temperature treatment.

dicot: a flowering plant that has two seed leaves.

diffraction: includes phenomena produced by the spreading of waves around and past obstacles that are similar in size to the wavelength.

diffraction gratings: a pattern of fine lines drawn close together on a transparent surface which can separate a mixture of wavelengths into a spectrum similar to that produced by a prism.

diffusion shell: an imaginary hemisphere above a pore in which molecules of diffusing gases radiate outwards.

DNA (deoxyribonucleic acid): a double-stranded, helical nucleic acid molecule capable of replicating and of determining the cell's proteins.

DNA photolyase: an enzyme that splits thymine dimers in DNA (see *photoreactivation*).

***doc* mutant:** dark overexpression of CAB (chlorophyll *a*/*b* binding protein).

dolomite: a mineral composed of double carbonate of calcium and magnesium.

dormancy: a condition in which growth of tissues such as buds, seeds, tubers and corms is temporarily arrested and they fail to grow even though they are provided with adequate moisture, oxygen and nutrition.

dose–response curves: plots of physiological responses in relation to varying levels of stimulus.

ecological niche: the sum total of an organism's utilization of the biotic and abiotic resources of its environment.

ecology: the study of how organisms interact with their environment.

ecosystem: a biological community together with its associated abiotic environment.

electric potential: the potential energy of charge separation across a membrane.

electrochemical gradient: the sum of the gradients of concentration and of electric charge of ions across a membrane.

electrogenic pump: an active transport protein that pumps ions against their concentration gradients across membranes; the main electrogenic pumps in plants are *proton pumps*.

electromagnetic spectrum: the distribution of electromagnetic energies arranged according to their wavelengths, frequencies or photon energies and ranging from cosmic rays (10^{-16} nm) to radio waves (10^4 nm).

electron carrier: a molecule that accepts electrons from donor molecules and passes them on to acceptor molecules in an electron chain.

endergonic reaction: a chemical reaction that absorbs energy, i.e. has a positive free-energy change; such reactions occur spontaneously.

end-of-day treatment: a phenomenon of mimicking shade-avoidance responses by giving plants a brief exposure to far-red light at the beginning of a daily dark period in order to reduce the amount of P_{fr} present in darkness.

endogenous oscillator: an internal pacemaker that regulates rhythmic fluctuations in plants – *a biological clock*.

endoplasmic reticulum: network of interconnected double membranes within the cytoplasm of eukaryote cells.

energy or photon flux: energy or photons per unit area per unit time expressed in watts or joules; the same as *irradiance* or *fluence rate* and often loosely called *intensity*.

enhancement effect: when light beams of two different wavelengths are superimposed on each other the photosynthetic rate obtained is greater than the sum of the individual rates.

entrainment: the synchronization of a biological rhythm by means of an external signal such as light or dark.

enzyme: a protein that functions as a catalyst.

epidermal strips: peels, a few layers of cells thick, from plant leaves.

epiphyte: an organism that is attached to another organism without parasitizing it.

epitope: the particular chemical group or groups within a macromolecule (antigen) to which a given antibody binds; an antigenic determinant.

erythrose: a four-carbon monosaccharide.

ethylene: a plant hormone responsible for fruit ripening.

etiolation: the general pale, spindly condition exhibited by dark-grown plants.

etioplast: proplastids of dark-grown or etiolated seedlings. They contain semi-crystalline arrays of tubular membranes known as prolamellar bodies; instead of chlorophyll etioplasts contain carotenoids.

eukaryote: an organism that has a membrane-bound nucleus and cell organelle.

excited-state electron: an electron that has absorbed a quantum and thereby moved to a higher energy level than its ground state.

exon: a coding region of a eukaryote gene that is expressed; exons are separated from each other by introns.

factor X: an unidentified external factor (not light or darkness) capable of resetting the biological clock.

FAD (flavin adenine dinucleotide): a coenzyme that carries electrons and hydrogen in a variety of biological oxidations and reductions.

ferredoxin: a membrane-associated, iron–sulphur protein that acts as an electron-acceptor in the light reactions of photosynthesis.

ferredoxin-NADP-reductase: a soluble flavoprotein that catalyses the reduction of $NADP^+$ to NADPH.

first messenger: a stimulus that brings the original 'message' of a signal transduction pathway to a cell.

flaccid cell: a cell without any turgor pressure.

flavonoid: phenylpropane derivative with a basic C_6—C_3—C_6 composition; major groups are *flavones*, *flavonols* and *anthocyanins*.

floral evocation: events at the meristem that result in the formation of floral primordia rather than leaves, following induction.

floral primordia: a small mass of cells that grows into a flower.

floral induction: the reorganization of meristem cells to produce floral primordia.

florigen: a hypothetical, unidentified, photoinduced flowering hormone thought to be present in plants.

fluorescence: the spontaneous emission of a quantum by an excited electron which allows the electron to return to its ground state.

free energy (G): a measure of the potential energy of a system to do work.

free radical: molecule with one or several unpaired electrons.

free-running rhythm: a rhythm that is allowed to run under constant environmental conditions (continuous darkness or dim light) so that its periodicity is close to but not exactly 24 hours.

frequency (n): the number of wave crests (peaks in energy) passing a given point in a given interval of time.

fusca **mutants:** a class of mutations (e.g. *det* and *cop*) that are characterized by short hypocotyls, open cotyledons and high concentrations of anthocyanins when grown in the dark. Cells exhibit differentiation that is typical of light-grown plants and their light-regulated genes are active in the dark.

gated ion channel: a specific ion channel that opens and closes to allow the cell to alter its membrane potential.

gene: a discrete unit of hereditary information consisting of a specific sequence of nucleotides in DNA.

genetic map: an ordered list of genetic loci along a chromosome.

genotype: the genetic make-up of an organism.

germination: resumption of growth of the embryo with consequent rupture of the seed coat and emergence of the new plant.

gibberellin: a plant hormone that stimulates stem elongation and leaf growth, triggers the germination of seeds and the breaking of bud dormancy.

girdling: removal of stem tissue external to the xylem.

gluconeogenesis: the biosynthesis of a carbohydrate from simpler, non-carbohydrate precursors such as oxaloacetate or pyruvate.

glycogen: a long, branched polymer composed exclusively of glucose units and found mainly in liver and muscle; sometimes termed 'animal starch'.

glycoprotein: a molecule in which one or more oligosaccharide chains are covalently linked to a protein; frequently found in the plasma membrane.

glycoside bond: the covalent linkage between two monosaccharides formed by a condensation reaction in which one carbon of one sugar reacts with a hydroxyl group on the second sugar with the loss of a water molecule.

Gotthaus–Draper principle: only light that is absorbed can be active in a photochemical reaction.

G-protein: a heterotrimeric guanine nucleotide-binding protein usually linked to a receptor molecule at the cell surface. Binding of a hormone to the receptor activates the molecule which, in turn, stimulates or inhibits an effector protein that generates a second messenger.

graft union: the physical joining of different species or cultivars of plant in order to combine the best qualities of both. The plant that provides the root system is called the *stock*; the twig grafted on to the stock is the *scion*.

granum: the stacked region of the thylakoid membrane within a chloroplast.

gravitropism: the downward growth of roots or stems in response to gravity.

greenhouse effect: the absorption of significant amounts of infrared radiation by the water vapour and carbon dioxide and other gases present in the earth's atmosphere.

guard cells: two kidney-shaped cells that border the pore of a stoma in a leaf and regulate the opening and closing of the pore.

GUS reporter gene (*β*-glucuronidase): a reporter gene that encodes a protein that is readily assayed.

half-life: the time required for the disappearance or decay of one-half of a given component in a system.

halophyte: a plant that grows in high-salt soils or salt marshes.

heat-shock protein: a protein that helps protect other proteins during heat stress.

herbicide: a weed-killer that selectively disrupts metabolic pathways in plants thereby causing the death of the plant.

Hill reaction: the evolution of oxygen and the photoreduction of an artificial electron acceptor by a chloroplast preparation in the absence of carbon dioxide.

HIR (high-irradiance response): light-dependent responses in which full expression of the response requires long exposure to high levels of irradiance with action spectra in the far-red and blue parts of the spectrum.

hydrophilic: refers to compounds that are polar or charged and are relatively soluble in water and other polar solvents and insoluble in lipids and other non-polar solvents (literally, *water-loving*).

hydrophobic: refers to compounds that are non-polar and are insoluble in water and other polar solvents and are soluble in lipids and other non-polar solvents (literally, *water-fearing*).

hyperpolarization: an electrical state whereby the inside of the cell is made more negative relative to the outside than at the resting membrane potential.

hypocotyl: the initial length of stem (located between the cotyledons and the radicle) that emerges from a germinating seed.

imbibition: water uptake by dry seeds.

induction: the ability of one group of embryonic cells to influence the development of other cells.

inductive resonance: the transfer of absorbed light energy across pigment molecules.

infradian rhythm: a period longer than 28 hours.

interference: a phenomenon caused by reinforcement when energy crests (waves) are superimposed on each other (in phase) or cancel or dampen each other (out of phase).

interference filters: a thin layer of a reflective medium on a glass surface of such a thickness that one wavelength (or multiples thereof) is strongly reinforced by passing through the filter, whereas other wavelengths are cancelled.

internal pacemaker: an endogenous oscillator (*biological clock*) that regulates biological rhythms.

internode: portion of a stem between nodes and where there are no leaves.

intron: a non-coding, intervening sequence within a eukarotic gene.

invertebrate: an animal that lacks a backbone.

in vitro: in an isolated cell-free extract, as in a test tube (literally, *in glass*).

in vivo: in an intact, living cell or organism (literally, *in life*).

ion channel: an integral membrane protein that allows for the regulated transport of selected ions across the membrane.

ion pump: any transmembrane ATPase that couples hydrolysis of ATP to the transport of a specific ion across a membrane against its electrochemical gradient.

IP$_3$ (inositol triphosphate): a second messenger that mediates an increase in cytoplasmic Ca^{2+} concentration.

isoprene: the basic five-carbon unit of terpenoid polymers.

isotope: one of several atomic forms of an element, each containing a different number of neutrons and thus differing in atomic mass.

juvenile phase: an early vegetative stage of growth when plants are unable to flower.

kerogen: sedimentary organic matter.

***Kranz* anatomy:** specialized leaf anatomy characteristic of C4 plants in which the vascular bundles are surrounded by a sheath (German: *Kranz*) of photosynthetic cells.

LFR (low-fluence response): inducible responses which cannot be initiated until fluence levels reach $1.0 \, \text{mmol m}^{-2}$; the fluence of red light required is in the range $1-10^3 \, \text{mmol m}^{-2}$.

lichen: a symbiotic association between a fungus which provides inorganic nutrients and an alga, which provides organic nutrients.

ligand: a small molecule, other than an enzyme substrate, that binds tightly and specifically to a macromolecule, usually a protein, forming a macromolecule–ligand complex, that generally signals some cellular response.

light: the visible portion of the electromagnetic spectrum.

light flux: the light energy falling on a surface for a given period of time.

light-harvesting complexes (LHCs): extended antenna systems for harvesting additional light energy during photosynthesis.

light reactions: the sum total of the light-dependent reactions that generate ATP and NADPH during photosynthesis.

light-sensitive seeds: certain seeds (e.g. Grand Rapids lettuce seeds) which require a pretreatment of light before they will germinate.

locus: the position of a gene on a chromosome (plural: *loci*).

long-day plant: a plant that will not flower unless the light period is longer than a critical minimum.

luciferase: an enzyme that catalyses the oxidation of luciferin with a resultant emission of light.

lutein: hydroxylated form of α-carotene.

lycopene: the principal pigment in tomato fruit; a carotene.

macromolecule: a large molecule of living matter composed of smaller molecules.

malate formers: plant species that use malate as the main vehicle for CO_2 transport to photosynthetic cells in the leaf.

membrane potential: charge difference across a membrane due to excess of positive ions (cations) on one side and negative ions (anions) on the other, resulting from the selective permeability of the membrane to different ions or from pumping of ions across the membrane by active transport.

mesophyll: parenchyma tissue between the epidermal layers (but excluding them) of the leaf; usually photosynthetic.

midpoint potential: the condition under which a redox couple is 50% reduced and is designated E_m^1.

mitosis: the process of nuclear division in eukaryotes whereby a cell divides to produce two identical daughter cells.

mole: the gram molecular weight of a substance.

monoclonal antibody: an antibody produced by a cloned hybridoma cell and is thus a homogeneous protein exhibiting a single-cell antigenicity.

monocot: a member of the group of flowering plants characterized by having one seed leaf (cotyledon) per seed.

mutation: a permanent transmissible genetic change in the nucleotide sequence of a chromosome, usually in a single gene, which leads to a change in or loss of its normal function.

NAD-ME: nicotinamide-adenine-dinucleotide-malic enzyme.

NADP (nicotinamide-adenine-dinucleotide-phosphate): a coenzyme that accepts two electrons and is thereby reduced to $NADPH_2$ during photosynthesis.

night break: the interruption of the long dark period in short-day plants by a flash of light that nullifies the dark period.

non-cyclic photophosphorylation: the light-driven flow of electrons from water to NADPH during photosynthesis and requires the cooperation of both PSI and PSII.

nucleolus: a structure within the nucleus where ribosomal RNA is synthesized and assembled in ribosomal subunits.

nucleoplasm: the substance located within the nucleus – DNA, RNA, histones, enzymes, nucleic acids.

nyctinasty: sleep movements of leaves in response to changes in the turgor of cells at the base of their petioles.

OEC (oxygen-evolving complex): a small complex of proteins associated with a cluster of four manganese ions and which is located on the luminal side of the thylakoid membrane; it is responsible for the splitting (oxidation) of water during the light reactions of photosynthesis.

organelle: one of several formed bodies with a specialized function, suspended in the cytoplasm of eukaryotic cells.

osmosis: the movement of water across a semi-permeable membrane from a region of higher to a region of lower water concentration.

osmotic potential: a component of water potential; a measure of the effect of solute concentration on the capacity of a solution to absorb or lose water; also called *solute potential.*

oxidation: loss of electrons or hydrogen from an atom or molecule.

oxidation–reduction: a chemical reaction in which one reactant loses electrons (oxidation) while the other gains electrons (reduction).

ozone layer: a form of oxygen (O_3) in the stratosphere which shields living organisms from intense UV-radiation.

palisade: a layer of elongated cells immediately below the leaf epidermis and containing chloroplasts.

PAR (photosynthetically active radiation): radiation in the range 400–700 nm used by plants in photosynthesis.

patch-clamp method: a technique in which a small 'patch' of membrane is isolated and has been adapted to study ion movement through single-ion channels in cell membranes.

PCA: photosynthetic carbon assimilation cycle.

PCK (phosphoenolpyruvate carboxykinase): an enzyme located in the cytosol that catalyses the decarboxylation of oxaloacetate to phosphoenolpyruvate and ADP and releases CO_2.

PCO (photosynthetic carbon oxidative) cycle: the enzymatic pathway of glycolate synthesis.

PCR (photosynthetic carbon reduction) cycle: the metabolic pathway leading to carbon assimilation; also called reductive pentose pathway (RPP), Calvin cycle, C3 cycle.

pentose: a five-carbon monosaccharide.

pentose phosphate pathway: a series of reactions interconnecting six-carbon (hexoses) and five-carbon (pentose) sugars and during which NADP is reduced to NADPH; also called the phosphogluconate pathway.

PEP: phosphoenolpyruvate.

PEPCO: phosphoenolpuruvate carboxylase.

peptide bond: the covalent bond between two amino acids in a polypeptide.

perennial: a plant whose life-cycle lasts for a number of years.

peroxisome: small organelle (microbody) in eukaryotic cells that contains enzymes that metabolize hydrogen peroxide and glycolic acid.

phase point: any point on the cycle of a rhythm recognizable by its relationship to the rest of the cycle, e.g. the maximum (peak) and minimum (trough) positions.

phase shift: a change in the phase of a rhythm achieved by moving the whole cycle forwards or backwards in time without altering the period.

phenotype: the physical, observable characteristics of an organism; an expression of the genetic make-up (*genotype*) of an organism.

pheophytin: a form of chlorophyll *a* in which the Mg^{2+} has been replaced by two H^+s; it is considered to be the primary electron-acceptor in PSII.

phospholipase C: an enzyme located on the cytoplasmic side of the plasma membrane and which catalyses the breakdown of phosphoinositol bisphosphate (PIP_2) to yield second messengers inositol triphosphate (IP_3) and diacylglycerol (DAG).

phospholipid bilayer: a symmetrical two-layer structure formed by phospholipids in aqueous solution; the basic structure of all biomembranes.

phosphorescence: an electron brought to a high-energy level due to the absorption of a quantum of light may have its spin reversed and be 'trapped' at that level in a *triplet state*; the electron may have its spin re-reversed and return to the ground state, giving off excess energy in a process termed phosphorescence.

photoassimilate: any photosynthetic product.

photoautotroph: an organism that uses light energy to drive the synthesis of organic compounds from carbon dioxide and water.

photobleaching: an oxygen-dependent destruction of chlorophyll molecules.

photochemical reaction: the 'light' reactions of photosynthesis which occur in the chloroplast and yield ATP and NADPH.

photoelectric emission: ejection of electrons from a conductor by light incident on its surface.

photoinductive cycle: the relative duration of light and darkness in a daily cycle required to induce flowering in a plant.

photolyases: a unique class of flavoproteins that use blue light to repair UV-induced damage to DNA.

photolysis: splitting of water molecules by light absorption in PSII.

photomorphogenesis: the light-controlled processes that regulate plant morphogenesis.

photon: a particle of visible light.

photon-fluence rate: the number of quanta impinging on a surface in a given time.

photoperiodism: the physiological response of a plant to relative lengths of light and dark periods.

photophile phase: the light-loving phase in a plant's photoperiod.

photophosphorylation: the production of ATP by means of the energy of light-excited electrons in photosynthesis.

photoprotection: protection of pigments by the absorption of excess of harmful wavelengths of light, e.g. the absorption of blue light by carotenoids protects chlorophyll from photo-oxidation.

photoreactivation: an enzyme repair mechanism in which a bacterial culture is quickly irradiated with UV-A or blue light after lethal UV irradiation, resulting in the reversal of the inactivation. This phenomenon is based on the existence of *DNA photolyase* whose catalytic action is sensitized by radiation in the range 300–500 nm.

photoreceptor: a light-sensitive pigment in a cell.

photorespiration: a wasteful metabolic pathway in which RubisCO/O uses oxygen (not CO_2) to produce phosphoglycolate and generates no ATP; generally occurs in hot, dry, bright days when stomata close and oxygen concentration in the leaf exceeds that of carbon dioxide.

photosensory pigments: pigments that absorb particular wavelengths of light.

photosynthesis: a complex series of *light* and *dark* reactions that occurs in the chloroplasts of higher plants using light energy to generate carbohydrates from CO_2 and H_2O and evolving O_2.

photosynthetic efficiency: the chemical-bond energy of the hexose produced during photosynthesis divided by the total photosynthetically active radiation (PAR).

photosystem: the light-harvesting unit in photosynthesis, located on the thylakoid membrane of the chloroplast and consisting of an antenna complex, the reaction-centre chlorophyll *a* and a primary electron-acceptor. There are two photosystems: PSI absorbs light of $l > 680$ nm and PSII absorbs light at $l < 680$ nm.

phototropism: growth of a plant part towards or away from light.

pH scale: a measure of hydrogen ion concentration equal to $-\log[H^+]$ and ranging in value from 0 to 14.

phycobilins: plant pigments with properties similar to bile pigments (*phyco* = plant).

phycobiliprotein: plant chromoprotein composed of a phycobilin (chromophore) and a protein portion (apoprotein).

phycobilisome: a large macromolecular complex composed of phycobilins and biliproteins.

phytochrome: a pigment involved in plant photomorphogenesis; it exists in two reversible forms – red-absorbing P_r and far-red-absorbing P_{fr}.

phytol: a 20-carbon alcohol tail attached to the porphyrin head of the chlorophyll molecule.

plasmadesmata: a membrane-lined channel between adjacent cells.

plasma membrane: a bilayer of phospholipid and protein that surrounds the cytoplasm of cells.

plastids: organelles within plant cells: *proplastids* – young plastids found in meristematic cells; *chloroplasts* – chlorophyll-rich plastids; *amyloplasts*, which store starch; *chromoplasts* – plastids that contain coloured pigments; *leucoplasts* – colourless plastids.

plastocyanin: a small peripheral copper-binding protein that can diffuse freely along the luminal surface of the thylakoid membrane and transport electrons.

plastoquinone: a lipid-soluble electron located in thylakoid membranes.

pleiotropy: a condition in which a single gene affects several traits.

polarization: functional or structural differences in distinct regions of a cell or cellular component resulting from an ion flow that creates a net negative charge inside a cell and a net positive charge outside.

polyene: a network of alternating single and double bonds that make pigment molecules efficient photoreceptors.

polypeptide: a polymer of amino acids linked together by peptide bonds.

polyploidy: a chromosomal alteration in which the organism possesses more than two complete sets of chromosomes.

porins: proteins that line transmembrane channels that permit the passage of chemicals having a molecular weight of between 1000 and 5000.

porphyrin: a cyclic tetrapyrrole made up of four nitrogen-containing pyrrole rings in a cyclic fashion; in the chlorophyll molecule the porphyrin contains Mg^{2+}.

primary action: some form of chemical transduction such as a conformational change in a protein or a redox reaction.

primary productivity: the rate at which light energy or inorganic chemical energy is converted to chemical energy of organic compounds by autotrophs in an ecosystem.

prokaryote: an organism that lacks a true membrane-limited nucleus and other organelles.

promoter: a specific nucleotide sequence in DNA that binds RNA polymerase and indicates where to start transcribing RNA.

prosthetic group: a tightly bound small molecule or metal that is required for catalytic activity of some enzymes; also called *coenzyme* or *cofactor*.

protein: a polymer of amino acids linked by peptide bonds.

protein kinase: an enzyme that phosphorylates certain amino acid residues in specific proteins.

protein phosphatase: an enzyme that removes phosphate groups from proteins thereby reversing the action of protein kinases.

proteolysis: degradative splitting of proteins.

proton-motive force: the energy equivalent of the proton concentration gradient and electric potential gradient across a membrane. In chloroplasts and mitochondria these gradients are generated by electron transport maintained by the thylakoid or inner mitochondrial membrane and drive ATP synthesis by ATP synthase.

proton pump: an active transport mechanism in cell membranes that uses ATP to force hydrogen ions out of a cell and in the process generates a membrane potential.

protoplast: the contents of a plant cell exclusive of the cell wall.

purine: a basic compound containing two fused heterocyclic rings that occurs in nucleic acids. The purines commonly found in DNA and RNA are adenine and guanine.

pyrimidine: a basic compound containing one heterocyclic ring that occurs in nucleic acids. The pyrimidines commonly found in DNA are cytosine and thymine; in RNA, uracil replaces thymine.

pyrophosphate (PP$_1$): the double phosphate molecule released from the enzymatic breakdown of ATP to AMP and PP$_i$.

pyruvate phosphate dikinase: an enzyme that catalyses two different phosphorylation reactions simultaneously: pyruvate is phosphorylated to phosphoenolpyruvate (PEP) and phosphate (P) is phosphorylated to pyrophosphate (PP$_1$).

Q$_{10}$ (temperature coefficient): a quantitative measure used to describe the effect of temperature on a reaction. Within the temperature range 5 °C to 30 °C a doubling of rate for every 10 °C rise in temperature (Q$_{10}$ = 2) is typical for enzyme reactions.

Q-cycle: a model for coupling electron transport from plastoquinol to cytochrome with the translocation of protons across the thylakoid membrane.

Q enzyme: a branching enzyme that catalyses the formation of α[1–6] linkages in the synthesis of amylopectin.

quantum: a unit of electromagnetic energy; a term used interchangeably with photon.

radiant energy (radiation): a form of energy that is emitted or propogated through space or a material medium.

ratiospec: an instrument that measures absorbance at two different wavelengths simultaneously.

reaction centre: a complex of four to six molecules of chlorophyll *a*, plus associated proteins and a primary electron-acceptor, within a photosystem.

recombinant DNA: a DNA molecule made *in vitro* with segments from different sources.

redox couple: an electron donor and its corresponding oxidized form, e.g. NADH and NAD$^+$; also called *redox pair*.

redox potential: the tendency of a molecule to accept or donate electrons during a chemical reaction.

redox reaction: a chemical reaction involving the transfer of one or more electrons from one reactant to another; also called *oxidation–reduction reaction*.

reduction: the gain of electrons or H$^+$ as a result of a chemical reaction.

refraction: the change in direction (bending) that occurs when a ray of light passes from one medium into another in which its velocity is different.

reporter genes: genes that encode proteins that are readily assayed and are used to replace a gene of interest, whose products are difficult to express, leaving the original flanking DNA sequences intact.

resonance transfer: electronic energy transfer between accessory pigments and chlorophyll.

rhythm: the repeated occurrence of some function.

ripeness to flower: the condition that a plant must achieve before it will flower in response to the environment (see *Blühreife*).

RNA splicing: the removal of non-coding portions (introns) of an RNA primary transcript.

rosette: a form to give a compact appearance of growth in which plant leaves or branches are clustered.

RubisCO: ribulose carboxylase, the enzyme that catalyses the first step (the addition of CO_2 to ribulose bisphosphate) of the Calvin cycle.

second messenger: a small, non-protein, water-soluble molecule or ion, such as Ca^{2+} or cAMP (an effector) produced in a cell in response to an external stimulus (first messenger).

semi-permeable membrane: a membrane that is permeable to water and some substances but relatively impermeable to others.

shade plants: plant native to shady habitats; these species exhibit lower photosynthetic rates and their photosynthetic responses are saturated at lower irradiances than plants grown in open areas.

short-day plant: a plant that flowers when the length of the dark period is greater than a critical minimum length.

signal transduction: a pathway of coupled intracellular events by which an external signal (stimulus) is received by a cell, amplified and converted to a cellular response.

singlet oxygen ($^1O_2$): electronically excited oxygen produced when excess light energy absorbed by chlorophyll is passed directly on to O_2.

skotomorphogenesis: plant development that occurs in darkness and is characterized by a pale, spindly appearance of the plant.

skotophile phase: dark-loving phase in the life-cycle of a plant (see photophile phase).

small nuclear ribonucleoprotein (snRNP): one of a variety of small particles in the cell nucleus, composed of RNA and protein, forming part of spliceosomes active in RNA splicing.

solarization: a light-dependent inhibition of photosynthesis which usually causes chlorosis and death of a plant.

solar time: time based on a 24-hour day.

solar tracking: a phenomenon in which leaf laminae follow the path of the sun during daylight hours.

Southern blotting: a DNA hybridization technique by which fragments of DNA are separated by gel electrophoresis, transferred to a filter and probed with DNA that is complementary to the fragment of interest.

spectrophotometer: an instrument that measures the proportions of light of different wavelengths absorbed and transmitted by a pigment solution.

spliceosome: a complex assembly of small nuclear ribonucleoproteins that functions in RNA splicing by releasing an intron and joining two adjacent exons.

Stokes shift: conservation of energy requires that the energy of a fluorescent photon be lower than that of the excitation photon, hence the shift to a longer wavelength; chlorophylls usually fluoresce in the red and the shift in wavelength is typically about ten nanometres.

stoma: a pore surrounded by two guard cells that occurs in the leaf epidermis and facilitates the exchange of gases with the external atmosphere.

stroma: the matrix surrounding the grana within a chloroplast; the site of the 'dark reactions' of photosynthesis.

subjective day: the phase of the free-running rhythm that corresponds to day in the previous entraining cycle; that which corresponds with the dark phase is called the *subjective night*.

succulent: a plant having thick, fleshy leaves or stems; these generally have sunken stomata in order to conserve water.

sunflecks: patches of full sunlight that pass through the canopy and travel along the shaded leaves as the sun moves; in a dense forest, sunflecks can, within seconds, increase ten-fold the photon-flux density impinging on a shaded leaf.

superoxide (O_2^-): a toxic form of oxygen known as a superoxide radical (a radical being a molecule with an unpaired electron) which is highly reactive and will oxidize, and thus destroy chlorophyll and most organic molecules within the cell.

superoxide dismutase (SOD): a ubiquitous enzyme that detoxifies superoxide radicals (O_2^-); it can scavange and inactivate superoxide radicals by forming hydrogen peroxide and molecular oxygen.

symbiosis: an ecological, mutually beneficial relationship between two organisms that live together in direct contact.

temperature compensation point: the temperature at which the amount of CO_2 fixed by photosynthesis equals the amount of CO_2 released by respiration.

thermal dissipation: the loss of energy by heat.

thylakoid: a sac-like membranous structure of the grana of the chloroplast containing the photosynthetic pigments.

tonic effect: the effect of light in decreasing the sensitivity of an organ to subsequent applied light.

tonoplast: the inner membrane of the cytoplasm; it surrounds the vacuole of a cell.

transcription: the synthesis of RNA from a DNA template.

transgenic plant: a plant that has been transformed by the insertion of a gene from another source.

translocation stream: the flow of organic material through the phloem in plants.

transplanting shock: transient wilting, bleaching and poor development expressed by young plants grown under UV-impermeable glass and then transplanted into the open.

triplet state of electrons: a metastable light-excited state of electrons from which energy may be lost by radiation-less decay or by phosphorescence; the triplet state is sufficiently long-lived to allow photochemical reactions to occur.

tropical rainforest: a complex community located near the equator where rainfall is abundant and that harbours a wide range of plant and animal species.

tropism: a growth response that results in the curvature of whole plant organs toward or away from stimuli due to differential rates of cell elongation.

ubiquitin: a small polypeptide, found in nearly all eukaryotes, that becomes covalently attached to proteins, thus tagging them for degradation by proteases.

ultradian rhythm: periods less than 20 hours.

UV stress: a form of radiation stress that occurs in plants growing at high altitudes; long-wave UV (295–400 nm) can cause photoinhibition of chloroplasts, resulting in photobleaching and death of leaves.

vacuole: a membrane-bound (tonoplast) space within a cell which stores water and water-soluble substances.

visible spectrum: the portion of the electromagnetic spectrum that is visible to the human eye.

voltage-gated channels: ion channels in a membrane that open and close in response to changes in membrane potential.

Warburg effect: the inhibition of photosynthesis in C3 plants by O_2.

water potential (ψ): the chemical potential of water; a measure of the capacity of a substance to absorb or release water to another substance.

wavelength (l): the distance between waves or crests of energy in electromagnetic radiation; wavelength is equal to velocity divided by the frequency: $l = c/n$.

winter annuals: plants such as wheat and rye that germinate in the autumn, overwinter as seedlings, and flower in the following spring.

xanthophyll: a yellow carotenoid; an oxygenated carotene.

xanthophyll cycle: the conversion of violaxanthin to zeaxanthin under conditions of excess light in a process of energy dissipation.

X-ray crystallography: a technique for determining the three-dimensional structure of macromolecules (particularly proteins and nucleic acids) by passing X-rays through a crystal of the purified molecules and analysing the diffraction pattern of the discrete spots that results.

***Zeitgebers*:** periodic environmental signals such as *dark-to-light* transition at dawn and *light-to-dark* transition at dusk (from the German for *time-giver*).

Z-scheme: a model that outlines the coupling of electron transport and vectorial H^+ transport in photosynthesis.

Index

Photobiology of Higher Plants. By M. S. Mc Donald.
© 2003 John Wiley & Sons, Ltd: ISBN 0 470 85522 3; ISBN 0 470 85523 1 (PB)